D0753173

SHALLOW-WATER GAMMARIDEAN AMPHIPODA OF NEW ENGLAND

Gammarus oceanicus Segerstråle   (magnification x 4)

$\mathcal{S}$HALLOW-WATER
GAMMARIDEAN AMPHIPODA OF NEW ENGLAND

Chief Zoologist
National Museum of Natural Sciences, National Museums of Canada
Ottawa, Ontario

Visiting Investigator, 1963–1970
Systematics-Ecology Program, Marine Biological Laboratory
Woods Hole, Massachusetts

COMSTOCK PUBLISHING ASSOCIATES a division of **CORNELL UNIVERSITY PRESS** Ithaca & London

First published 1973 by Cornell University Press. Published in the United Kingdom by Cornell University Press Ltd., 2-4 Brook Street, London W1Y 1AA.

International Standard Book Number 0-8014-0726-5 Library of Congress Catalog Card Number 72-4636

Printed in the United States of America by Vail-Ballou Press, Inc.

Librarians: Library of Congress cataloging information appears on the last page of the book.

*In memory of **Ruth Von Arx***
whose artistic skill and dedication
made this work possible

Contents

ILLUSTRATIONS

FOREWORD

On September 1, 1962, the Marine Biological Laboratory inaugurated the Systematics-Ecology Program and embarked on a ten-year basic, year-round program of research and training. One of the major objectives of the Program was extension of the knowledge of estuarine and coastal marine organisms by investigation of important problems in the context of the New England area.

High on our list of priorities was the preparation of systematic monographs. Not only do these advance systematic biology, but they provide a base for the extensive research and training now going on in marine biology, biological oceanography, and applied environmental sciences. The deterioration of the coastal marine environment italicizes the need for baseline systematic information in the assessment of coastal degradation.

It was thus a great pleasure to have Edward L. Bousfield join the Program as a part-time Senior Visiting Investigator in August 1963 to commence field work and synthesis of information for a monographic revision of the shallow-water gammaridean amphipod crustaceans of New England. In the intervening years, somehow sandwiched into his heavy schedule at the National Museum of Natural Sciences in Ottawa, he has been able to take time to visit the Marine Biological Laboratory a number of times to carry out field investigations, confer with associates in Woods Hole and elsewhere in New England, and guide the work of Mrs. Ruth von Arx, who so ably illustrated the major part of the volume.

The monograph has been written for the informed layman, student biologist, and fisheries biologist, as well as the professional carcinologist.

Approximately 200 species of coastal marine intertidal, estuarine, and semiterrestrial species are included in the keys, of which 125 species have been fully figured and diagnosed, and descriptions of geographic range, ecology, life history, and behavior have been provided to the limit of existing information. Also included are sections on the general biology of the Amphipoda, a glossary of terms, and an index.

Approximately 25 percent of the species fully treated in the monograph have been described as new herewith or in separate publications by the author and colleagues since 1963, all or partly based on specimens obtained during the field work in New England.

Our association with Dr. Bousfield has been most pleasant and stimulating. We take pride in having had a small part in the preparation of this important contribution to systematic biology.

October 1972   Melbourne R. Carriker
                                        Past Director, Systematics-Ecology Program
                                        Woods Hole, Massachusetts

$\mathscr{A}$CKNOWLEDGMENTS

Preparation of this guidebook has been made possible through the help of many agencies and interested persons. Most of the work was conducted during the period 1963–1970 with the support of the Systematics-Ecology Program (SEP), Marine Biological Laboratory, Woods Hole, Massachusetts (through a grant from the Ford Foundation, and grants GB-7387, GB-8264, and GB-13250 from the National Science Foundation) and of the National Museum of Natural Sciences, Ottawa.

Valuable study material and ecological data were generously made available through the Gray Museum and by other SEP biologists, notably by R. H. Parker and J. S. Nagle from the Hadley Harbor region; by V. A. Zullo from the Woods Hole region and Cape Cod Bay; by J. B. Pearce from Vineyard Sound and Quick's Hole; by D. C. Grant and A. D. Michael from the Cape Cod Bay Census, and by L. R. McCloskey and J. L. Simon from Vineyard Sound. Personal field work in the Cape Cod region was capably assisted by J. C. H. Carter, P. E. Schwamb, P. J. Oldham, and at times by D. Burnham and T. Williams. Valuable illustrative material, especially from offshore and deeper localities, was generously made available by R. L. Wigley, U.S. Bureau of Commercial Fisheries Biological Laboratory, Woods Hole, and by T. Chess, Sandy Hook, New Jersey. Useful comparative material was supplied by T. E. Bowman and J. L. Barnard, Museum of Natural History, U.S. National Museum; L. Watling, University of Delaware; R. A. Croker and M. F. Gable, University of New Hampshire; E. L. Mills, Dalhousie University; D. H. Steele, Memorial University, Newfoundland; and Pierre Brunel, University of Montreal. The author benefited greatly from professional advice and services rendered on numerous occasions by H. L. Sanders, R. R. Hessler, D. J. Zinn, Nelson Marshall, Diana Laubitz, and many others. For all materials and for valuable critical suggestions I am most grateful. Any errors in text or plates are, however, attributable to myself.

The rendition of the line drawings, almost all of which are original, is a tribute to the artistic skill, ingenuity, and long-suffering dedication of the late Mrs. Ruth Von Arx, senior artist, Systematics-Ecology Program. She was capably assisted, particularly in drawings of mouthparts and in some of the whole-mount drawings, by Mary Lou Florian, Ottawa; Ann Stamper, Lewes, England; and Charles H. Douglas and Donald R. Pentz, National Museum of Natural Sciences, Ottawa. I made an attempt to impress uniformity upon their diverse artistic styles by personally composing the plates and adding minor finishing touches. The color frontispiece shows a living specimen photographed by M. Patricia Morse, Edwards Marine Laboratory, Nahant.

The section on general amphipod biology is based largely on *Invertebrate Zoology* (New York, 1967–1970), Vol. III, as translated and adapted by H. W. and L. R. Levi, of Alfred Kaestner's *Lehrbuch der Speziellen Zoologie,* with the kind concurrence of Herbert Levi, of Harvard University, and Interscience Publishers, by permission of John Wiley and Sons, Inc.

Finally, to Melbourne R. Carriker, Director, Systematics-Ecology Program, I am deeply grateful for support, encouragement and many personal kindnesses that made this study not only possible but a most pleasant and stimulating experience.

E. L. B.

Ottawa

SHALLOW-WATER
GAMMARIDEAN AMPHIPODA OF NEW ENGLAND

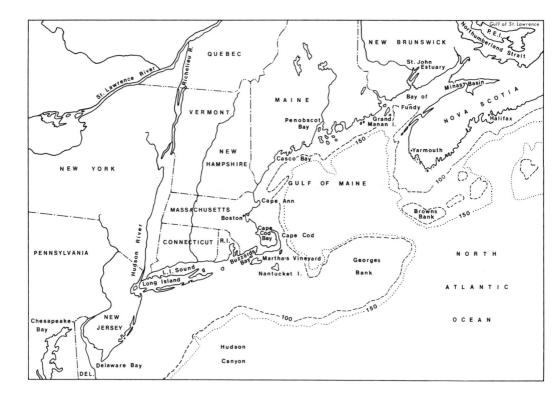

The New England Coastal Marine Region

The present volume is intended as a guide to the gammaridean amphipod crustaceans that commonly inhabit waters of less than one hundred feet (about 30 meters) in depth in the New England coastal region. Geographically, the New England region extends from the northern part of the Gulf of Maine southward to Long Island, as shown in the accompanying map. Many of the littoral marine amphipods of this region also occur predominantly throughout the eastern Canadian region northward to the St. Lawrence estuary and Strait of Belle Isle. "Southern" faunistic elements are characteristic also of the Middle Atlantic States southward to and including Chesapeake Bay. Ecologically, the book also treats semiterrestrial, brackish-water, and epigean fresh-water amphipods that occur in streams, rivers and heads of estuaries within a few miles of the coast. Morphologically, the study is restricted to the gammaridean amphipods—as outlined in the next section—which comprise nearly 90 percent of all amphipod species and perhaps over 95 percent of the amphipod biomass of this region. Of the three other regional suborders, the Caprellidea are represented by about eight shallow-water species (see McCain, 1968), the Hyperiidea (essentially pelagic) by about six coastal species, and the Ingolfiellidea (abyssal-benthic) are rare.

The need for a comprehensive, illustrated, fully keyed guide to the regional amphipod fauna has long been felt. Existing comprehensive manuals treat only a fraction of the known species, even where amply illustrated (e.g., Sars, 1895; Holmes, 1905). Others are limited regionally or faunistically (e.g., Kunkel, 1918; Miner, 1950). About half the regional gammaridean species have been described since the work of Holmes (1905) and about 25 percent were newly discovered during the past decade. The quality of pertinent taxonomic work has increased markedly during and since the era of the late C. R. Shoemaker, but the papers of Shoemaker and other recent authors are widely scattered in the literature. During the past thirty years, systematic monographic treatments of family, subfamily, and generic groups by Shoemaker, Barnard, Mills, Bousfield, and others have led to further studies on the ecology, physiology, and life histories of component species by Sameoto (1969a, b, c), Dexter (1967, 1971), Croker (1967a, b, 1968), and others. A large body of unpublished information on the systematics, distribution, and ecology of the regional species has resulted from recent surveys of the Woods Hole region, Cape Cod Bay, and various New England shores by collaborators in the Systematics-Ecology Program, the U.S. Fish and Wildlife Service, by staff members of regional universities, and by the author.

The collation of all this information in a readily accessible, accurate, readable, and up-to-date form is therefore the aim of this book.

Development of knowledge of the systematics of New England amphipods commenced with the work of Thomas Say (1818). His descriptions of common species such as *Gammarus fasciatus, G. mucronatus, Talorchestia longicornis, Cerapus tubularis,* and *Lepidactylus dytiscus* are recognizable today and his remarks on the ecology and behavior of these species are keenly perceptive. Some of his type specimens are still extant (see Bousfield, 1958b). By contrast, the contemporary works of C. F. Rafinesque, mainly on the fresh-water amphipod species, fared less well and his names are now largely unrecognized (Bousfield and Holthuis, 1969). Gould (1841) and DeKay (1844) included a few amphipod species in their lists of marine fauna of New England and New York, respectively. The first definitive systematic account of New England regional amphipods was that of Stimpson (1853), who not only introduced several new taxa, but identified many of the northern New England species with those of northern Europe described earlier by Krøyer (1842), Leach (1814), Montagu (1813), Pallas (1766), and Linnaeus himself (1758).

S. I. Smith (1873) significantly extended knowledge of the Cape Cod fauna with descriptions of several new amphipod taxa, particularly those with southern or warm-water affinities. Most notable of these were *Gammarus annulatus, Byblis serrata, Melita nitida, Elasmopus levis,* and three species of Amphithoidae. He added further regional records of amphi-Atlantic species described by Krøyer, Bate (1862), Boeck (1871), and contemporary European systematists but failed to identify some of his new species (e.g., *Orchestia agilis, O. palustris*) with previously described forms. The remarkable and beautifully illustrated work on Norwegian amphipods by G. O. Sars (1895) included about thirty New England species, mostly of boreal affinities, and set quality standards for nineteenth-century publications that are difficult to surpass even today.

The most comprehensive previous treatment of New England Amphipoda was that of S. J. Holmes (1905). Among approximately seventy gammarid species, most at least partially illustrated, twelve species and two genera were newly described. Four of these species were subsequently synonymized. Eleven other species have since been reassigned different names. Holmes evidently overlooked several distinct new species in his material, especially in the Haustoriidae, that have only recently been described. Holmes' study included many of the species listed by Whiteaves (1901) from eastern Canada and by Paulmeier (1905) from New York City. Data on the distribution and breeding seasons of Cape Cod amphipods were compiled by Sumner, Osborn, and Cole (1911).

During the next fifteen years, emphasis on marine research in New England shifted away from systematics, relieved partially by the useful publication of Kunkel (1918) on the amphipods of Connecticut. During the thirty years following 1925, C. R. Shoemaker single-handedly accounted for most of the progress of regional amphipod systematics through his successive monographic treatments of Bateidae, *Photis, Unciola,* Pontoporeiinae, *Leptocheirus, Corophium,* and comprehensive regional studies such as the Amphipoda of the Cheticamp Expedition (1930). Also contributing usefully during this period were Blake (1929), Dunbar (1954) and a few European revisors such as Sexton and Spooner (1940), Dahl (1938), and Segerstråle (1947). The major European publications of Chevreux and Fage (1925), Schellenberg (1942), Bulycheva (1957), and Gurjanova (1951, 1962) gave updated descriptions, keys, and illustrations that included a significant fraction of the New England fauna.

The past decade or so has been marked by a renewed interest in taxonomic work by American students. Their studies have treated mostly infaunal amphipod groups such as the Phoxocephalidae (e.g., Barnard, 1960), Oedicerotidae (e.g., Mills, 1962), Ampeliscidae (e.g., Barnard 1958; Mills, 1964, 1967), Haustoriidae (e.g., Bousfield, 1965), Lysianassidae (e.g., Steele and Brunel, 1968), tubicolous and/or commensal groups such as Liljeborgiidae (e.g., Mills, 1963; Wigley, 1966), estuarine groups such as the Gammaridae (e.g., Bousfield, 1969), and the semiterrestrial Talitridae (e.g., Bousfield, 1958a). A number of regional faunistic studies (e.g., Bousfield and Leim, 1962; Brunel, 1970; Bousfield, 1960; Mills, 1964; Michael, 1972) added new distributional and ecological information. The species of marine and estuarine environments of the Cape Cod region have been usefully summarized by Yentsch et al. (1966). In essence, all these studies have revealed a wealth of new amphipod taxa, at both specific and generic levels, that was previously overlooked because of unrefined taxonomic methods and uncritical adherence to old systematic criteria. New species and new distributional records are still being published and further new taxonomic discoveries are predicted, particularly in parasitic, commensal, and deeper-water amphipods. However, the number of remaining regionally undescribed taxa is fast diminishing, and further systematic studies will probably emphasize problems of subspeciation, ecophenotypic variation, and infraspecific units. Two recent comprehensive papers by Barnard (1958, 1969) provide useful bases on which the New England amphipod fauna can be identified, catalogued, and related to the worldwide amphipod fauna.

EXTERNAL MORPHOLOGY

Amphipods are generally small to medium-sized peracaridan crustaceans. At maturity, the smallest known species are about 1 mm and the largest about 28 cm in body length. Peracaridans (to which amphipods, mysids, cumaceans, isopods, and tanaids belong) differ from other malacostracan crustaceans (such as euphausiids, decapods, stomatopods) in having only one true thoracic segment fused to the head (e.g., in having only one pair of maxillipeds), in having eggs borne in a thoracic brood pouch, and which hatch into a juvenile rather than a true larval stage. Amphipoda differ from other peracaridans in the following combination of characters: laterally compressed body (typical) with coxal plates directed ventrally; sessile eyes; seven pairs of uniramous peraeonal walking limbs; absence of a carapace; and lack of antennal squame. Amphipods are unique in possessing coxal gills at the inner bases of walking limbs; the first four pairs of walking limbs are directed forwards, the last three pairs backwards; and the abdomen bears three anterior pairs of multisegmented swimmerets (pleopods) and three posterior pairs of few-segmented uropods.

From the standpoint of morphology, life cycle, and low subordinal diversity, the amphipods may be considered the most modern and the most recently evolved of the higher crustacean ordinal groups. The amphipod fossil record is meager, extending back only to the Baltic Amber (Eocene), and all known fossils are identifiable as family Gammaridae (sens. lat.). This record contrasts with that of the Isopoda, known from the Triassic (or Permian),

5

and of the Decapoda, which is well represented from the Carboniferous on-
ward (see Hessler, personal communication).

Within recent Amphipoda, about 5500 species have been described to
date of which nearly 85 percent are Gammaridea, 9 percent Hyperiidea, 6
percent Caprellidea, and much less than 1 percent are Ingolfiellidea. The
Gammaridea may be considered to be undergoing evolutionary "explosion,"
as to some extent are the more ecologically restricted hyperiids and caprellids. Existing ingolfiellids are possibly highly modified relicts of a more primitive group that survive in interstitial or hypogean habitats—not yet fully ex-
ploited by more modern gammaridean counterparts.

The Gammaridea are distinguished from the other three amphipod
suborders by characters given in the "Key to Suborders of Amphipoda" (p.
40), and illustrated in Figures 1 and 2. Gammarideans are basically more
"primitive" or conservative than hyperiids or caprellids. The Hyperiidea
have probably evolved from a lysiannasidlike ancestor, whereas the Caprellidea are directly linked with the gammaridean family Podoceridae via the
family Caprogammaridae (McCain, 1968). Most gammarideans have well-developed coxal plates that are characteristically small in hyperiids and very
minute or lacking in the caprellids and ingolfiellids.

The detailed external morphology of a typical gammaridean is indicated
in Figures 1 and 3, and closely parallels that of Barnard (1969, Fig. 1). In
keeping with correct modern convention, the appendages of the peraeon
are numbered from 1 to 7, corresponding to the segment number; thus
"gnathopods" 1 and 2 are also peraeopods 1 and 2 and the posterior five
pairs of peraeopods are numbered consecutively from 3 to 7.

The head or cephalon bears two pairs of antennae, a pair of mandibles,
two pairs of biramous maxillae, and one pair of basally fused biramous maxillipeds (Figs. 4A, 9). The mandibles, first maxillae, and maxillipeds each
bear a palp (Fig. 5). The first antenna (or antennule) has a peduncle of three
segments and a multisegmented flagellum, and an accessory flagellum that
is usually short and may be vestigial or lacking (Fig. 4A). The second antenna has a five-segmented peduncle; the first segment is short, deep, and
may form the frontal or frontolateral part of the "face"; the second segment
is also very short and bears an outwardly directed "gland cone" through
which the antennal gland excretes. The third, fourth, and fifth peduncular
segments combined are often longer and usually stouter than the multisegmented flagellum. The eyes are paired, lateral, but may be located at the
base of the rostrum or on the anterior head lobes. The upper lip and epistome are immediately anterior to the mouth, the lower lip (possessing inner lobes in more specialized species) immediately posterior.

The peraeon (thoracic region, mesosome) is 7-segmented, each segment bearing a pair of 7-segmented uniramous limbs (Fig. 4C). The coxal
segments (limb segment 1) bear ventrally directed extensions or outer
plates that protect the gills and brood plates. Coxal plates 5 and 6 are bilobed to allow the corresponding limb some lateral action. Limb segment 2

(basis), especially of peraeopods 5–7, may be expanded posteriorly to form further platelike protection for the ventral body processes (Fig. 11C). Segment 3 is usually shortest. Distal segments are spinose and/or setose. Dactyl (seventh segment) may be simple and unguiform, or duct-bearing and lacking nail. The first two pairs of peraeopods (referred to as gnathopods, gamopods) are usually subchelate, often powerfully so in males where they function as precopulatory grasping appendages (Fig. 4B). The gnathopods may be small, weakly subchelate, or simple and may serve as accessory mouthparts in food gathering, burrowing, tube construction, sound production, or may be essentially tactile in function (Fig. 10).

The epimeral plates of the anterior three abdominal segments (pleon, pleosome) are true body outgrowths, as in the isopods, but are directed ventrally, not laterally (Fig. 12). The three pairs of pleopods are biramous, subequal in size, and the rami are multisegmented (Fig. 4D). Tube-dwelling fossorial, or pelagic species may have powerfully developed peduncles and rami (Figs. 13B–D). The peduncles of each pair of pleopods are hooked together by means of special small spines at the distal inner margins, ensuring a smooth metachronal beat simultaneously on both sides. The urosome bears two anterior pairs of uropods that are similar, with stiffly articulated, one-segmented rami (Fig. 4E). The uropods function in saltation, swimming, or burrowing. Uropod rami (especially uropod 2) may be modified for copulatory function (as yet undescribed) in Lysianassidae, Crangonycidae, Talitridae. The third uropods are more flexible and more loosely articulated; the rami are often foliaceous (as in pleopods); the outer ramus is usually two-segmented and longer than the inner (Fig. 4F). The telson, usually bilobed, is a terminal plate of the sixth abdominal segment, dorsal to the anus (Figs. 5G, 8I–N).

The coxal gills are paired leaf-shaped epipodites, directed inwardly, on thoracic limbs 2–7, but may be lacking on 7, and on 2 (as in Corophiidae). Coxal gills are essentially thin-walled respiratory sacs that may be stalked, foliate, or dendritic (Fig. 4B). They are particularly large and convoluted in terrestrial "leaf-mold" species to compensate for loss of surface respiration where the general body cuticle is hardened and "waxy" to prevent water loss. Fingerlike accessory gills and sternal gills may be found in some brackish- and fresh-water Gammaridae, Crangonycidae, Hyalellidae, and Pontoporeiinae. Gills and other permeable regions of the body may also be important in ionic regulation.

The penes are short, slender, paired papillae projecting ventromedially from the venter of peraeon 7. The marsupium is formed by spoon-shaped, marginally setose plates attached to the margin of coxae 2–5 (Fig. 4B). Oostegite primordia appear long before maturity, but only in the last molt stage are the margins setose. In some mature females the setae of the oostegites are lost after brooding (the "resting" stage) and are regained before the next brooding.

The eyes are usually paired, multifaceted, variously pigmented, and usually located laterally on the head (Figs. 6F–I). They are never stalked but may be located on or at the tips of the anterior head lobes or rostrum. Pigmented eyes are frequently lacking in subterranean, deep-sea, or burrowing species. The eyes are usually larger in males than in females of free-living species. Each eye may be divided into two parts (*Quadrivisio,* some Hyperiidea) or into three parts (many Ampeliscidae) wherein each anterior eye may be covered by a corneal lens (Fig. 6J).

Aesthetascs are clublike chemoreceptors on flagellar segments of the first antennae. Calceoli are club-, plate-, or cup-shaped organs, presumably sensitive to chemical or vibrational stimuli, attached to the anterior margins of flagellar and distal peduncular segments of the 2nd male antennae (Fig. 4A). Statocysts occur in the dorsal head region of some Gammaridae and Hyperiidea. Pressure receptors are presumed to be present in *Synchelidium,* an intertidal sand-burrowing genus. Sensory setae of various types occur generally over the body surface, epimeral plates and appendages.

INTERNAL MORPHOLOGY

The digestive system is composed of a foregut, midgut, and hindgut. The foregut consists of a cardiac or chewing stomach located in the anterior peraeon connected to the mouth by a short esophagus. From the anterior end of the midgut, directed posteriorly, are two (or four) elongate, tube-shaped digestive caeca and a short anteriorly directed caecum. The hindgut is a short straight tube emptying via the anus.

The excretory system consists of (1) paired caeca arising from the posterior end of the midgut and surrounded by the posterior aorta, and (2) antennal glands that open through the gland cone of segment 2 of antenna 2. The duct may be especially prominent in brackish- and fresh-water species but is very short or vestigial in terrestrial species.

In the circulatory system, the heart is an elongate segmented tube dorsally situated in the peraeon and surrounded by a pericardial sinus. Blood reaches the pericardial sinus from gills and appendages and passes through the ostia into the heart. From there the blood is pumped back into the body by an anterior and posterior aorta and paired lateral arteries. There is no subneural artery.

In reproductively mature males, the paired testes are cylindrical tubes extending from the third to the sixth peraeonal segments and continue as vas deferens into the seventh. They open through short muscular ejaculatory ducts at the tips of the paired penis papillae. The anterior pleopods do not function as gonopods, except possibly in certain fresh-water amphipods of the families Hyalellidae and Crangonycidae. In mature females, the ovarian tubes open on the venter of the fifth peraeonal segment into the marsupium.

BODY PROCESSES

Gammarideans are of two basic feeding types, macrophagous and microphagous. Macrophagous animals feed on coarse, solid food. These include the predaceous, carnivorous species (Eusiridae, some Gammaridae, most Crangonycidae) and the carrion or handicapped-animal feeders (many Lysianassidae) which tend to have a strong mandibular incisor and reduced molar process. Macrophages with well-developed mandibular molars include the macroalgae and detritus feeders (domicolous species), omnivores, and wood-boring species (*Chelura*).

Among the microphagous feeders are the sand-cleaning or sand-licking haustoriids that scrape microfloral and interstitial organisms from sand grain surfaces by means of finely setose mouthparts and anterior peraeopods. The true filter-feeders strain suspended organic materials, microalgae, etc., from the aquatic medium. Some groups (e.g., *Leptocheirus, Cheirocrates*) utilize filter setae of the anterior peraeopods from which food is transferred by maxilliped palps to the mouth. Others (e.g., Haustoriinae) utilize various mouthparts, especially maxilla 2, to set up a filter current that directs food particles onto mouthpart setae and thence toward the mouth. Deposit feeders, such as many Ampeliscidae and Corophiidae, scrape food material from surrounding surfaces by means of setose antennae. *Dulichia* (Podoceridae) extends its long setose antennae onto which floating detritus may settle and from there be transferred via mandibular palps and maxillipeds to the mouth.

Some amphipods are parasitic or semiparasitic and have rather specialized mouthparts characterized by unusual development of the mandibular incisor and reduction or loss of the molar process. Truly parasitic species include Lafystiidae and some Lysianassidae (e.g., *Opisa*) on fish, certain Hyalidae on turtles, Cyamidae on whales, and some Hyperiidea in large jellyfish, salps, etc. Many amphipods live commensally in or with other invertebrates and exhibit remarkable adaptive modifications of body appendages and mouthparts; e.g., Liljeborgiidae in tubes of polychaete worms, ghost shrimps, etc.; *Leucothoe* in sponges, tunicates; *Colomastix* in sponges; *Dulichia* on sea urchins; and *Polycheria* on colonial tunicates.

Digestion involves mastication of solid particulate matter in the cardiac stomach; enzymatic action in the midgut facilitates digestion of carbohydrates and mucilages. A weak cellulase is present in some talitrids. The cells of the midgut glands secrete digestive fluids, and those of the gut wall absorb digested food.

Respiration takes place partly through the body surface but mainly through the thin walls of the coxal gills. Action of the pleopods maintains a steady circulation of oxygenated water over gill surfaces. The current flows from anterior to posterior along the ventral channel formed by the body as a ceiling and the coxal plates of the peraeopods and epimeres of the abdomen as walls.

The reproductive process is diverse. The sexes are separate and sexual dimorphism is pronounced in less specialized or "primitive" amphipods. Males are typically larger than females; their larger, more strongly subchelate gnathopods enable the male to grasp and hold the female by the dorsal margins of the thoracic segments, prior to copulation. In certain essentially fossorial amphipods (e.g., Phoxocephalidae, Pontoporeiinae, Ampeliscidae), the mature male form emerges in abrupt metamorphosis from a femalelike penultimate stage. These males possess very elongate calceolate antennae, large eyes, very setose appendages, powerful pleopods, and a well-developed feathery or foliaceous uropod 3. There is no precopulatory "carrying" of the female, and mating takes place freely in the water column. In other burrowing forms (Haustoriinae), sexual dimorphism is suppressed, males are usually smaller than females, and copulation may take place in the substratum (mechanism as yet unknown). In other families (e.g., some Lysianassidae, Crangonycidae), one ramus of uropod 2 may be modified and probably assists in holding the female during actual copulation.

In *Gammarus,* copulation follows immediately after the female maturation molt. The male turns the female venter up. By bending his abdomen, the male pushes his three anterior folded pleopods several times (at half-second intervals) between the posterior brood plates of the marsupium. The sperm, issuing in long strings from the male genital papillae, is transported to the female gonopores. Egg deposition and fertilization takes place in the marsupium during mating. The marsupium is ventilated by action of the pleopods and stretching of the oostegites. There is no nutrient fluid. The duration of development and hatching of eggs varies considerably but depends largely on temperature. The young may first molt within the marsupium and leave a few days after hatching. Except in certain parasitic hyperiids, newly hatched amphipods more or less resemble the parents and do not undergo a "larval" metamorphosis. In arctic-boreal species of *Gammarus* (e.g., *G. oceanicus*), the female may enter a "resting" stage between main breeding periods in which the oostegites lack marginal setae.

HABITS

The body shape and type of appendages determine the locomotory ability and habits of gammaridean amphipods. Pelagic species are usually elongate, with short swollen trunk and large powerful pleosome and pleopods, and often possess flotation devices such as spinose body processes and internal oil globules. Species that crawl in vegetation, hide in crevices, or swim freely usually have an elongate, parallel-sided, subcylindrical body (in dorsal view) (Fig. 3B). Species that cling tightly to the substratum, including external parasites, have shortened, broad, flattened body segments and short powerful curved claws. Tube-dwelling or domicolous species may have cylindrical shallow bodies with powerfully broadened pleopods that can maintain a steady water current through the tube. Free-living burrowing species have fusiform or spindle-shaped bodies, narrowing to either end (in dorsal view), slender in some Phoxocephalidae and most Pontoporeiinae (Fig. 3A), but broad-fusiform and anteriorly truncate in a few Haustoriinae and some Pontoporeiinae (*Urothoe, Priscillina*) (Fig. 3C).

The body color of most epifaunal amphipods resembles the color of the substratum on which the animal normally rests; green, red, or brown in the case of algal dwellers; whitish or cream colored in the case of sand burrowers or subterranean species. Some amphipod species (i.e., in *Caprella*) are able to change color to match changing substratum coloration.

Swimming is accomplished by metachronal beats of the three pairs of pleopods. During their backward movement, the plumose setae of the pleopod rami spread out, increasing their surface. During forward movement water pressure pushes the setae backwards and the pleopod surface is reduced. The swimming speed may be jerkily increased by flexing or unflexing the abdomen and thereby bringing the powerful tail fan of uropods and telson into play. Most amphipods swim dorsal side up, but haustoriid amphipods swim with venter up.

In normal ambulatory motion, *Gammarus* "sidles" along the bottom with the legs of one side pushing while those of the other side are held tightly against the body. Occasionally the posterior uropods and spines of the urosome on one side hook into the bottom and push the body forward when the abdomen unflexes or beats backward. Many amphipods (e.g., *Crangonyx, Corophium, Orchestia, Haustorius*) walk dorsum upright, like most other crustaceans. *Corophium* hooks the tips of its stout second antennae into the substratum and flexes the powerful peduncle, pulling the body forward. Talitrids pull with peraeopods 3 and 4 and push with peraeopods 6 and 7, using the sixth (or seventh) for lateral support to prevent falling over. The fifth peraeopods are used for turning or backing. These animals can jump several times their body length (as can *Hyale*) by suddenly unflexing the abdomen and pushing hard with uropods and telson. Caprellidae (and some Podoceridae) crawl like a measuring worm, using gnathopods and peraeopods alternately and arching the trunk.

Amphipods burrow head first. In Talitridae, peraeopods 1, 3, and 4 pull, and the uropods push sand back to be thrown out of the burrow by sudden unflexing of the abdomen and spinose tail fan. Haustoriids literally swim through the sand by means of a hydraulic tunneling action. The broad body and wide deep peraeopods form a ventral cylindrical tube in which a current is set up by the powerful pleopods. Sand grains are passed rapidly rearward in this current and prevented from returning by the posterior "plug" formed by the urosome and appendages.

Tube-dwelling amphipods, such as Ampeliscidae and some Corophiidae, may construct solid tubes in the sand whose inner walls are solidified by glandular secretions from the third and fourth peraeopods. The tubes of *Ampelisca* are shallow thin-walled sacs, open at the upper end from which the head and antenna are extruded. The tube of *Corophium volutator* is U-shaped, normally extends 4–8 cm into the substratum, but in winter may extend downwards about 20 cm. Cement is secreted by one-celled glands of segments 2, 3, 4 of peraeopods 3 and 4, glands which open at the tip of the dactyl (Fig. 11F). *Microdeutopus* constructs a tube between algae, open at both ends. *Ampithoe* cements together bits of algae and detritus, to form an open-ended tube or nest in which it can rapidly turn around. *Cerapus* and *Siphonoecetes* construct portable tubes, those of the former genus being square-sided. *Leptocheirus* constructs a tube with a domed ceiling by cementing mud and algal particles. *Unciola* occupies and modifies burrows of other small animals in sand. *Microprotopus* burrows in sand with peraeopods, then turns in all directions, spinning sand grains together with third and fourth peraeopods to form a sand tube.

Directional orientation may be a complicated process. In beach hoppers (Talitridae), the sun, moon, and distribution of polarized light serve as a compass. Certain sand-hoppers can compensate for movement of the sun during the course of the day. An animal trying to reach the tide line early in the morning would take a different angle to the sun from that during the afternoon. Compensation depends upon an internal clock, independent of sensory perception, and is temperature-influenced.

ECOLOGY

Amphipod crustaceans occur in virtually all marine habitats and many permanent fresh-water habitats. Most amphipods occur in the sea, from the supratidal level to the greatest recorded depths. Free-living gammaridean amphipods may be free-swimming in coastal waters, sometimes termed tychoplanktonic. These include *Calliopius, Pontogeneia, Gammarellus, Gammarus annulatus, G. daiberi,* the pelagic terminal adult mating stages of many infaunal species, and the young or juvenile (dispersal) stages of nearly all species, particularly infaunal or tubicolous amphipods such as *Jassa,* aorids, and corophiids. Epifaunal species such as *Dexamine, Hyale, Gammarus, Melita, Batea* may clamber about or cling to algae, rocks, eelgrass, detritus, etc. Free-living infaunal species burrow into soft unstable substrata, usually staying within a few centimeters of the surface, or burrow into wood (e.g., *Chelura*). Tubicolous or domicolous amphipods construct nests or tubes on or in the substratum. In dense concentrations, these animals become important fouling organisms on ships and aids-to-navigation. Infaunal tube-builders (Ampeliscidae, *Leptocheirus, Unciola*) construct tubes in the substratum with one end open at the surface. A few very small, elongate, vermiform amphipods (some *Melita,* Crangonycidae, Bogidiellidae, Ingolfiellidea) are interstitial. Several family groups are commensal in the tubes of worms and crustaceans (e.g., Liljeborgiidae), and may be host-specific. Interspecies associations have been noted between hydroids such as *Tubularia* and members of the Pleustidae, Bateidae, Stenothoidae, and Caprellidea. At least one member of the Pleustidae is known to mimic the shape and coloration of small gastropods.

New England shallow-water amphipods are subjected to wide seasonal and sometimes wide daily fluctuations in temperature. Most species are therefore mainly eurythermal. Boreal and subarctic species reaching their southern limit in the Gulf of Maine region are reproductively cold-stenothermal and breed only in the coldest months of the year. On the other hand, warmer-temperate and Virginian species that reach northern limits at Cape Cod or in isolated summer-warm pockets to the north are reproductively warm-stenothermal and propagate only during the summer months. A few species are both vegetatively and reproductively stenothermal and occur only in oceanic areas with relatively uniform year-round temperatures (e.g., *Orchestia gammarella, Corophium volutator, Gammarellus angulosus*).

A large number of New England amphipods are strictly marine and seldom occur in salinities of less than about 28 o/oo. Such characteristically stenohaline groups are the Lysianassidae, Pleustidae, Stenothoidae, Phoxocephalidae, Atylidae, Podoceridae, as well as the Hyperiidea and Caprellidea, but a few brackish-water exceptions occur in nearly all these groups. Many of the shore-dwelling and intertidal species are widely euryhaline or otherwise can tolerate considerable lowering of salinities for extended periods, as in tidal pools exposed to rain storms, or in fresh-water seeps, or in estuaries. Examples are most species of *Gammarus* (sens. lat.), *Orchestia*, several *Corophium, Microdeutopus gryllotalpa, Ampelisca abdita, Cymadusa compta,* and *Melita nitida.* Obligate oligohaline-brackish amphipods are rarely found in full marine salinities—for example, *Gammarus tigrinus, G. daiberi, Leptocheirus plumulosus,* and *Corophium lacustre.*

Most marine amphipods occur subtidally or penetrate landward only in the infralittoral zone. Their toleration of desiccation and of extremes of temperature, salinity, light, and other physical variables is relatively low. In less than two dozen (about 15 percent) of the New England shallow-water amphipod species is the bulk of the population found between the tide marks in summer (Table 1). In winter, offshore migration into deeper water may take place. Paradoxically, this is a relatively high percentage for comparable coastal marine areas in other parts of the world. Compared to the American-Pacific coast, the climate of New England is relatively severe with wide extremes of temperature in summer and winter, especially in rocky shore areas. Amphipods are apparently better adapted to such harsh physical conditions than are other macrocrustaceans. Amphipods are the dominant intertidal macrocrustaceans in New England, outnumbering the isopod species (including the terrestrial Oniscoidea) and the crabs of which only *Carcinus maenas* and some xanthids might be considered significantly intertidal in this region. In winter-mild temperate coasts, or in the tropics, the reverse domination is observed.

Obligate intertidal amphipods of exposed or stony rocky shores include

Orchestia gammarella and *O. palustris, Marinogammarus* spp., *Gammarus duebeni, G. oceanicus,* and *Hyale nilssoni* (Table 1); on sandy shores *Talorchestia* spp., *Haustorius canadensis, Neohaustorius* spp., and in protected bays, estuaries, and salt marshes *Orchestia grillus, O. uhleri, Gammarus palustris, G. mucronatus, Hyale plumulosa,* and *Corophium volutator.* In all these species part of the population is subtidal, if only during the dispersal of juveniles in the water column, or may occur in other habitats at the same vertical level. Many intertidal species such as *Orchestia platensis, Gammarus oceanicus,* and *G. lawrencianus* are sufficiently euryokous or eurytopic that they can be found in significant numbers in almost all habitats categorized in Table 1. The epifaunal species tend to be more eurytopic than the infaunal species, occurring on or in a much wider range of substrata and wider range of salinities.

Strictly fresh-water amphipods include the Crangonycidae, Hyalellidae, and *Gammarus fasciatus,* but these may be washed into estuaries or even marine habitats during heavy runoff and lead to anomalous situations. Thus, Holmes' *"Melita parvimana"* (1905) was probably a male specimen of *Crangonyx pseudogracilis* that had been washed into Newport Harbor, R.I. Its identity has not been verified in the apparent absence of the type specimen.

TABLE 1   **ECOLOGICAL "PREFERENCE" OF SOME COMMON NEW ENGLAND INTERTIDAL GAMMARIDEAN AMPHIPODA**

MAINLY OUTER COAST (surf- or wave-exposed)			
Rock or Stone Substratum with Fucoids		**Sandy Substratum**	
MAINLY INTERTIDAL	INFRALITTORAL AND SUBTIDAL	MAINLY INTERTIDAL	INFRALITTORAL AND SUBTIDAL
Orchestia gammarella	Ampithoe rubricata	Talorchestia megalophthalma	Gammarus annulatus
O. platensis	Jassa falcata	Haustorius canadensis	Amphiporeia lawrenciana
Hyale nilssoni	Ischyrocerus anguipes	Neohaustorius schmitzi	A. virginiana
Gammarus duebeni	Dexamine thea		Parahaustorius longimerus
G. oceanicus	Pleusymtes glaber		Acanthohaustorius spinosus
Marinogammarus finmarchicus	Gammarellus angulosus		Psammonyx nobilis
M. obtusatus	Calliopius laeviusculus		Trichophoxus epistomus
M. stoerensis	Pontogeneia inermis		

MAINLY INNER EMBAYED COASTAL OR ESTUARINE (surf-protected)		
Sandy or Sandy Mud	Mainly Mud and/ or Salt Marsh	Miscellaneous (stones, eelgrass, fucoids)
MAINLY INFRALITTORAL	MAINLY INTERTIDAL	INFRALITTORAL AND SUBTIDAL
Talorchestia longicornis	Gammarus mucronatus	Hyale plumulosa
Gammarus lawrencianus	G. palustris	Elasmopus laevis
Monoculodes edwardsi	Orchestia grillus	Lysianopsis alba
Neohaustorius biarticulatus	O. uhleri	Chelura terebrans
Acanthohaustorius millsi	G. setosus	Corophium acutum
Protohaustorius deichmannae	G. tigrinus	C. acherusicum
Phoxocephalus holbolli	Melita nitida	Microdeutopus gryllotalpa
Ampelisca abdita	Corophium volutator	Ampithoe longimana
A. vadorum	C. insidiosum	Cymadusa compta
	C. lacustre	Stenothoe minuta
	Ampithoe valida	
	Leptocheirus plumulosus	

ZOOGEOGRAPHIC CONSIDERATIONS

The New England coastal marine region embraces two major life zones that overlap rather sharply in the Cape Cod region. To the north is the subarctic-boreal or cold-temperate zone, where summer surface temperatures usually do not exceed about 15°C except in isolated warm pockets such as Minas Basin and Northumberland Strait. To the south is the temperate, warm-temperate, or Virginian life zone where summer surface temperatures are consistently 18°C or higher. Since rocky, sandy, and muddy habitats are present on both sides of Cape Cod, the amphipod fauna exhibits a remarkable range of habitat specificity and zoogeographical affinity. Thus the 120 (approx.) shallow-water marine amphipod species found within a 75-mile radius of Cape Cod greatly exceed the number in equivalent regions either to the north or to the south anywhere on the American Atlantic coast. Of these 120 species, more than two-thirds are "southern" species limited northward, and less than one-third are "northern" species limited southward. We may note that of approximately 85 "southern" species, 46 do not extend north of Cape Cod Bay, 11 other species "cut out" in the Casco Bay region, and only 28 species reach Canada. On the other hand, of 34 "northern" Cape Cod species, 15 do not extend south of Cape Cod Bay, an additional 13 terminate in the Long Island Sound region, and the remaining 6 extend about to Delaware Bay.

In effect, all the northern species reach a southern limit that coincides with the southern limit of igneous bedrock on the American Atlantic coast at Connecticut. All rocky-shore intertidal species of amphipods in the New England region are amphi-Atlantic in distribution; none are American-endemic. Pleistocene continental glaciation extended south of existing rocky coast. Glacial action would probably have eradicated pre-Pleistocene endemic rocky-shore species by severely altering the substratum and the thermal and salinity regimes along the exposed continental margins. On the other hand, the obligate-estuarine and sand-burrowing faunas are almost entirely American-endemic. The 85 "southern" species (above), are almost entirely estuarine or sand-burrowing, and only 9 (about 10 percent) are amphi-Atlantic. This anomaly is attributable mainly to the unique geological history of the North Atlantic ocean. During the past 100 million years, as a result of continental drift, this basin has been formed as a gradually widening "estuary." The American Atlantic coast south of Connecticut has throughout been a continuous "sandbar" estuarine coastline. Herein major new taxa have radiated (e.g., Haustoriinae), essentially cut off from eastern Atlantic counterparts by immense distance, and by temperature barriers to the north and south. The arctic-boreal and cold-temperate species (except for *G. lawrencianus*) are almost entirely amphi-Atlantic, whether open-coast, rocky- or sand-shore, or estuarine, and have a natural stepping-stone contact with European populations via Labrador, Greenland, Iceland, Faroes, and the British Isles.

The postglacial sequence of events that resulted in isolated pockets of "warm-water" marine invertebrates in the southwestern region of the Gulf of St. Lawrence, and pockets of "subarctic" invertebrates in the northern region of the Gulf of Maine, has been described in detail by Bousfield and Thomas (1973). At the close of Pleistocene glaciation, continental ice covered the entire coastal region, except for large archipelagos near the continental slope exposed by lowered sea levels. In the subsequent "hypsithermal" or warm period (10,000–6000 b.p.) the climate was significantly warmer than at present, the ice retreated from the entire region, and the warm shallow waters of the Scotian archipelago provided northward dispersal routes into the Gulf of St. Lawrence. The subsequent deepening and cooling of the coastal seas resulting from melting continental glaciers and cooling climate obliterated the coastal archipelagos and their warm faunas. Remnants of the warm fauna were isolated in the summer-warm southwestern Gulf of St. Lawrence and Minas Basin regions, and elements of the cold-water fauna dispersed southward once again into the summer-cold northern Gulf of Maine.

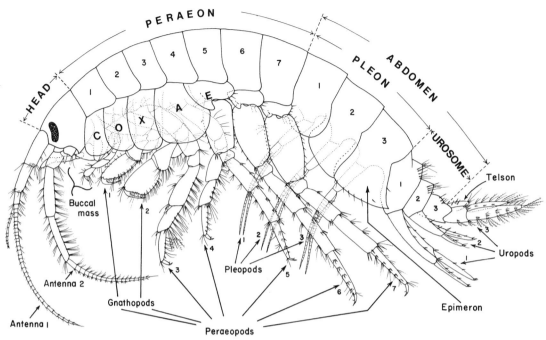

Basic gammaridean amphipod (lateral view)

Figure 1

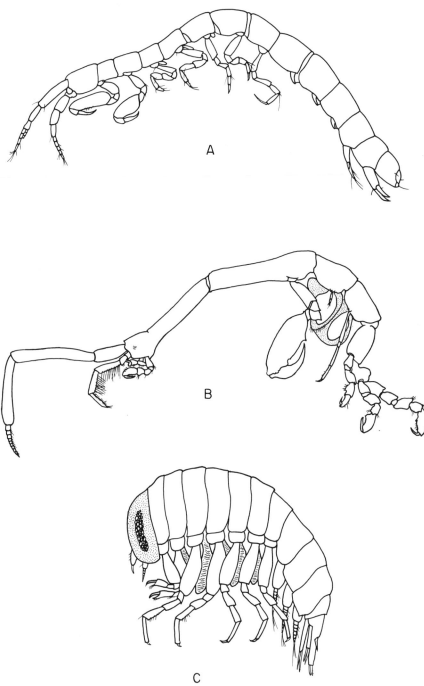

Figure 2   Other amphipod suborders (lateral view)
A—Ingolfiellidea.   B—Caprellidea.   C—Hyperiidea.

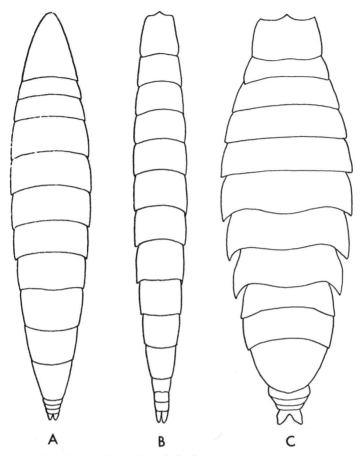

Basic gammaridean body outlines (dorsal view)
A—Fusiform.   B—Subcylindrical.   C—Broad (truncate) fusiform.

Figure 3

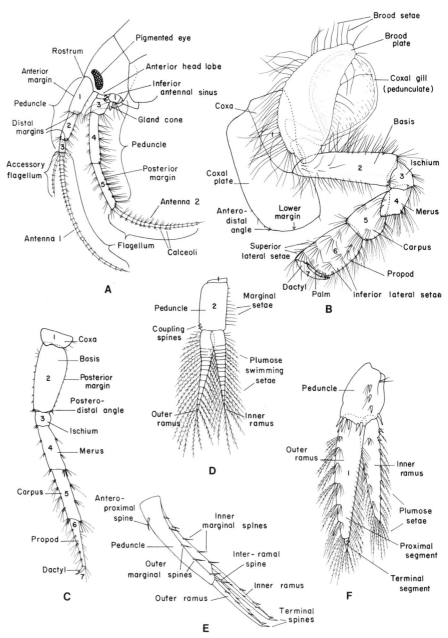

Figure 4   **Basic gammaridean body appendages (not to scale)**
            A—Head region.   **B**—Gnathopod 2.   **C**—Peraeopod 7.   **D**—Pleopod.   **E**—
            Uropod 1.   **F**—Uropod 3.

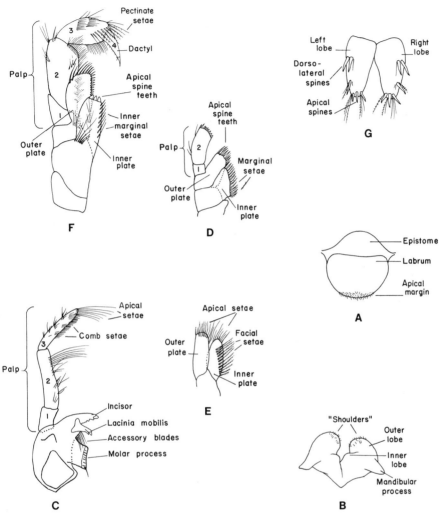

Basic gammaridean mouthparts and telson
A—Upper lip.   B—Lower lip.   C—Mandible.   D—Maxilla 1.   E—Maxilla 2.   F
—Maxilliped.   G—Telson.

Figure 5

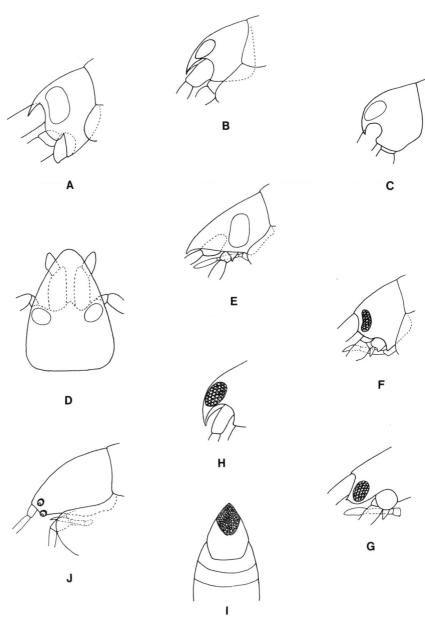

Figure 6   **Characteristics of rostrum (A–E), eye position and shape (F–I)**
A—Acute.   **B**—Falcate.   **C**—Decurved.   **D**—Hooded (dorsal view).   **E**—Hooded
(lateral view).   **F**—Lateral, reniform.   **G**—Anterior, ovate.   **H**—Rostral, fused
(lateral view).   **I**—Rostral, fused (dorsal view).   **J**—Divided with corneal lens.

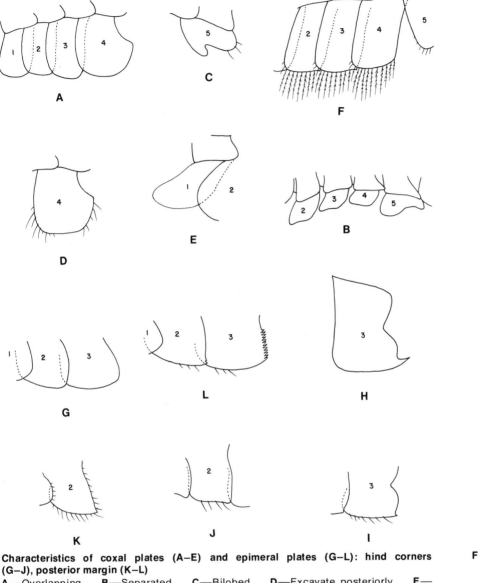

Characteristics of coxal plates (A–E) and epimeral plates (G–L): hind corners (G–J), posterior margin (K–L)

A—Overlapping.   B—Separated.   C—Bilobed.   D—Excavate posteriorly.   E—Anterodistally expanded.   F—Marginally setose.   G—Hind corners, rounded.   H—Toothed, produced.   I—Mucronate.   J—Subquadrate.   K—Setose.   L—Serrate.   H–I—Sinuous.

Figure 7

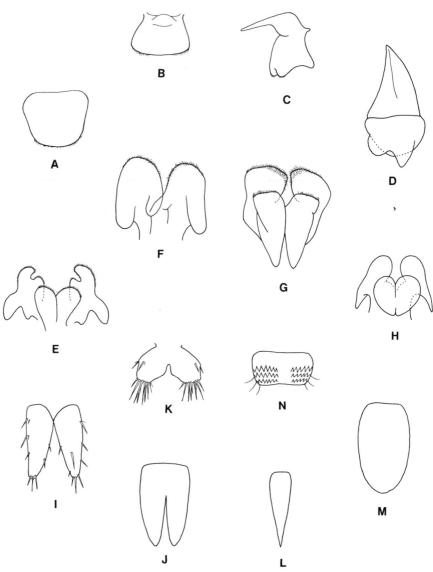

Figure 8   **Characteristics of upper lip (A–D), lower lip (E–H), and telson (I–N)**
A—rounded.   **B**—Broad.   **C**—Bilobed, epistome produced (lateral view).   **D**—
Asymmetrically bilobed.   **E**—"Shoulders" notched.   **F**—Inner lobes lacking.
G—Mandibular processes lacking.   **H**—Mandibular processes elongate.   **I**—
Cleft to base.   **J**—Deeply cleft.   **K**—Broad, notched.   **L**—Entire, acute.   **M**—
Entire, linguiform.   **N**—Dorsally uncinate.

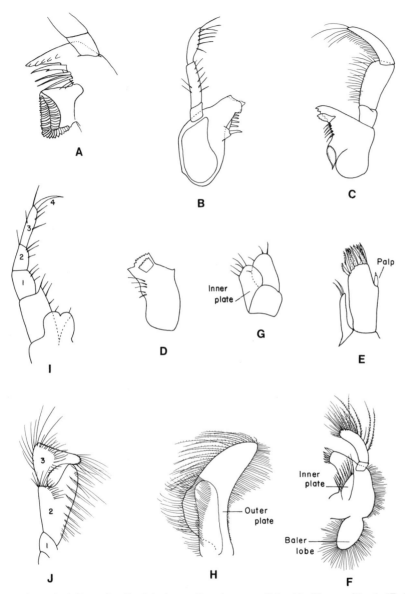

Some characteristics of articulated mouthparts: mandible (A–D), maxilla 1 (E–F), **Figure 9**
maxilla 2 (G–H), maxilliped (I–J)
A—Molar strong, triturating.   **B**—Molar vestigial.   **C**—Palp strong, falcate.   **D**—
Palp lacking.   **E**—Palp vestigial.   **F**—Basal baler lobe large.   **G**—Plates small,
weakly armed.   **H**—Outer plate large, lunate.   **I**—Inner plates fused, palp dactylate.   **J**—Palp geniculate.

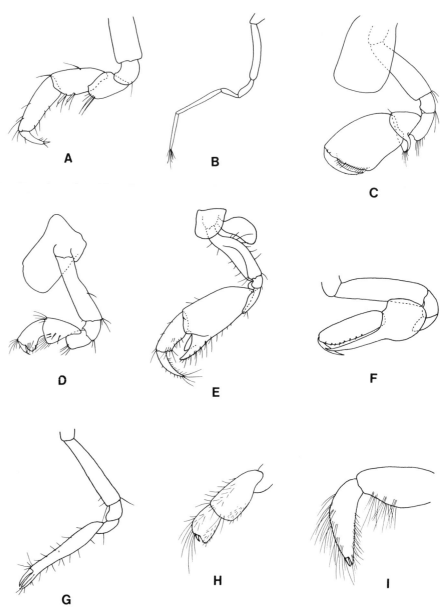

Figure 10   Characteristics of gnathopods
A—Simple, normal.   B—Simple, filiform.   C—Subchelate, powerful.   D—
Subchelate, weak.   E—Complexly subchelate.   F—Carpochelate.   G—Chelate.
H—Minutely subchelate.   I—Minutely chelate.

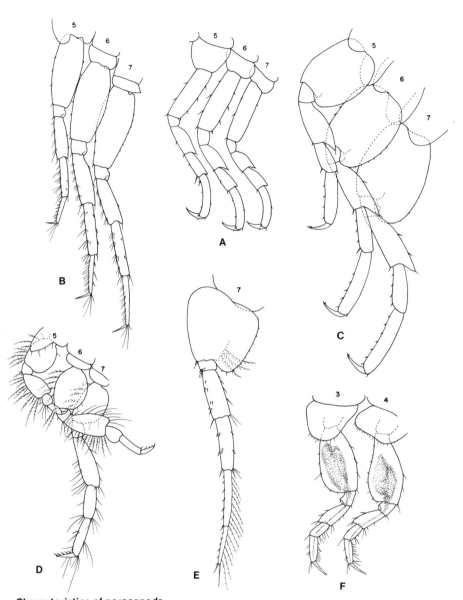

Characteristics of peraeopods
A—Basis linear.   **B**—Basis slightly expanded.   **C**—Basis broadly expanded.   **D**
—Posteriorly directed (5, 6).   **E**—Elongate, linear distal segments.   **F**—Basis glandular.

Figure 11

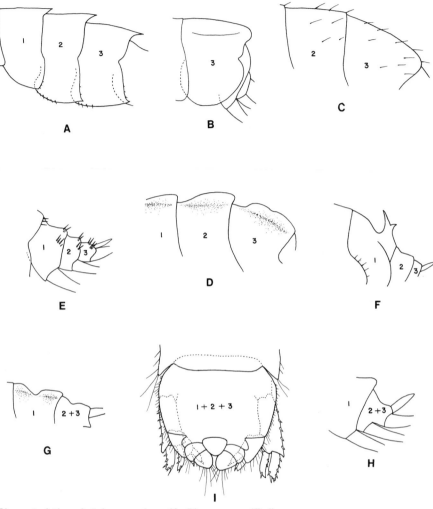

Figure 12   Characteristics of abdomen: pleon (A–D), urosome (E–I)
A—Mucronate.   B—Posteriorly recurved overhanging urosome.   C—Setose, hirsute.   D—Carinate.   E—Spinose.   F—Toothed.   G—Bicarinate.   H—Posteriorly carinate.   I—Segments fused (dorsal view).

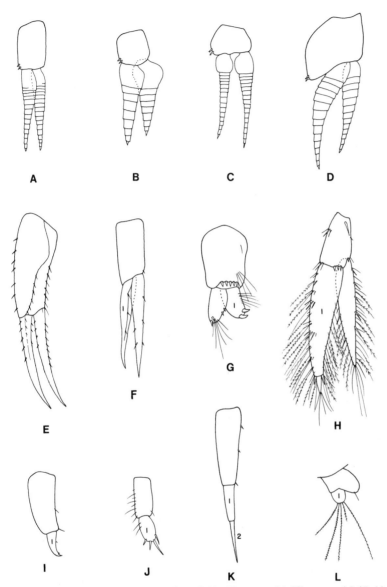

Characteristics of pleopods (setae not shown) (A–D), uropod 1 (E), uropod 3 (F–L).   **Figure 13**
A—Normal.   B—Rami proximally expanded.   C—Peduncle broad.   D—Peduncle
expanded medially.   E—Rami falciform.   F—Rami lanceolate.   G—Outer ramus
with hook spines.   H—Rami foliaceous.   I—Uniramous, uncinate.   J—Uniramous,
spatulate.   K—Uniramous, styliform.   L—Uniramous, peduncle medially lobate.

$\mathcal{H}$OW TO COLLECT AND PREPARE AMPHIPODS FOR STUDY

COLLECTION AND PRESERVATION

The secret of successful amphipod collecting is knowing the habitats of the species. The most probable habitats for common intertidal and shallow-water species, in which collecting success is virtually guaranteed, are summarized in Table 1. The collector must also know the proper collecting methods according to the habitat.

For work along the shore, the collector should wear a pair of hip-length waders over which may be pulled a pair of rubberized pants to protect clothing against wave splash. Collecting gear should include a plastic bucket, a set of large plastic vials containing preservatives, a pair of curve-tipped forceps, and fine-meshed hand nets. Intertidal species may be obtained by turning over rocks, parting the attached algae, and transferring the exposed amphipods directly by hand or forceps to a vial. Drift-line debris, fucoids, eelgrass, etc., can be shaken over a cleared spot or over a large bucket, and the released amphipods picked up individually. Beach fleas (*Orchestia*), entrapped in high-water drift debris that is being dunked in sea water, will often swim out freely and climb up onto the collector's legs for easy "picking." A small kitchen strainer can be used to sweep through algae in tidal pools, up into grassy tidal marsh banks, or set in seeps and small tidal streams to catch material and cryptozoic species dislodged up-current.

For work in hip-boot depth along shore or along steep rocky shores, a long-handled dip net with replaceable nylon mesh bag will prove more efficient than the kitchen strainer. If a round-mouth net is used, a steel cutting bar may be attached to the outer rim. This net can be swept through algae, kelp, or eelgrass, especially from steep rocky ledges, and can be used to cut algae, sponges, hydroids, and encrusting materials from rock faces, automatically entrapping the attached amphipods in the nylon bag. The amphipods and debris can be deposited directly in a pail partially filled with water to be concentrated, preserved, and sorted later. Net sweeps through kelp and fucoids may capture free-swimming, clinging, and nest-building species. Net sweeps through *Spartina* grasses of tidal marshes, flooded at high tide, may yield *Orchestia uhleri* and *O. grillus* that have climbed up onto the stems.

In working on sandy bottom the collector may push the net forward at an angle through the sand until partly filled. The net is then lifted to the surface where the sand is easily washed through the bag, leaving the amphipods. The collector may also pull the strainer along the bottom, simultaneously kicking sand into the mouth of the net, until several "bootfuls" have been strained. Well-oxygenated "ripple" sand is usually rich in haustoriids, phoxocephalids, and oedicerotids; grey or black "reducing" sand usually contains little amphipod life. A small, wooden-sided, wire-screen strainer is stronger and better for sampling amphipods that lie buried in the top few centimeters of intertidal beach sand. Several shovelfuls can be strained at one time, and thereby provide a semiquantitative picture of the distribution and abundance of species between tide marks. Beach hoppers (*Talorchestia*) can be dug out of the sand at the drift-line level or found under nearby boards and damp wave-cast algae. Shallow, steep-walled pans can be set flush with the ground surface at various levels on the beach and left out overnight. The pans should contain an inch-deep layer of 10 percent formalin into which the nocturnal beach hoppers may fall by chance and be unable to escape. A shovel is useful for digging out shallow burrows and tubes of mud and sandy mud-dwelling amphipods (e.g., Ampeliscidae, *Corophium volutator*) or the much deeper burrows of polychaete worms, gephyreans, callianassid shrimps, etc., in which some animals are commensal or may be temporarily sheltering (e.g., Liljeborgiidae, *Melita nitida*).

For subtidal and deeper-water sampling, ship-operated benthic dredges, sleds, trawls, and various types of grab samplers may be utilized.* The first three collecting devices are basically wide-mouth rectangular frames behind which are attached fine-meshed collecting bags. As the dredge with frame vertical is towed slowly across the bottom at the end of a long wire or rope, the lower bar of the frame cuts into the bottom and dislodges into the bag whatever material may be on or in the surface layer of substratum. The loaded dredge is winched up onto the deck and dumped onto a series of large sieves. The substratum is then washed down, leaving

---

* See N. A. Holme, Methods of sampling the benthos, *Adv. Mar. Biol., 2* (1966): 171–260, for detailed review of benthic samplers.

the animals on the screens. The animals and residue may be bottled, preserved, labeled, and stored for later sorting. Dredges usually collect a mixture of epifaunal and infaunal species. Grab samplers such as the Ekman, Van Veen, and Smith-McIntyre employ various jaw mechanisms to take a "bite" of fixed size from the substratum. These samplers work best in pure sediments wherein the jaws can close tightly and prevent washing out of the contents as the bucket is being winched to the surface. Grab samplers collect mainly infaunal species.

Epibiotic or "hitchhiking" species may be strained from floating kelp or Sargassum weed. Pelagic species may be collected by coarse-meshed plankton nets towed at various depths or held in tidal currents from a dock or fixed station. A night light held over the dockside will invariably attract nocturnal amphipods and other free-swimming animals to the surface, where they can be readily dip-netted. By collecting during fall, winter, and early spring, the reader may obtain different assemblages of amphipods or the sexually mature stages of winter-breeding, arctic-boreal species. In summer, the proportion of juveniles and subadult stages of both planktonic and benthic amphipods is exceptionally high.

The avid collector should be on the lookout for unusual habitats that might harbor amphipods. Thus the interior cavities of tunicates, sponges, coelenterates, the bases of sea anemones, and the holdfasts of kelps frequently contain rich populations of commensal, parasitic, or iniquilinous species. *Chelura* bores into wooden stakes and pilings previously riddled by gribbles and teredos. Tubicolous animals (*Jassa, Erichthonius, Corophium*) may occur in dense mats under the surface of ships, buoys, and on wharf pilings, and may be collected at low tide or when these objects are hauled ashore for periodic cleaning. The stomach contents of fishes, sea birds, and marine mammals often yield remarkable amphipod species, albeit not in the best of condition, that might not be obtainable in any other way. Thus the California grey whale is known to feed on essentially *infaunal* amphipods (Lysianassidae, etc.), which it dislodges by roiling up the bottom and sieving on its baleen plates. A few species (*Lafystius, Opisa*) are ectoparasites on cod and other fishes; some (*Hyale* spp.) may occur on marine turtles; others (*Cyamidae*) occur on whales, and many small species (Stenothoidae, Photidae, etc.) may be found on the carapaces or around the eye depressions of large decapod crustaceans that have not recently molted. Fresh-water amphipods are essentially epifaunal, although *Hyalella* and *Crangonyx* may burrow into shallow sand and bottom ooze.

Marine species are best preserved in a 5 percent formalin in seawater. Body colors fade relatively slowly in formalin, and penetration is more rapid and more complete than in alcohols. Preservatives should be added as soon after collecting as possible to prevent loss or damage from specimens' feeding upon one another or threshing about in the container. Formalin-preserved specimens should later be transferred to 70 percent ethanol or to 40 percent isopropyl alcohol, fluids that are less effective penetrants but are much easier to work with in the laboratory.

PREPARATION OF ANIMALS FOR STUDY

Amphipods should be sorted from the field containers and field matrix as soon as possible. The bulk field material is placed in a shallow white-bottomed enamel pan. Macrospecimens (larger than 5 mm) can be picked out by eye, using curve-tipped forceps for easy handling. A low-power stereobinocular microscope with large visual field will facilitate detection of the very small amphipods and immature instars. Handpicking of individual specimens can be a tedious task, particularly where several hundred specimens occur in fine algal or eelgrass detritus. The process can be speeded up by decanting most of the tray water into a beaker, allowing amphipods at the surface of the residue to partially dry. If fresh water is slowly and carefully added to one end of the pan, surface tension will float up the drying amphipods. Special fluids of heavier density (sugar solutions) may achieve the same effect. The floating specimens can readily be decanted into a beaker and depressed to the bottom. The amphipod concentrate can then be placed in a vial, labeled, stoppered, and stored for future use.

Identification of amphipod specimens may require partial or complete dissection of the body appendages, particularly of the mouthparts. These appendages can best be studied, under relatively high-power magnification, by preparation of temporary or permanent slide mounts. An excellent account of the proper dissection of specimens and preparation of glass slide mounts is given in Barnard (1969, pp. 505–510). Fine forceps, fine needles, and a sharp scalpel are required. The body of the specimen, covered by twice its depth of preserving fluid in a shallow dish, is held by one pair of forceps or speared by a strong needle. The appendages of one side are removed by gently pulling on the coxal segment (or peduncle, as the case may be) with another pair of forceps. The peraeopods usually come away easily (with muscular attachments) but may have to be cut away with the scalpel. If dissected carefully, the coxal gills and brood plates (where present) will remain attached to the detached limb. The larger body appendages (antennae, gnathopods, peraeopods, uropods, pleopods) are ideally mounted on separate slides. A depression slide should be used if appendages have appreciable thickness, otherwise downward pressure of the cover slip may distort the appendages. Or, as the mounting medium dries (over a long period), air bubbles may leak in around the edges and obscure or damage the appendages. Glycerin or 7 percent ethylene glycol is used as a temporary mounting medium. Parts may be transferred directly into the latter from 70 percent ethanol. Other viscous fluids with clearing properties such as CMC-10 (clear or dye-stained), Euparol, and Canada Balsam are useful in preparing semipermanent and permanent slide mounts. CMC-10 medium (and derivatives) permit direct transfer of the appendages from water and most preservatives, whereas Canada Balsam requires dehydration through laborious transfer through alcohols and xylols to the pure resin mount. The former medium is semipermanent, but will make a reasonably permanent slide mount if the specimen is small and the cover slip is carefully ringed. Old, partially dried, or damaged CMC-10 mounts can be redissolved in water over a 48-hour period, the appendages picked out, and remounted in fresh medium. The addition of a small amount of stain such as methylene blue to the mounting medium facilitates the location of very small appendages, especially mouthparts and telson, on the slide.

It is often advantageous to mount body appendages under one cover slip for easier study and efficiency of storage. This practice is recommended only for medium- to small-sized appendages. Some practice and experience are required to arrange limbs in a counterclockwise ring around the drop of mounting fluid so that they will retain their relative positions when the cover slip is set in place and the mounting medium is pressed out to the edges. The mouthparts may be dissected out individually from the buccal mass in large specimens and dropped separately into the mounting medium. In small specimens it is advisable to remove the buccal mass as a unit and dissect it in the drop of mounting medium in order to minimize loss of pieces such as palps. The mouthparts are arranged in any suitable way and the cover slip is set as above. Gills, brood plates, epimeral plates, telson, and other parts, may be mounted separately, or with other appendages, depending on their thickness and convenience of the operation. In sexually dimorphic species, the appendages of ♂ and ♀ specimens should be mounted on separate slides. The body, with appendages on one side intact, may be mounted in a depression slide or placed in a separate labeled and stoppered vial for future reference. All slide mounts should be accompanied by a permanent label, glued to the righthand side of the slide, on which is typed or written in permanent ink the species name, locality data, and reference number.

This guidebook is intended as a comprehensive systematic treatment and as a means of rapid and reliable identification of the amphipod fauna of the New England region. To serve both purposes, every shallow-water species has been illustrated, mostly in full, all have been diagnosed in detail, and all are keyed. Explanation of technical terminology, unavoidable in a topic of this complexity, is provided in the glossary at the end of the book and in Figs. 1 through 13. The "Guide to Pronunciation and Root Meanings of Scientific Names" attempts to standardize pronunciation of Latin names of species, since common English names are not in use for any of them. Basic information on the general morphology and biology of amphipods is provided in the preceding sections.

The whole-body illustrations are of mature male or female animals, viewed from the left side, anterior end to the left. Dorsal views of body regions showing diagnostic features are provided for some species groups (e.g., Plates XXXI–XXXII, LXI–LXVIII). Illustrations of species are grouped by genus and family, closely equivalent to their order in the text. Selected mouthparts and other appendages, not visible in lateral view or without dissection, are illustrated at higher magnification immediately adjacent to the whole animal drawing. The actual size (average maximum at maturity) of the animal is indicated by the short horizontal line above each drawing. The length of the animal is measured from the tip of the rostrum to the base of the telson, with body in a straightened or unflexed position. The width (thickness) is the greatest distance between lateral body margins when viewed from above (see Fig. 4). Depth is the distance from dorsal to ventral margin of a segment when viewed laterally.

The keys are intended to provide a reliable means of species identification that is independent of sex or instar. Where possible only the most conspicuous of reliable taxonomic characters are utilized in order to facilitate rapid identification of large lots of mixed species. Characters that require dissection of the specimen or preparation of slide mounts of appendages, or sex-linked characters, are used only as alternatives or where other reliable characters have not been found.

The key is a combination of phylogenetic (or "natural") and artificial characters. The former are utilized mainly in keys to families and genera where it is desirable to show group similarities as well as differences. Artificial characters are utilized mainly in separating species, where differences are the paramount consideration.

Specimens in hand, properly prepared for study, may be identified in one of two ways: by quick visual comparison with illustrations, or by use of the keys in combination with illustrations. The first method is advocated in cases where the animal is of striking morphology and is unlikely to be mistaken for any other regional species (e.g., *Rhachotropis, Chelura*) or where its ecology or habits are so restricted that it can be identified by exclusion of all others. Thus, strand-line amphipods (beach fleas) are almost certainly Talitridae; fresh-water species must be either *Hyalella, Crangonyx,* or *Gammarus;* white-bodied, free-living sand burrowers are likely to belong in Haustoriidae, Phoxocephalidae, or Lysianassidae; sponge or sea-squirt iniquilines are likely to be either *Leucothoe* or *Colomastix;* and external fish parasites are probably *Lafystius* or *Opisa.*

The keys are dichotomous; that is, each stage of progress (couplet) offers a clear choice between one of two alternatives. The reader proceeds from one couplet to the next according to the alternative with which the specimen better agrees.

In using this key, the reader may encounter difficulties through limitations of the key and/or his material. The key is regional and treats only 30 of approximately 60 world families, and slightly under 200 of approximately 5500 world gammaridean species. The incidence of morphological convergence or parallel evolution in component species of the various families and genera of Amphipoda is suprisingly high. Good all-inclusive characters that rigorously define higher taxa are therefore difficult or even impossible to find. The words "usually," "often," "typically," "may have," and so on are necessitated by such exceptions. For instance, most nondomicolous families are characterized by a proximally excavate posterior margin of coxal plate 4, but this character is not apparent in species of Colomastigidae, Stenothoidae, and some genera and species of other families in which the character is typically prominent.

The use of multiple-character couplets attempts to provide a more reliable "best-fit" criterion for keying exceptional species. There are, however, critical instances in the key to families and genera where only two-character couplets are used, and/or which involve "difficult" characters such as presence or absence of an accessory flagellum. Where such a character may be vestigial, or not obvious to the viewer, or where the pertinent appendage is missing in the specimen at hand, the reader may be obliged to follow through both succeeding branches of the key, eventually eliminating the wrong choice by obvious impasses in subsequent couplets or by "visual" comparison of the specimen with pertinent illustrations.

The reader must make allowance for imperfections in the material being keyed. Specimens may show abnormalities in growth or structure of appendages, and some body parts may be damaged or lost entirely. Inter-sexes and interspecific crosses are not infrequent in some species com-plexes, especially in *Gammarus*. Newly molted or emerging individuals usually do not have well-developed armature of spines and setae, and rela-tive lengths of segments and appendages may be distorted. Similarly, very young instars and juveniles may not have diagnostic characters sufficiently well developed to be "keyable."

When a specimen has been keyed to a family name, its validity should then be checked quickly against illustrations of component species of that family. If identification is obviously erroneous, the species should again be "run through" the key, taking the other alternative where there may have been previous doubt, until the correct family identification is obtained.

The reader then proceeds to the key to genera of that family (the page is indicated). The above procedure is followed until the correct name is ob-tained. The procedure is repeated in the key to the genus until the correct species name is obtained. The reader should verify the identification by comparison with the pertinent illustrations, paying particular attention to characters marked by arrows. The reader should also consult the text to check all diagnostic features of the species, and also to satisfy himself that his collecting locality and associated data are consistent with the known in-formation on the ecology, life history, and distribution of the species. After thorough checking, material that does not conform to keys, descriptions, or illustrations of this manual may be new to the region or possibly new to sci-ence. Where "taxonomic" and "ecological" identifications do not coincide, the species may have been misidentified, or, if correctly identified, new in-formation on the regional distribution and ecology may have been discov-ered. In all such cases, the opinion of an amphipod systematist should be enlisted. Museum specialists are usually willing to provide this service, pro-vided the request is made before the material is sent in for identification.

BASIC INFORMATION

The systematics of gammaridean amphipods herein is modeled on that of J. L. Barnard (1958, 1969). That author, however, has pointed out the existing unsatisfactory arrangement of genera within families and families within superfamilies, particularly in cases where higher taxa are inadequately defined, or where species and genera that apparently "bridge" existing higher taxa have recently been discovered. These difficulties are still unresolved in the Calliopiidae-Pleustidae, Eusiridae-Pontogeneidae-Bateidae, and Photidae-Aoridae-Corophiidae assemblages, and in the Talitroidea. The present level of taxonomic refinement in the Amphipoda is unsophisticated relative to that of more intensively studied crustacean groups such as the Decapoda, Isopoda, Copepoda, Ostracoda, other invertebrate groups such as the Insecta and Mollusca, and the vertebrate classes. The history of taxonomic development in these groups suggests that a real improvement would result from further subdivision of existing families and genera in the Amphipoda. The creation of smaller, more "natural" and less unwieldy systematic units would result from application of comparable taxonomic criteria at comparable levels. In my view, the taxonomy of terrestrial Talitridae would be unquestionably improved by refining generic and subgeneric groupings to the level of sophistication in oniscoidean isopods. The Gammaridae has hitherto been an unrealistic and virtually "unkeyable" family concept that has long masked the inclusion of discrete family units. I do not view the discovery of an "intermediate" genus or species as an excuse for automatic fusion of two related families or genera. Such discovery might just as readily result in confirmation of the applicability of existing family or generic criteria, a realignment of existing criteria, or creation of superfamily and/or subfamily and subgeneric units.

Whereas many of the systematic and nomenclatural changes proposed by Barnard (1969) from older reference works as Gurjanova (1951) and Stebbing (1906) are herewith accepted, I have departed from these authors in areas where more satisfactory systematic treatment is desirable. These include the separation of new families Crangonycidae and Melitidae from Gammaridae, the re-division of Eusiridae and Pontogeneiidae, realignment of genera within Aoridae and Corophiidae based on new taxonomic considerations, and interim retention of Photidae in preference to Isaeidae. I have continued recognition of the family arrangement in Talitroidea proposed by Bulycheva (1957) and subfamily arrangement in Haustoriidae proposed by myself (1965). I have also proposed the new genera *Psammonyx* and *Pseudunciola* and resurrected *Trichophoxus* K. H. Barnard 1930 for highly distinctive species and/or species complexes. The genus *Rozinante* is transferred to family *Calliopiidae*. Six new species *Synchelidium americanum, Proboloides holmesi, Colomastix halichondriae, Amphiporeia gigantea, Bathyporeia parkeri* and *Rudilemboides naglei* are described, and *Stenopleustes inermis* Shoemaker is elevated to full species status. Other changes, mostly in the direction of generic subdivision, seem desirable but must await further study of material on a broader basis.

The families are arranged in two groups, more or less in phylogenetic sequence. Recent amphipod families do not lend themselves to a linear phylogenetic arrangement because of intense recent radiation on the one hand and obvious convergence on the other. Common ancestral types and lines of phylogeny are difficult to reconstruct. The analysis of Barnard (1969, Graph 1) is a good attempt to do so. In families herewith considered, the first group consists of the nondomicolous families, commencing with the probably most primitive family, the Gammaridae, and related families. Radiating from this stock are (1) the fossorial Haustoriidae-Phoxocephalidae, (2) the Eusiridae-Pontogeneiidae-Bateidae-Calliopiidae-Pleustidae complex, (3) the Atylidae-Dexaminidae-Ampeliscidae group, (4) the Stenothoidae-Amphilochidae-Leucothoidae-Cressidae assemblage, (5) the Liljeborgiidae (possibly from melitoid ancestry leading to Sebidae), and fossorial Oedicerotidae, (6) the Talitroidea, and (7) families of uncertain affinity such as the Lysianassidae, Colomastigidae and Lafystiidae. The second group, the domicolous families, combines two closely related offshoots from Gammaridae; the Aoridae-Ampithoidae group, and the Photidae-Ischyroceridae-Corophiidae-Cheluridae-Podoceridae line, which is also believed ancestral to the Caprellidea. In my opinion there is practical taxonomic advantage in formal recognition of at least some of the aforementioned groups (especially numbers 3 and 4, and Gammaridae family groups) at the superfamily level, a move already initiated in amphipod taxonomy by Bulycheva (1957) with the Talitroidea. To this end, more detailed study of component families on a worldwide basis would be recommended, and indeed required.

KEY TO SUBORDERS OF AMPHIPODA

1. Head and first peraeon (second thoracic) segment variously fused; coxal plates lacking or vestigial; abdomen of no more than 5 segments and appendages usually vestigial; brood plates and gills on thoracic segments 3 and 4, rarely gills on 2 . . . . CAPRELLIDEA.
Peraeon of seven distinct segments; peraeon appendages with coxal plates; abdomen strong, of six segments, with appendages; brood plates and coxal gills on three or more segments . . . . 2.
2. Body vermiform; eyes lacking; pleopods vestigial; two pairs of biramous uropods; coxal gills on three segments . . . . INGOLFIELLIDEA.
Body shrimplike; eyes usually present; pleopods usually strong, biramous; three pairs of uropods; coxal gills usually on 4 or more segments . . . . 3.
3. Eyes usually of large size, covering most of head; maxilliped palp lacking; coxal plates small or indistinct; uropods sometimes without rami . . . HYPERIIDEA.
Eyes of normal size, usually occupying less than half of head; maxilliped with palp; coxal plates present, usually large; uropod 1 always biramous . . . . . .
. . . . GAMMARIDEA (p. 40).

KEY TO FAMILIES OF GAMMARIDEAN
AMPHIPODA OF THE NEW ENGLAND REGION

(* Component species marked with an asterisk are seldom encountered in depths of less than 30 meters and are not treated in the Systematic Section herein.)
1. Body usually shallow or depressed, dorsally smooth (except some Cheluridae); coxal plates usually medium or small, never large, 4th not excavate behind (Fig. 7D); eyes medium to small, usually located on anterior head lobes (Fig. 6G); rostrum usually weak or lacking; dactyls of peraeopods 3 and 4 with gland ducts (Fig. 11F); mouthparts normal or basic; telson entire; animals usually domicolous
. . . . 31.
Body usually moderately to strongly compressed, often dorsally carinate or toothed; coxal plates usually deep, sometimes very large, 4th usually excavate behind; eyes usually medium to large, lateral, or at base of rostrum; rostrum often large; dactyls of peraeopods 3 and 4 simple, without gland ducts (except Ampeliscidae); mouthparts and telson variable; animals usually free-living, sometimes commensal or parasitic . . . . 2.
2. Urosome segments 2 and 3 fused (Fig. 12H); peraeopods 5–7 with dactyls turned backwards (Fig. 11D5, 6) . . . . 3.
Urosome segments 2 and 3 separate; some or all peraeopods 5–7, dactyls turned forwards (except Melphidippidae) (Fig. 11A–C) . . . . 5.
3. Body (esp. pleon) carinate dorsally; rostrum acute; peraeopod 3 longer than 4; uropod 2 much shorter than 1, rami unequal; 1 pair compound eyes . . . . . 4.
Body smooth, or carinate only on urosome; rostrum lacking; peraeopod 4 longer than 3; uropod 2 slightly shorter than 1, rami subequal; eyes simple, usually two pairs (Fig. 6J) . . . . AMPELISCIDAE (p. 132).

4. Body very compressed; rostrum strong; peraeopods 3 and 4, segment 5 very short; mouthparts about normal . . . . ATYLIDAE (p. 131).
   Body moderately or little compressed; rostrum weak; peraeopods 3 and 4, segment 5 slightly smaller than segment 6; mandible lacking palp . . . .
   . . . . DEXAMINIDAE (p. 129).
5. Peraeopods specialized for burrowing, spinose; peraeopod 6 dissimilar to (and usually longer than) peraeopod 7; body smooth dorsally, often broad; antennae short (except in pelagic males), accessory flagellum present, often prominent; mandible with palp . . . . 6.
   Combination not so; mostly nonburrowing; accessory flagellum often minute or lacking; body narrow or fusiform; peraeopod 7 longest (usually) . . . . 7.
6. Head usually with large hooded rostrum (Fig. 6D); peraeopod 6 usually distinctly longer than 7 (except *Playtyischnopus* group); peraeopod dactyls strong; mandible with very weak molar . . . . PHOXOCEPHALIDAE (p. 122).
   Rostrum short or lacking, not hoodlike; peraeopod 6 slightly longer than peraeopod 7; peraeopod dactyls weak or lacking; mandible, molar strong . . . . . .
   . . . . HAUSTORIIDAE (p. 99).
7. Antenna 1, accessory flagellum usually prominent, of one or more distinct segments; antenna 2, peduncle well developed, usually equal to or longer than its flagellum (♀), and/or longer than peduncle of antenna 1 (except in Lysianassidae, Stegocephalidae) . . . . 8.
   Antenna 1, accessory flagellum vestigial or lacking; antenna 2, peduncle short, usually much shorter than its flagellum and/or little longer than peduncle of antenna 1 (except in Oedicerotidae, Stenothoidae family group, Talitroidea) . . 18.
8. Antenna 1, peduncular segment 1 very stout, segments 2 and 3 usually very short, broad; body and coxal plates deep, broad, only distal parts of peraeopods protruding below; gnathopods weak, usually unlike; mouthparts abnormal, mandibular molar and palp often weak or lacking . . . . 9.
   Antenna 1, peduncular segment 1 not unusually inflated or longer than 2 and 3 combined; body and coxal plates not abnormally deep, appendages largely visible below; gnathopods usually strongly subchelate, alike; mandibular palp strong, molar usually strong (except Pardaliscidae, Liljeborgiidae) . . . . 10.
9. Coxal plate 4 very large; peraeopod 5, basis not expanded; peraeopod 7 not longer than 6; gnathopod 2 simple, segment 3 short; mandible without palp . . .
   . . . . STEGOCEPHALIDAE.*
   Coxal plate 4 not much broader than coxae 1–3; peraeopod 5 basis broadly expanded; peraeopod 7 usually longer than 6; gnathopod 2 minutely chelate or subchelate (Fig. 10H); mandible with palp . . . . . . LYSIANASSIDAE (p. 142).
10. Coxal plates 2 and 3 much smaller than 1 and 4; gnathopods simple, segment 5 elongate; peraeopod dactyls small, at right angle to segment 6; eyes of a few paired peripheral facets . . . . ARGISSIDAE (p. 121).
    Coxal plates 2 and 3 usually about as deep as 1 and 4; gnathopods usually subchelate, strongly so; peraeopod dactyls and eyes various, not as above . . . 11.
11. Peraeopods 3–7 very slender and elongate, bases narrow; dactyls of peraeopods 6 and 7 turned backwards; coxal plates very shallow, wider than deep . . . . .
    . . . . MELPHIDIPPIDAE.*
    Peraeopods not extremely elongate, bases of 5–7 variously expanded; dactyls of peraeopods 6 and 7 usually as in 5, usually turned forward; coxal plates not wider than deep. . . . . 12.

12. Telson elongate, narrowly cleft distally; uropod 3, rami lanceolate, subequal, outer ramus 1-segmented; accessory flagellum 1 ½-segmented; gnathopods very powerful, alike . . . . EUSIRIDAE (p. 77).
    Characters not combined . . . . 13.
13. Coxal plates shallow, 4th not distinctly excavate behind; gnathopods 1 and 2, segment 5 longer and stouter than segment 6; mandibular molar vestigial; maxilliped inner plate vestigial . . . . PARDALISCIDAE.*
    Coxal plates deeper than wide, 4th distinctly excavate behind; gnathopods 1 and 2, segment 6 longer and stouter than segment 5; mandibular molar usually strong; maxilliped inner plate developed . . . . 14.
14. Peraeopod 7 usually distinctly stronger than peraeopod 6; coxa 1 expanded anteriorly (Fig. 7E); mandibular palp, segment 1 elongate, molar very reduced or vestigial . . . . LILJEBORGIIDAE (p. 70).
    Peraeopod 7 usually little longer (or shorter) than 6; coxa 1 not expanded distally; mandibular palp, segment 1 short, molar strong . . . . 15.
15. Compound eyes nearly fused at base of strong rostrum; coxal plate 4 often less deep and broad than coxa 3; antenna 1 shorter than 2; gnathopods slender, weakly subchelate or simple; upper lip bilobed; pleosome very powerful . . . .
    . . . . TIRONIDAE.*
    Compound eyes separated; coxal plate 4 usually deepest and broadest; antenna 1 usually longer than 2; gnathopods usually strongly subchelate (esp. in ♂); upper lip rounded; pleosome not exceptionally powerful . . . . 16.
16. Antenna 1, accessory flagellum 2-segmented, terminal segment very small; peraeopod 6 usually longer than 7; urosome dorsally smooth; uropod 1, peduncle smooth anteriorly; head lacking inferior antennal sinus.   CRANGONYCIDAE (p. 67).
    Antenna 1, accessory flagellum usually with 3 or more segments; peraeopod 7 usually longest; urosome dorsally carinate, mucronate, or spinose; uropod 1, peduncle with proximal anterior spine(s) (Fig. 4E); head, inferior antennal sinus distinct . . . . 17.
17. Eye nearly round, small; gnathopod 1 distinctly smaller and weaker than gnathopod 2 (esp. in ♂); some urosome segments mucronate or with small spine clusters; uropod 3, rami narrow, margins spinose . . . . MELITIDAE (p. 61).
    Eye reniform, large; gnathopod 1 nearly as large as or larger than gnathopod 2; all urosome segments with dorsal and dorsolateral groups of spines (except *Gammarellus*); uropod 3, rami usually broad, foliaceous. . GAMMARIDAE (p. 47).
18. Eyes fused or approximated dorsally on or at base of rostrum (Fig. 6H); peraeopod 7 much longer than 5 and 6; basis very expanded; gnathopods strongly subchelate (or chelate), carpal lobe usually prolonged behind . . . .
    . . . . OEDICEROTIDAE (p. 94).
    Eyes lateral, separated; peraeopod 7 little longer (or shorter) than 6; gnathopods variously subchelate or simple . . . . 19.
19. Body thick, integument hard, strongly toothed or spined (esp. abdominally) dorsally and laterally; coxal plates 1–3 generally narrow, acute, successively deeper, 4 acutely produced behind; antennal peduncle short; telson notched at apex . 20.
    Body carination (if present) dorsal abdominal; anterior coxal plates uniformly broad and/or deep; antennal peduncle various; telson cleft or entire . . . . 21.
20. Gnathopods similar, weakly subchelate; peraeopod 5, basis not expanded; mandible with molar; maxilliped palp 4-segmented . . . . . PARAMPHITHOIDAE.*

Gnathopod 1 very weak, minutely chelate; gnathopod 2 variously subchelate; peraeopod 5, basis expanded, coxa bilobed, posterior lobe deeper; mandibular molar lacking; maxilliped palp 3-segmented . . . ACANTHONOTOZOMATIDAE *

21. Body more or less depressed, broad; peraeopods 3–7 about equal in length; coxal plates medium, shallow, subequal; antenna short, subequal . . . . . 22.
Body laterally compressed; coxal plates usually deep, large; peraeopods 3 and 4 usually shorter than peraeopods 5–7; antenna 1 and 2 variable; flagellum of one or other elongate, multisegmented . . . . 23.

22. Rostrum strong, overhanging peduncle of antenna 1; peraeopods 5–7, bases expanded, dactyls strong, hooklike; antenna 1 and 2, flagella distinct, more than 3-segmented; gnathopod 1 not filiform . . . . LAFYSTIIDAE (p. 141).
Rostrum short, weak; peraeopods 5–7, basis narrow, dactyls normal; antenna 1 and 2, peduncular segments tubular, flagella minute; gnathopod 1 filiform. . . .
. . . . COLOMASTIGIDAE (p. 139).

23. Uropod 3 very short, usually uniramous, ramus 1-segmented; antenna 1 usually much shorter than antenna 2; mandible lacking palp; maxilla 1, palp reduced or vestigial . . . . TALITROIDEA (p. 152).
Uropod 3 strong, biramous (2-segmented uniramous in some stenothoids); antenna 1 and 2 not greatly unequal; mandible usually with palp; maxilla 1, palp about equal to outer plate . . . . 24.

24. Coxal plate 4 (often 2 and 3 also) very large, deep, covering peraeopods (except Leucothoidae); antenna 2 usually longer than antenna 1, peduncle long; upper lip asymmetrically bilobed; telson entire; body usually dorsally smooth . . . . 26.
Coxal plates not markedly unlike or enlarged, peraeopods exposed; antenna 1 usually longer than 2, peduncles short; upper lip rounded; telson cleft or entire; body (pleon) often strongly carinate dorsally . . . . 25.

25. Telson entire or apically notched . . . . 29.
Telson narrowly cleft . . . . 30.

26. Uropod 3 biramous; uropod 2 short, extending little beyond telson; telson elongate, apex acute; lower lip, inner lobes lacking, outer lobes closely approximating . . . . 27.
Uropod 3 uniramous; uropod 2 extending well beyond telson; telson rounded; lower lip, inner lobes fused, outer lobes widely separated. . . . . 28.

27. Rostrum strong, deflexed over peduncle of antenna 1; coxa 1 very small; gnathopods 1 and 2 small, alike, usually subchelate; mandible, molar process strong; maxilliped, outer plate developed, small . . . . AMPHILOCHIDAE.*
Rostrum short; coxa 1 nearly as large as 2; gnathopods very large and unlike; mandibular molar lacking; maxilliped, outer plate very small . . . .
. . . . LEUCOTHOIDAE (p. 92).

28. Telson and urosome 3 fused; peraeopod 5, basis broadly expanded; coxa 4 distinctly excavate behind . . . . CRESSIDAE.*
Telson and urosome 3 separate; peraeopod 5, basis narrow; coxa 4 shieldlike, not distinctly excavate behind . . . . STENOTHOIDAE (p. 85).

29. Rostrum generally strong; uropod 3, rami unequal, sublinear, spinose; upper lip bilobed . . . . PLEUSTIDAE (p. 81).
Rostrum weak or lacking; uropod 3, rami subequal, lanceolate, foliaceous or partly so; upper lip entire . . . . CALLIOPIIDAE (p. 79).

30. Coxal plate 1 and gnathopod 1 vestigial; peraeopod 5, basis not expanded as in

peraeopods 6 and 7 . . . . BATEIDAE (p. 75).
Coxal plate 1 and gnathopod 1 developed, normal; peraeopod 5, basis expanded
about as in peraeopods 6 and 7 . . . . PONTOGENEIIDAE (p. 73).

31. Gnathopod 1 larger and more powerful than gnathopod 2 (esp. in ♂), often complexly subchelate; accessory flagellum always present; uropod 3 biramous or uniramous, terminal spines simple, not hooked . . . . 32.
Gnathopod 2 larger and more powerful than gnathopod 1 (esp. in ♂), often complexly subchelate; accessory flagellum (when present) short; uropod 3, rami
variously reduced, outer ramus may be hook-tipped. . . . . 33.

32. Pleon 3 with large dorsal spine; uropods 1 and 2 markedly unlike; uropod 3,
outer ramus longer than peduncle, broad; antenna 2, flagellum clavate · · · · ·
. . . . CHELURIDAE (p. 206).
Pleon dorsally smooth; uropods 1 and 2 usually alike, rami slender; antenna 2,
flagellum filiform, multisegmented; lower lip, mandibular processes prominent
(Fig. 8H) . . . . AORIDAE (p. 165).

33. Antenna 2 longer and more powerful than antenna 1 (esp. in ♂); accessory flagellum not present; coxal gills lacking on peraeon segment 2; body depressed,
broad; pleopods variously expanded or modified . . . COROPHIIDAE (p. 192).
Antenna 2 about equal to antenna 1; accessory flagellum usually present, short;
coxal gills usually present on peraeon segment 2; body usually compressed laterally; pleopods about normal. . . . . 34.

34. Urosome segment 3 and appendages very small or lacking; coxal plates small,
separated basally; peraeopod 5, dactyl normally turned forward . . . .
. . . . PODOCERIDAE (p. 207).
Urosome segment 3 and appendages distinct, uropod 3 usually biramous; coxal
plates medium deep, overlapping at base; peraeopod 5, dactyl often turned backward (Fig. 11D) . . . . 35.

35. Head, anterior lobe blunt; eye usually at its base; antennal peduncles weakly setose behind; uropod 3, outer ramus with 2 strong apical hooks; lower lip, outer
lobes notched at shoulders (Fig. 8E); mandible, palp weakly setose or lacking. . .
. . . . AMPITHOIDAE (p. 178).
Head, anterior lobe usually acute, bearing eye; antennal peduncles usually
strongly setose behind; uropod 3, outer ramus hook spines very fine or lacking;
lower lip, outer lobes rounded distally; mandibular palp present, strongly setose . .
. . . . 36.

36. Uropod 3, outer ramus about as long as peduncle, apical spines simple; antenna
2, peduncle not stout, flagellum slender, multisegmented . . PHOTIDAE (p. 184).
Uropod 3, outer ramus much shorter than peduncle, uncinate, with hooked tip,
hooked spines or denticles at tip; antenna 2, peduncle stout, flagellum short, segments few or fused . . . . ISCHYROCERIDAE (p. 189).

$\mathcal{S}$HALLOW-WATER
GAMMARIDEAN AMPHIPODA OF NEW ENGLAND

Primitive, sexually dimorphic, mostly free-living amphipods with compressed, anteriorly smooth bodies, pleosome smooth or carinate, mucronate, urosome dorsally carinate, or spinose; coxal plates moderately deep, 4th excavate behind, head with rostrum variously developed; antenna strong, flagellate, accessory flagellum present; mouthparts with plates and palps normally present, setose; mandibular molar strong; gnathopods 1 and 2 subchelate, usually strongly; peraeopods normally ambulatory, dactyls nonglandular; uropods normally biramous, last pair typically foliaceous; telson usually deeply cleft.

Family GAMMARIDAE sens. str.

Head, rostrum short, eyes lateral; coxal plate 1 unmodified. Abdominal segments usually distinct, urosome typically with dorsal clusters of spines and/or setae. Antennae strong, accessory flagellum usually prominent. Mouthparts normal; lower lip inner lobes present or variously reduced. Gnathopods 1 and 2 subchelate, strongly in ♂, usually similar, 2nd usually larger. Peraeopods 3–5 successively increasing in length, bases not excessively expanded. Uropod 1, peduncle with proximal anterior spine(s). Uropod 3, rami usually foliaceous, extending beyond uropods 1 and 2. Telson usually cleft, usually deeply. Coxal gills pedunculate, present on mesome segments 2–7, accessory and sternal gills sometimes present. Brood lamellae large, with many long simple marginal setae, present on segments 2–5 inclusive.

KEY TO GENERA OF GAMMARIDAE

1. Abdomen carinate dorsally; urosome lacking dorsal groups of spines; telson entire; antenna 2, peduncle short, about equal to that of antenna 1 . . . .
. . . . *Gammarellus* (p. 59).
Abdomen not carinate (may be mucronate); urosome with dorsal clusters of spines and/or setae; telson cleft to base; antenna 2, peduncle longer and stronger than that of antenna 1 . . . . 2.
2. Uropod 3, inner ramus more than one-half length of outer ramus (in mature animals); peraeopod 6, posterior margin of basis usually lacking free posterodistal

lobe (Fig. 4C); accessory gills lacking on thoracic segments; antenna 2 often calceolate in male. *Gammarus* (p. 48).
Uropod 3, inner ramus less than one-half outer ramus; peraeopod 6, posterior margin of basis with free distal lobe; accessory gills usually present on some thoracic segments; antennae noncalceolate. . . . . *Marinogammarus* (p. 57).

Genus *Gammarus* L. 1768 *   (găm′âr ŭs)

Urosome with prominent dorsal and dorsolateral groups of spines and setae. Head, rostrum very short or lacking; inferior antennal sinus deeply concave; eyes lateral, reniform. Accessory flagellum well developed. Antenna 2, peduncle strongly setose posteriorly, flagellum calceolate or not. Upper lip slightly notched; maxilla 1, inner plate with many marginal setae. Gnathopod 2 larger than gnathopod 1, both powerfully subchelate in ♂, carpus shorter than propod. Peraeopods 6 and 7, basis often lacks posterodistal free lobe. Uropods 1 and 2 usually extend beyond peduncle of uropod 3; rami slender, with marginal spines; uropod 2, outer ramus shorter than inner. Uropod 3, rami foliaceous, setose, inner shorter; outer ramus with short terminal segment. Telson cleft to base, lobes with dorsolateral and apical spines.

KEY TO AMERICAN ATLANTIC SPECIES OF *GAMMARUS* (SENS. LAT.)

1. Antenna 1, peduncular segment 3 elongate, posterior margin with 4–6 groups of setae; arctic and subarctic only . . . . *G. wilkitzki* Birula.
   Antenna 1, peduncular segment 3 not elongate, length not more than three times width; posterior margin with 0–3 groups of setae . . . . 2.
2. Posterodistal margin of basis (segment 2) of peraeopod 6 (and 7 also) forming a distinct free lobe, lacking stout spines at junction with segment proper; superior antennal angle of head broadly rounded . . . . 3.
   Posterodistal margin of basis of peraeopod 6 merging directly with segment proper, junction marked by one or more spines; superior antennal angle of head usually angular or acuminate, sometimes sharply rounded . . . . 4.
3. Antenna 1, penduncular segment 1, posterior margin with 3–4 groups of setae distally; peraeopods 5–7, distal segments richly adorned with clusters of long simple setae; urosome dorsal spine clusters containing several long simple setae; in brackish spray pools. . . . . *G. duebeni* (p. 56).
   Antenna 1, penduncular segment 1 lacking groups of posterior marginal setae (except terminally); peraeopods 5–7, distal segments spinose, lightly setose; urosome dorsal spine clusters with spines only; in strictly fresh-water bodies . . . .
   . . . . *G. pseudolimnaeus* Bousfield.
4. Coxal plates 1–4, lower margins armed with several long setae (some plumose); urosome 1, dorsolateral spine-cluster lacking; antenna 1, peduncular segment 1 relatively stout, segment 3 short; antenna 1 distinctly shorter than 2. . . . . . 5.
   Coxal plates 1–4, middle of lower margins usually unarmed, setae present at an-

---

* Includes *Rivulogammarus* Karaman 1931, of which *G. pulex* is the type.

teroventral angle or along anterior margin; urosome 1, dorsolateral cluster of spines present; antenna 1, peduncular segments distinctly longer than deep; antenna 1 not shorter than 2. . . . . 6.

5. Antenna 1 with long plumose setae on lower (posterior) margins of peduncular segments extending onto basal flagellar segments; peraeopods 3–7, segment 6 richly armed with curly setae ($\male$); animal mainly pelagic . . *G. annulatus* (p. 53).
   Antenna 1 lacking long plumose setae on peduncular or basal flagellar segments in either sex; peraeopods 3–7, segment 6 with few curly setae ($\male$); animal mainly benthic (estuaries), or pelagic; mainly north of Cape Cod. *G. lawrencianus* (p. 54).

6. Coxal plate 1 with several (5–8) setae at anteroventral angle; peraeopod 7, basis usually armed posteriorly with several long setae; antenna 2 ($\male$) sometimes with curly setae, never with calceoli . . . . 7.
   Coxal plate 1 with a few short (1–5) setae at anteroventral angles, or with several short setae lining anterior coxal margin; peraeopod 7, basis usually with only a few relatively short posterior marginal setae; antenna 2 ($\male$) not bearing curly setae, calceolate in some species . . . . 9.

7. Urosome segments with distinct midsegmental dorsal "hump"; lateral spines of urosome in clusters of 2–3; abdominal side plates 2 and 3, lower margins setose; antenna 1, peduncular segment 2 with one major cluster of posterior marginal setae. *G. fasciatus* (p. 53).
   Urosome segments with small distal dorsal elevation; lateral spines usually single or in clusters of two; abdominal side plates submarginally spinose; antenna 1, peduncular segment 2 with 3–5 groups of posterior marginal setae; antenna 2 and peraeopods armed with clusters of curly setae ($\male$) . . . . 8.

8. Antenna 1, basal flagellar segments with alternate posterior setae longer than twice width of respective segments; antenna 2, peduncular segments 4–5 each with 4–7 clusters of long stiff setae, 5–6 setae per cluster; urosome spine clusters weak, located at posterior segmental margins; pelagic (South of New England) . . . . *G. daiberi* (p. 52).
   Antenna 1, basal flagellar segments with alternate posterior setae short, scarcely exceeding width of segment; antenna 2, peduncular segments 4–5 each with 3 (occ. 4) clusters of long setae, about 3 setae per cluster; urosome spine clusters located anterad of margin; benthic and pelagic . . . . . . *G. tigrinus* (p. 51).

9. Dorsal abdominal spine clusters relatively weak, 1–2 per cluster, nearly lacking accessory setae; antenna 1, accessory flagellum usually 4–5 segmented; antenna 2, flagellum setose, lacking calceoli ($\male$); animals small (8–12 mm); body strongly banded or mottled in life. 10.
   Dorsal abdominal spine clusters strong, usually 3 spines per cluster, interspersed with setae; antenna 1, accessory flagellum usually 6–10 segmented; antenna 2, flagellum calceolate ($\male$); animals large (15–35 mm); body not mottled or banded, usually unicolorous in life . . . . 11.

10. Abdominal segments 1–3 usually dorsally mucronate, posterior dorsal margin produced backward in sharp process; urosome segments with dorsal elevations; urosome dorsolateral spines tall, in clusters of two per group; eye normally reniform . . . . *G. mucronatus* (p. 55).
    Abdominal segments 1–3 dorsally smooth; urosome segments flat, lacking dorsal elevations; dorsolateral spines small, single; eye narrowly reniform . . . .
    . . . . *G. palustris* (p. 55).

11. Gnathopod 2 ($\male$), median palmar spine flask-shaped, having swollen basal portion

and truncate tip; mandibular palp segment 3 armed on inner (ventral) margin with row of comblike spinules, smallest posteriorly; European exclusively . . . .
. . . . *G. locusta* group.*
Gnathopod 2 (♂), median palmar spine simple, not flask-shaped; mandibular palp segment 3, comb setae uniform, palp segment 1 may be setose; amphi-Atlantic . .
. . . . 12.

12. Antenna 1, peduncular segment 1 with 2–4 clusters of setae distally on posterior margin; peraeopod 7, basis with numerous longish posterior marginal setae; some plumose setae present on peraeopods, urosome, and telson. . . . . . .
. . . . *G. setosus* (p. 50).
Antenna 1, peduncular segment 1 with 1–3 clusters of setae centrally on posterior margin; peraeopod 7, basis with short posterior marginal setae; simple setae only on peraeopods, urosome, and telson . . . . *G. oceanicus* (p. 50).

Gammarus oceanicus Segerstråle 1947 Plate I.1 and Frontispiece
(ō shē ăn′ ĭk ŭs)

L. 15–22 mm. Head, upper angle acute, slightly produced. Eye normally reniform black. Antenna 1, peduncle 3 $\geqq$ one-half peduncle 2, peduncle 1 with two groups of posterior marginal setae placed centrally. Coxal plates 1–2, anterodistal angles rounded, bearing 2–4 short setae. Gnathopod 2 (♂), propod not swollen, posterior margin nearly straight. Peraeopods 6 and 7, posterior margin with about 10 short setae. Abdominal side plate 3, hind corner acute, produced, posterior margin with single seta. Urosome segments 1 and 2 dorsally elevated or "humped," 2–4 spines per cluster. Setae on peraeopods, dorsal spine clusters, and telson are simple, not plumose.

Distribution: Amphi-Atlantic, southward in Europe to Northern France; in America regularly south to Long Island Sound, sporadically to Chesapeake Bay.
Ecology: A very common, dominant intertidal and shallow-water species (esp. in winter) to depths of 100 ft, mainly along rocky shores, but also regularly in estuaries to salinities as low as 6 parts per thousand (o/oo). Under stones, among algae, in tide pools, in debris in the wave zone, from subtidally to about half-tide level or higher.
Life Cycle: Ovigerous females mainly from March to July in New England. Two broods, occasionally three per year. Adult may live two years.

Gammarus setosus Dementieva 1931 Plate I.2   (sē tōs′ ŭs)

L. 15–25 mm. Head, upper angle acute, slightly produced. Eye black, reniform, slightly broader and less deep than in *oceanicus*. Antenna 1, peduncle 3 = one-half peduncle 2, peduncle 1 with 3–4 groups of distal posterior

---

* As defined by Stock (1967).

marginal setae. Antenna 2 calceolate in ♂. Coxal plates 1–3, anterodistal angles rounded, with 3–6 short setae. Gnathopod 2 (♂), propod somewhat swollen, posterior margin convex; in ♀, propods of gnathopods 1 and 2 stouter than in *G. oceanicus*. Peraeopods 6 and 7, posterior margin of basis with about 12 evenly spaced moderately long setae. Abdominal side plates 2 and 3, posterior corners acute, produced, posterior margins usually with several short setae. Urosome segments not strongly elevated dorsally. Some setae on peraeopods, dorsal spine clusters, and telson are plumose.

Distribution: Circumpolar, subarctic-boreal; in western Atlantic region south in cold-water areas to Nova Scotia, Bay of Fundy, and Maine (Penobscot Bay).
Ecology: Uncommon, occurring intertidally and in shallows of cold-water, somewhat muddy or silty habitats, in stream outflows and beach seeps, and in estuaries down to salinities of about 3 o/oo. Occurs under stones, in *Enteromorpha,* and under fucoids up to about mid-tide level; also in tide pools on rocky shores.
Life Cycle: Ovigerous females from September to May, peak release January–February. One brood per year at southern limit of range, breeding again in second year, cycle affected by photoperiod.

Gammarus tigrinus Sexton 1939 Plate IV.1   (tī'grĭn ŭs)

L. 10–14 mm. Head, upper angle acute, occ. slightly notched. Eye large, broad-reniform, black. Antenna 1, peduncular segments 1, 2, and 3 with 1–2, 2–4, and 2 posterior marginal groups of setae; proximal flagellar setae short, not longer than width of respective segment. Antenna 2 noncalceolate in ♂; peduncle strong, posterior margin bearing curly setae in ♂. Antenna 1 ≦ antenna 2. Mandibular palp segment 3, distal row of pectinate setae uneven in length. Coxal plates 1–3, anterior distal angle with 4–6 relatively long setae. Peraeopods 3–7, distal segments bearing curly setae in ♂; peraeopod 7 basis, posterior margin with numerous long setae. Abdominal side plates 2 and 3, hind corners acute, slightly produced, hind margins lightly setose, lower margins spinose. Urosome segments 2 and 3 not conspicuously elevated, spine clusters set back from posterior margin of segment. Uropod 3, terminal segment of outer ramus without lateral marginal setae.

Distribution: American endemic; from southern Labrador and north shore of the Gulf of St. Lawrence (but not Newfoundland) south to Chesapeake Bay and, as a variety, sporadically south to Florida. Introduced into the British North Sea region, where it was originally described, and from where it is now spreading rapidly in brackish areas bordering the North Sea.
Ecology: A common and dominant, essentially benthic, species of the upper portions of estuaries, in salinities 1–25 o/oo. Occurring in debris, *Enteromorpha,* and among hydroids (*Cordylophora*), on stakes and pilings, and the

low intertidal, but essentially subtidal. Intertidal on rocky shores in beach seeps and stream outflows.

Life Cycle: Ovigerous females from April to October; annual life cycle; female has several broods during summer in southern or "warm-water" parts of the range and may reproduce in its first summer; overwinters in the resting stage.

Gammarus daiberi Bousfield 1969 Plate IV.2 (dā′bĕr ī)

L. 8–12.5 mm. Head, upper angle subacute, not broadly rounded. Eye short-reniform, purplish-brown (in preservative). Antennae 1 and 2 subequal. Antenna 1, peduncular segments 1, 2, and 3 with 1–2, 3–5, and 2–3 groups of posterior marginal setae; proximal flagellar setae longer, each about 2–3 times width of respective segment; accessory flagellum 4–5 segmented. Antenna 2, peduncle not stoutly expanded, posterior margins of segments 4–5 each with 4–7 clusters of long stiff setae, 5–6 setae per cluster; flagellum 13–16 segmented, posterior margins bearing whorls of simple and (in ♂) curly setae, but lacking calceoli.

Mandibular palp segment 3, distal two-thirds of posterior margin lined with about 16 short, stiff, unequal, finely pectinate setae and 4–5 long apical setae.

Coxal plates 1–4 with 3–5 moderately long stiff setae at anteroventral angle. Peraeopods 3–7 (in ♂), distal segments with posterior marginal clusters of simple and curly setae, especially on peraeopods 6 and 7. Peraeopod 7, posterior margin of basis gently convex, with 10–12 medium-length, well-spaced setae and two slender spines at distal angle.

Abdominal side plates 2 and 3 with acute, slightly produced posterior angles, posterior margins lined with 4–5 short setae; plate 3 with a few slender setae and 3–4 submarginal spines.

Urosome segments with prominent middorsal paired spines and a few setae set on low distal elevations, dorsal and dorsolateral clusters set close to posterior margin of segments. Uropod 3, terminal segment of outer ramus relatively long, occasionally bearing 1–2 lateral marginal setae.

Distribution: American endemic—Delaware and Chesapeake estuarine systems to estuaries of South Carolina; possibly occurring northward (via canals) to the Hudson and Long Island Sound.

Ecology: Reaches most dense concentrations in the upper estuary, especially in spring and summer, in salinities of 1–5 o/oo and in mid- to near-bottom depths but occurs seaward to about 15 o/oo, largely pelagically. Co-occurs with *G. fasciatus* and *G. tigrinus* in tidal fresh and slightly brackish waters.

Life Cycle: Annual; ovigerous females from March to October; several broods per female.

Gammarus fasciatus Say 1818 Plate V.2   (făs ē ăt′ ŭs)

L. 8–13 mm. Head, upper angle acute, occ. notched. Eye large, broad-reniform. Urosome segments 1 and 2 dorsally elevated or "humped," spine clusters anterior to segmental posterior margin. Coxal plates 1–3, anterior distal
angles with 4–8 relatively long setae. Antenna 1 ≦ antenna 2 (shorter in ♂).
Antenna 1, peduncular segments 1, 2, and 3 with 1, 1 (3) and 1 posterior
marginal groups of setae respectively; proximal flagellar setae short. Antenna 2 noncalceolate, peduncle with whorls of simple (nonplumose) setae
in male. Peraeopods 3–7, distal segments with numerous simple setae in ♂;
peraeopods 6 and 7, basis, posterior margin with several long setae. Abdominal side plates 2 and 3, hind corners acute, 3 produced, posterior margins setose, lower margins setose (not spinose). Uropod 3, terminal segment
of outer ramus (esp. in ♂) with 1–2 plumose lateral marginal setae.

Distribution: American Atlantic coastal drainages, from the south side of
Cape Cod and southern New England south to Chesapeake Bay tributaries.
Also in lakes and larger rivers of the St. Lawrence drainage system, Lake
Superior to tide water.
Ecology: Essentially fresh-water, benthic, and semipelagic, mainly in large,
summer-warm, turbid river systems, downstream into the upper parts of es-
tuaries (salinities of 1–3 o/oo), where it barely co-occurs with *G. tigrinus*
and *G. daiberi*. The latter has not yet been recorded in New England.
Life Cycle: Annual; ovigerous females from April to September; female with
several broods per summer, may reproduce the first year.

Gammarus annulatus Smith 1873 Plate II.1   (ăn ūl āt′ ŭs)

L. 12–17 mm. Head, upper angle acute, apex frequently notched. Eye black,
moderately large, short reniform. Body relatively long and slender and very
setose. Coxal plates 1–3 deep, 1st broad distally, lower margins lined with
long plumose setae. Urosome segments moderately elevated, urosome segment 1 lacking dorsolateral spine cluster. Antenna 1 distinctly shorter than
antenna 2 (both sexes). Antenna 1, peduncular segment 1 large, somewhat
inflated, with 3–4 close-set posterodistal setae groups; segment 3, short.
Most peduncular and proximal flagellar setae are plumose (both sexes). Antenna 2 noncalceolate, peduncle strong, longer than antenna 1 (in ♂) bearing
whorls of setae, some curly (in ♂). Mandibular palp, segment 3 elongate (>
seg. 2), with about 6 clusters of lateral facial setae. Maxilliped small, slightly
larger than maxilla 1. Peraeopods 1 and 2 slender, posterior margins
densely setose (esp. in ♂). Peraeopods 5–7 slender, moderately elongate;
distal segments, anterior margins richly setose. Peraeopods 6 and 7, posterior margin of basis with numerous long setae. Abdomen large; pleopods
large and powerful; side plates 2 and 3, hind corners acute, produced, posterior margins bare, lower margins slender-spinose. Uropod 3, rami sub-

equal, very elongate (in ♂), terminal segment of outer ramus bearing 2–4 lateral marginal plumose setae. Telson lobes with two lateral marginal spine groups and 3–5 plumose setae.

Body banded with dark brown at intersegmental areas.

Distribution: Southern New England, New Hampshire, Cape Cod, and Long Island Sound; also Sable Island, off Nova Scotia.
Ecology: An essentially pelagic species, occurring mainly in open coastal waters or along sandy shores in the surf zone, but in relatively high salinities. Juveniles fossorial in sandy bottoms.
Life Cycle: Presumably annual; ovigerous females from June–September, presumably bearing several broods per summer.

Gammarus lawrencianus Bousfield 1956 Plate II.2   (lâr ĕnts ĭ ăn' ŭs)

L. 8–13 mm. Body compact (esp. in ♀). Head, upper angle acute. Eye short, broad-reniform, black. Coxal plates 1–3 deep, 1st slightly broader distally, lower margins lined with long setae (some plumose). Urosome segments 1 and 2, slightly elevated. Urosome segment 1, lacking dorsolateral spine cluster. Antenna 1 longer than antenna 2 (both sexes). Antenna 1, peduncle short, segment 1 inflated, posterior setae (including flagellum) simple nonplumose. Antenna 2 noncalceolate, peduncle shorter than antenna 1 (both sexes), bearing whorls of setae, some of which are curly (♂).

Mandibular palp, segment 3 ≧ segment 2, with 3–4 groups of facial setae. Peraeopod 3 stouter and longer than 4; segment 4 strong, posterior margins thickly setose (in ♂). Peraeopods 6–7, posterior margin of basis with numerous medium-long setae; distal segments richly setose (in ♂). Abdominal side plates 2 and 3, hind corners acute, slightly produced, posterior margins distally with a few setae, lower margins slender-spinose. Uropod 3, rami not unusually long and slender, terminal segment of outer ramus lacking marginal plumose setae. Telson with 2 lateral marginal spine groups, setae simple.

Body color—light green, with brown bands at intersegmental areas.

Distribution: American Atlantic coast, from Labrador (Nain) and Newfoundland south to Connecticut and Long Island Sound.
Ecology: A very common, essentially benthic species of sandy and sandy mud bottoms, in lotic waters; mainly estuarine, in salinities as low as 3–4 o/oo, but also on sandy surf coasts in the offing of estuarine outflows. In flat rock pools affected by stream outflow and beach seeps, under stones, in eelgrass, among fucoids.
Life Cycle: Annual; ovigerous females from March to October, breeding in first season and carrying several broods per summer. Potential fecundity high.

Gammarus mucronatus Say 1818 * Plate III.1   (mū krŏ nat' ŭs)

L. 9–13 mm. Head, upper angle sharply rounded or blunt, not acute. Eye large, reniform. Pleosome segments 1, 2, and 3 typically with sharp posteriorly directed dorsal mucronation (not carination), variously reduced or even lacking (esp. in ♀) in material from southern outer coastal localities. Coxal plates 1–2 medium deep, anterodistal angle with 4–8 short marginal setae. Antenna 1 ≧ antenna 2, peduncular segments 1, 2, and 3 with 1, 3, and 1 groups of posterior marginal setae, respectively. Antenna 2, noncalceolate peduncle relatively slender, with whorls of simple setae extending onto flagellum (in ♂). Mandibular palp segment 3 shorter than 2, widest medially. Maxilliped palp segment 2 narrow, segment 4 short, not swollen.

Gnathopods 1 and 2 large; propod of 2, palm with medium concavity and blunt spine. Peraeopods 3 and 4, posterior margins of segments 4–6 richly setose (in ♂). Peraeopods 6–7, posterior margin of basis lined with 6–10 short setae; distal segments with simple setae (esp. in ♂). Abdominal side plates 2 and 3, hind corners acute, not produced; posterior margins with a few very short setae; hind margins sparsely short-spinose. Urosome segments 1 and 2 slightly elevated; spine clusters strong, tall. Uropod 3 normal, rami slender, terminal segment of outer ramus lacking lateral marginal setae. Telson, lobes relatively long, divergent, with spines on inner and outer margins.

Body light green, spotted and maculated with red, brown; antennae and appendages banded.

Distribution: Gulf of St. Lawrence (Chaleur Bay and southwestern Newfoundland) south to Florida and the Gulf states.
Ecology: A very common and dominant species of salt-marsh pools and estuaries, low-water level and subtidally, esp. in brackish (down to 4 o/oo) waters, among eelgrass, in muddy areas, in *Cladophora* mats, sometimes in *Enteromorpha* clumps, mainly benthic but also pelagic. Replaces *G. duebeni* in spray pools of rocky shores south of Cape Cod.
Life Cycle: Annual; ovigerous females from April–September, may breed in first summer and bear several broods.

Gammarus palustris Bousfield 1969 Plate III.2   (pă lŭs' trĭs)

L. 9–14 mm. Head, upper angle rounded, not acute. Eye black, narrow, deep-reniform. Pleosome dorsally smooth. Urosome segments dorsally flat, spines very short, singly inserted. Coxal plates 1–2 relatively shallow, lower margins rounded, anterodistal angles with 5–10 very short setae. Antenna 1 > antenna 2; peduncle segments 1, 2, and 3 with 1, 3, and 2 posterior marginal setae, respectively. Antenna 2 noncalceolate; peduncle relatively short, with

* Barnard and Gray (1968) have erected the subgenus *Mucrogammarus* based on this species.

whorls of simple setae extending onto flagellum (in ♂). Mandibular palp, segment 3 much shorter than 2. Maxilliped palp, segment 3 slender.

Gnathopods 1 and 2 (in ♂) relatively large, palm of gnathopod 2 oblique, with median concavity and blunt spine tooth. Peraeopod 3 distinctly larger than peraeopod 4, segment 4 strong. Peraeopod 5, segment 4 stout, longer than segment 5. Peraeopods 6 and 7, posterior margin of basis with a few short weak setae. Abdominal side plates 2 and 3, hind corners acute, not produced, hind margin with a few short setae, lower margin with short spines. Uropod 3 relatively short and weakly setose, esp. in ♀, outer ramus less than twice peduncle, terminal segment very short. Telson short, lobes each with one lateral group of spines and setae.

Body greenish brown, mottled with red, brown.

Distribution: American Atlantic coast, from New Hampshire (Piscataqua estuary) south to northern Florida.

Ecology: An estuarine and salt-marsh euryhaline species, mainly benthic, on muddy, grassy, and boulder-mud beaches, from about half tide level to mean high-water neaps, where it remains hidden in the culms of *Spartina alterniflora,* and under damp debris, boards, stones, and the like, when the tide is out. Optimal salinity range from about 5 to 20 parts per thousand but survives short periods of fresh water (at low temperatures) and full sea salt (at summer temperatures).

Life Cycle: Annual; ovigerous females from March until September; overwintering generation breeds in spring; young mature and breed in summer with 3-5 broods per female, fewest in northern part of range.

Gammarus duebeni Liljeborg 1851 Plate V.1   (dū′bĕn ī)

L. 14–23 mm ♂, 12–18 mm ♀. Eye normally reniform, with anterior middle notch, black. Coxal plates moderately large, rounded below, anterodistal angles with 2–5 short setae. Antenna 1, peduncular segment 2 with 2–4 posterior distal groups of setae; segment 3 with 2 posterior groups of setae. Antenna 2, flagellum stout, longer than peduncle, calceolate (in ♂). Lower lip, inner lobes obsolescent. Maxilliped palp, segment 2 broad; segment 3 large, broadened distally.

Peraeopods 5–7, distal segments with bundles of long simple setae among spine groups (esp. in ♂); peraeopod 7, basis with several long setae proximally on posterior margin, distally bare, a few at extreme hind lobe. Abdominal side plates 2 and 3, hind corners acute, 3 slightly produced, posterior margin with about 5 setae, lower margin sparsely spinose. Urosome spine clusters with tall spines and slender simple setae; uropod 1, peduncle with one proximal anterior spine, rami with 1–2 lateral marginal spines. Uropod 3 relatively large, terminal segment of outer ramus without marginal setae. Telson short, wide, with long simple setae among spine clusters.

Body uniformly dull greenish.

Distribution: Boreal amphi-Atlantic, south to Biscay coast of France; Iceland, Greenland; southern Labrador to Nahant, Mass.

Ecology: Primarily in brackish spray pools of bedrock coasts, at and above mean high water level; also subtidally in short estuaries (northern areas) and among cobbles in beach seeps; salinity range usually 2–32 o/oo; in fresh-water lakes and streams of western Europe (esp. Ireland) exposed to salt spray and salt rain, where sodium- and chloride-ion content is above 23 ppm.

Life Cycle: Annual. Ovigerous females from November to August, with a sequence of broods per year, female usually not breeding in first summer. Brood development 6–18°C.

Genus *Marinogammarus* Sexton and Spooner 1940   (măr ēn ō găm′âr ŭs)

Head, anterior lobe, upper angle rounded; eyes deep-reniform. Male usually larger than ♀. Antenna 2 noncalceolate in both. Lower lip lacking inner lobes. Coxal gills simple, pedunculate on peraeon segments 2–7; accessory gills often present on 6 and 7.

Gnathopods 1 and 2 subchelate, powerfully in ♂. Gnathopod 1 may be the larger, but gnathopod 2 is usually larger than 1. Peraeopods 5–7 successively longer, distal segments spinose, with few or no setae. Peraeopod 6 (and 7), basis with posterodistal free lobe. Urosome segments not dorsally elevated, spine clusters usually lacking supernumerary setae. Uropod 3, inner ramus short, less than one-third the outer ramus. Outer ramus narrow, spinose or foliaceous, terminal segment small or lacking. Telson lobes broad, apex spinose.

KEY TO AMERICAN ATLANTIC SPECIES OF *MARINOGAMMARUS*

1. Antenna 1, peduncle segment 1, posterior margin with one or no setal groups; uropod 3, outer ramus broad, foliaceous, 1-segmented; inner ramus about one-third outer; abdominal side plates 2 and 3, hind corner acute, slightly produced . .
. . . . *M. finmarchicus* (p. 58).
Antenna 1, peduncle 1, posterior margin with 3–5 groups of setae; uropod 3, outer ramus narrow, subcylindrical, spinose, with distinct terminal segment; inner ramus less than one-quarter outer; abdominal side plates 2–3, hind corners quadrate, blunt or obtuse . . . . 2.
2. Antenna 1, peduncle segment 2 shorter than 1, with 3–4 posterior marginal groups of setae; male smaller than female; gnathopod 2 larger than 1 (esp. in ♂); urosome dorsal spines very short; uropod 3, ramus with spines only . . . .
. . . . *M. stoerensis* (p. 59).
Antenna 1, peduncle segment 2 not shorter than 1, with 5–6 posterior marginal groups of setae; male larger than female; gnathopod 1 larger than 2 (in ♂); urosome dorsal spines prominent; uropod 3, ramus with spines and setae . . . . . .
. . . . *M. obtusatus* * (p. 58).

---

* Pinkster and Stock (1970) have placed *M. obtusatus* in the genus *Eulimnogammarus* along with *E. anisocheirus* Ruffo, *E. monocarpus* Stock 1969, and *E. toletanus* n. sp. from

Marinogammarus finmarchicus Dahl 1936 * Plate VII.1   (fĭn mârch′ī kŭs)

L. 14–24 mm. Male larger than female. Gnathopod 2 stronger than 1. Coxal plates 1–3 rounded below, anterior angle with 3–5 short setae. Eye deep, narrow reniform. Antenna 1 distinctly longer than antenna 2; peduncle segment 1 with 0–1, and segment 2 with 2–3 short setal groups on posterior margin. Mandibular palp widest about one-third from proximal end. Maxilliped, palp segment 3 very large, nearly as long as 2, dactyl strong.

Peraeopods 3–7, distal segments with clusters of spines and setae. Peraeopod 7, posterior margin with several very short setae. Abdominal side plates, hind corners acute, slightly produced, lower margins weakly spinose. Urosome segments with tall dorsal spines and a few setae. Uropod 1, outer ramus with inner and outer marginal spines. Uropod 3, outer ramus large, foliaceous, tapering, spinose and setose, 1-segmented; inner ramus about one-third length of outer. Telson lobes each with one lateral group of spines only.

Distribution: Amphi-Atlantic; northwestern Europe from Norway to Normandy, France, western Atlantic coasts, from eastern Nova Scotia (Cape Breton Island) to Cape Cod Bay, and Rhode Island.
Ecology: A dominant species of tide pools along bedrock surf coasts, at about mean high water level, at top of barnacle (*Balanus*) and *Fucus* zone; also in salt-marsh pools and their outflows at about the high water level and below; occ. among *Spartina* roots along salt-marsh embankments. In salinities down to less than 10 o/oo.
Life Cycle: Annual; females ovigerous from April–July.

Marinogammarus obtusatus Dahl 1936 Plate VI.1   (ŏb tūs āt′ŭs)

L. 12–18 mm. Male larger than female. Gnathopod 1 larger than gnathopod 2 in male. Coxal plates normal, rounded below, anterior distal angles with 3–5 very short setae. Antenna 1 longer than 2; peduncle segments 1, 2, and 3 with 4–5, 5–6, and 2–3 posterior marginal clusters of setae respectively. Antenna 2, flagellum setose, noncalceolate. Mandibular palp, segment 3 slender. Maxilliped, palp segment 3 shorter than segment 2.

Gnathopod 2 small, similar in ♂ and ♀; segment 5 longer than 6. Peraeopods 3–7, segments 5 and 6 with spine clusters and few simple setae. Peraeopod 7, posterior margin of basis with 6–8 short setae. Abdominal side plate 3, hind corner obtuse, posterior margin weakly setose, lower margin with a few short spines. Uropod segments with moderate dorsal elevations, lateral spine clusters with four strong spines but lacking setae. Uropod 3, outer ramus narrow, subcylindrical, with distinct terminal segment, and

---

Western Europe, based on relative sizes of gnathopods 1 and 2, and the fact that carpus of gnathopod 2 is elongate, not triangular.
* Stock (1970) places this species in the genus *Gammarus* (sens. lat.)

bearing marginal spines and setae; inner ramus short, setose and spinose, less than one-fourth outer ramus. Telson short, each lobe with single proximal outer marginal spine.

Distribution: Amphi-Atlantic; northern Norway south to the British Isles, English Channel; American Atlantic coast from Newfoundland (Avalon peninsula and Gulf coasts) and winter-mild protected localities along outer Nova Scotia south to Long Island Sound.

Ecology: Dominant in middle third and tidal zone; under stones and in pools, in somewhat lotic water. Also in salt marshes, under *Spartina* roots, among *Fucus* and *Ascophyllum* (esp. at Cape Ann, Mass.).

Life Cycle: Annual; ovigerous females November–September, bearing 2–3 broods of few large eggs; fecundity low.

Marinogammarus stoerensis (Reid) 1940 Plate VI.2   (stôr ĕn' sĭs)

L. 6–10 mm ♀, 5–7 mm ♂. Female larger than male; gnathopod 2 larger than 1. Head, eye deep reniform, narrow. Antennae relatively short, weak. Antenna 1 longer than 2; peduncle 1 and 2 with 3–4 clusters of posterior marginal setae. Antenna 2, peduncle weak, relatively few posterior marginal setae. Maxilliped, palp segment 3 short, not expanded.

Gnathopods 1 and 2 small, weak (♀); segments 5 and 6 of gnathopod 2 slender, strongly subchelate in male. Peraeopods 3–7, segments 5–6 with a few clusters of short spines, lacking setae. Peraeopod 7, posterior margin of basis with a few small hairs. Abdominal side plates 2 and 3, hind corners quadrate, lower margin spinose. Urosome segments dorsally flat, spines very short and weak. Uropod 3, outer ramus narrow, thick, with marginal spines only; terminal segment short; inner ramus very short, scalelike, with a single apical spine. Telson short, lobes broad, each with single lateral spine.

Distribution: Amphi-Atlantic; northwestern Europe from southern Norway and Great Britain to western France; in western Atlantic from eastern Nova Scotia to Cape Cod Bay, Mass.

Ecology: Occurring under stones and pebbles, intertidally from about mid-tide level down, in areas of beach seeps, small streams, and other lotic fresh waters at low tide, along semiprotected rocky and stony shores.

Life Cycle: Annual; females ovigerous from May until September; probably more than one brood per summer.

Genus *Gammarellus* Herbst 1793   (găm âr ĕl' ŭs)

Body deep, broad, middorsally carinate. Coxal plates relatively shallow; coxa 5, anterior lobe the larger. Urosome carinate, without spine clusters. Antennae short and stout, calceolate. Lower lip lacking inner lobes. Mandibular palp falcate. Maxilliped, outer plate short, palp very strong. Gnathopods powerfully subchelate, subequal (both sexes); propod oval, carpus short, deep. Peraeopods 3 and 4 stout, segment 5 short. Peraeopods 5–7 short, subequal.

Uropod 3, rami broad-lanceolate, foliaceous. Telson entire, or slightly emarginate apically.

Coxal gills folded, or plicate; on peraeon segments 2–7.

Brood plates large, foliaceous.

Remarks: Except for the well-developed accessory flagellum, this genus would more properly be placed in the family Calliopiidae. It is remote from all other Gammaridae (sens. str.).

KEY TO AMERICAN ATLANTIC SPECIES OF *GAMMARELLUS*

1. Dorsal carinations not produced posteriorly; eye very large, nearly contiguous middorsally; telson short, length = width . . . . *G. angulosus* (p. 60).
   Dorsal carinations large, produced posteriorly; eye large, lateral, not nearly contiguous dorsally; telson length twice width . . . . *G. homari* (Fabr.) *.

Gammarellus angulosus (Rathke) 1843 † Plate VII.2   (ăng ūl ōs′ŭs)

L. 13mm. Body stout, deep, heavy. Posterior peraeon segments and abdominal segment 1–4 with low middorsal carina not overhanging succeeding segment. Eye very large, black, subquadrate, nearly meeting middorsally (in ♂). Antenna 1 slightly shorter than 2, peduncle segments with few posterior marginal setae; accessory flagellum with 3–4 segments. Antenna 2, peduncle segments 4 and 5 short, little exceeding peduncle of antenna 1. Inferior antennal sinus shallow.

Coxa 1 produced anterodistally, corner blunt; coxa 2 broadly rounded in front. Gnathopods 1 and 2 closely subequal, palm very oblique, proportionately slightly larger and stouter in ♂. Coxa 5, anterior lobe deeper than posterior. Peraeopod 7, basis, posterior margin angled medially, weakly crenulate.

Pleon side plates 2 and 3, hind corners subquadrate, not acute. Uropod 3, rami broadly lanceolate, barely exceeding uropod 2, outer margin of outer ramus not foliaceous. Telson short, apically slightly emarginate.

Body whitish, mottled with red, green, brown, and blue.

Distribution: An amphi-Atlantic boreal species, extending in North America from Newfoundland and the Gaspé Penninsula (where it overlaps with *G. homari*) to the Gulf of Maine and the rockly coast of Connecticut. Also in Iceland and northwestern Europe.

Ecology: Low intertidal to about 20 m, on surf-exposed, algal-coated, rocky

---

* Northern, not recorded south of Newfoundland and Gaspé Peninsula.
† This species is often confused with *Calliopius laeviusculus* with which it frequently occurs on surf coasts. *Gammarellus* is distinguished readily by its true accessory flagellum and by the short segment 5 of peraeopods 3–7.

coasts; commonly clinging to algae or freely swimming in the swash zone at low water level.

Life Cycle: Annual; ovigerous females from April–July; one brood per year.

Family MELITIDAE n. fam.   (mĕl ĭt′ ĭ dē)

Body smooth anteriorly, generally slender. Coxal plates shallow, tending to separation at base. Sexual dimorphism of gnathopods marked, secondarily in antennae and uropod 3. Head, eye rounded, small; rostrum lacking; anterior lobe round, antennal sinus sharply incised. Abdominal segments distinct, urosome mucronate, dentate, spinose, or smooth. Mandibular palp slender, reduced, or vestigial, 1–3 segmented. Lower lip, inner lobes large, distinct. Maxilla 1, inner plate with few apical setae; outer plate with 7 apical spine teeth.

Gnathopod 2 larger than 1 (both sexes). Peraeopod 7 longest. Uropod 1, peduncle with anterior proximal spine. Uropod 3 large, rami spinose, not foliaceous, inner ramus tending to reduction. Telson deeply cleft, lobes diverging, apices acute, with spine(s).

Brood plates narrow, sublinear, with relatively few marginal setae. Coxal gills present on peraeon segments 2–6 only, pedunculate; no accessory gills.

Component Genera

Essentially warm temperate-tropical coastal brackish- to fresh-water group from which tropical fresh-water gammarids evolved. American Atlantic genera include *Melita, Maera, Elasmopus, Casco.* Tropical American genera include:

Metaniphargus	*Quadrivisio*
Weckelia	*Paramelita*
Paraweckelia	*Ceradocus*
Alloweckelia	*Maerella*, etc.

KEY TO AMERICAN ATLANTIC GENERA OF MELITIDAE

1. Antenna 1 distinctly longer than antenna 2; inferior head angle acute but not prolonged; coxal plates 3 and 4 usually subequal to or deeper than 1 and 2 . . . . 2.
   Antenna 1 shorter than antenna 2; inferior head angle strongly produced; coxal plates 1 and 2 distinctly deeper than 3 and 4 . . . . *Casco* (p. 62).
2. Uropod 3, rami subequal; abdomen without dorsal teeth or spines . . . . . 3.
   Uropod 3, rami markedly unequal, inner ramus scalelike; some or all abdominal segments with dorsal teeth and/or spines. . . . . *Melita* (p. 64).
3. Uropod 3 greatly exceeding uropods 1 and 2; coxal plates shallow, about as wide as deep; accessory flagellum long; maxilla 1, inner plate with 3 or more setae . 4.

Uropod 3 short, not greatly exceeding uropods 1 and 2; coxal plates deeper than broad; accessory flagellum 2–3 segmented; maxilla 1, inner plate with 2 setae . . . . *Elasmopus* (p. 63).

4. Pleon side plate 3, posterior margin sharply serrate; maxilla 1, inner plate with numerous marginal setae . . . . *Ceradocus*.
Pleon side plate 3 with tooth at hind corner only; maxilla 1, inner plate with 3 setae . . . . *Maera* (p. 66).

Genus *Casco* Shoemaker 1930   (kăs′ kō)

Body slender. Urosome 1 with middorsal tooth. Coxal plates 3 and 4 shallow, 1st produced anteriorly, 5th with deep anterior lobe. Head, rostrum very short; eye small, rounded; anterior head lobe blunt; inferior antennal sinus shallow, inferior angle strongly produced anteriorly. Antenna 1 shorter than antenna 2; accessory flagellum short.

Upper lip rounded. Lower lip with prominent inner lobes. Mandibular palp elongate, falcate, segment 2 longest. Maxilla 1, inner plate broad, setose; outer plate palp, distal segment medially broad. Maxilliped basal segments very broad, palp segment 2 strong.

Gnathopod 1 subchelate, palm very short; segment 5 longer than 6 (both sexes). Gnathopod 2 more powerfully subchelate (esp. in ♂), segment 6 longer than 5. Peraeopods 3 and 4 slender, segment 4 elongate, dactyls linear. Peraeopods 5–7 long, slender.

Uropods 1 and 2, rami long, spinose, subequal. Uropod 3 large, rami lanceolate, margins spinose. Telson deeply cleft, lobes diverging, acute.

Taxonomic Affinities: Although closely allied to *Cheirocratus,* this genus is also related to *Megaluropus,* members of which are sand-burrowing. Points of similarity are the short antenna 1, reduced coxae 3 and 4, form of gnathopod 1 and peraeopods 3 and 4, and the large subequal rami of uropod 3. Only one species is known to date.

Casco bigelowi (Blake) 1929 Plate XI.1   (bĭg ĕl ō′ĭ)

L. 25 mm. Urosome segments 2 and 3 each with pair of short middorsal spines. Eye weakly pigmented. Antenna 1, flagellum extending just beyond peduncle of antenna 2; accessory flagellum 2-segmented.

Mandible, palp segment 2 heavily long-setose. Maxilliped, basal segment with 3 facial groups of long curved setae.

Gnathopod 1, palm of segment 2 very short (esp. in ♀) greatly exceeded by weakly serrate dactyl. Gnathopod 2, (♂) segment 6 long-ovate, palm very oblique, irregularly toothed; in ♀, segment 6 narrow-ovate, little longer than 5, palm smooth, continuous with posterior margin. Interior and posterior margins of segments 5 and 6 of both gnathopods thickly setose.

Peraeopods 5–7 slender, basis little expanded, hind margin distally fusing almost directly with segment proper; segments 5 and 6, anterior margins setose. Coxal plate 7, posterior lobe acute.

Pleon side plate 3, lower corner acuminate. Uropod 3, outer ramus with very small terminal segment; peduncle longer than one-half rami. Telson cleft to base, subapical spine strong.

Distribution: Gulf of St. Lawrence, Chaleur Bay, Cape Breton Island, and Gulf of Maine south to Long Island Sound and off New Jersey. New England localities: Casco Bay (type), Cape Cod Bay—common, off Newport, R.I.
Ecology: Muddy and stony bottoms in cold-water areas, from extreme low water level to more than 50 m; often pelagic.

Genus *Elasmopus* Costa 1853   (ĕ lăs mō′pŭs)

Body smooth, not compressed. Coxal plates moderately deep, 4th largest, 4th and 6th anterior lobe deeper. Head, anterior lobe rounded, bearing eye basally; inferior antennal sinus short, sharply incised. Antenna 1 longer than 2; peduncle 3 elongate, accessory flagellum short. Antenna 2, peduncle slender.

Mandible, palp strong, segment 3 subfalciform. Maxilla 1, inner plate small, 2 apical setae. Maxilla 2, inner plate margin bare.

Gnathopod 1 small, segments 5 and 6 subequal (both sexes). Gnathopod 2 powerfully subchelate (esp. in ♂); segment 6 very large, segment 5 short, posterior lobe narrow. Peraeopods 5–7, basis broad, segments 4 and 5 expanded, stout, margins setose.

Uropods 1 and 2, outer ramus shorter. Uropod 3, rami short, stout, spinose; inner often shorter. Telson lobes approximate.

KEY TO AMERICAN ATLANTIC SPECIES OF *ELASMOPUS*

1. Peraeopod 6, posterodistal lobe of basis incised, lined with short spines . . . .
   . . . . *E. pectinicrus* Shoemaker.
   Peraeopod 6, posterodistal lobe of basis normally rounded. . . . . 2.
2. Gnathopod 2 (♂), segment 6 not hollowed medially to receive tip of dactyl; uropod 3, rami subequal . . . . *E. rapax* Costa.
   Gnathopod 2 (♂), segment 6 hollowed distally; uropod 3 inner ramus shorter. . 3.
3. Abdominal side plate 3 finely crenulate posteriorly; telson lobes rounded . . . .
   . . . . *E. pocillimanus* (Bate).
   Abdominal side plate 3 smooth posteriorly; telson apices acute . . *E. levis* (p. 63).

Elasmopus levis Smith 1873 Plate X.2   (lēv′ ĭs)

L. 12 mm ♂, 9 mm ♀. Antenna 1, accessory flagellum 2–3 segmented. Antenna 2 shorter, more slender; peduncle and 8-segmented flagellum with whorls of setae.

Gnathopod 2 (♂), segment 6 ovate, inner distal face excavate for reception of short, strongly curved dactyl; stout tooth at posterior proximal edge of excavation; in ♀, propod much smaller, dactyl closes normally on short convex palm.

Peraeopods 5–7, basis strongly expanded, posterodistal lobes produced, sharply rounding; in ♂ segments 4 and 5 strongly expanded; width nearly equal to length; distal lobes slightly produced; margins of segments 4–6 strongly setose; in ♀ segments are only slightly expanded and much less setose.

Pleon segment 3, hind corner produced in slightly upturned tooth, posterior margin smooth, not crenulate. Uropod 3, inner ramus distinctly shorter than outer, both longer than peduncle. Telson, lobes apically acute; outer subapical group of 3 spines, center spine longest.

Distribution: Cape Cod, from east side of Cape Cod Bay and Vineyard Sound, south to Georgia and northern Florida. Common along rocky or boulder shores of Vineyard Sound, Woods Hole, Buzzards Bay, and coast of Rhode Island and Connecticut.
Ecology: Intertidal and shallow water; under algae, stones (to mid-water level), among eelgrass clumps.
Life Cycle: Annual; ovigerous females May–September, several broods.

Genus *Melita* Leach 1813–14   (měl′ ĭt ă)

Body slender. Pleosome dorsally mucronate, dentate, or spinose. Head, rostrum very short, anterior lobe short, broad; inferior antennal sinus sharply incised; eye small, lateral. Coxal plates moderately deep, smooth below, 4th largest, 5th anterior lobe deeper. Antenna 1 longer than antenna 2, peduncle elongate; accessory flagellum prominent.

Mandible, palp slender, weakly setose, tending to reduction of segments. Maxilla 1, inner plate margin setose; outer plate with 7 apical spine teeth. Maxilliped, inner and outer plates well developed. Lower lip, inner lobes strong.

Gnathopod 1 small, slender; segment 5 longer than 6 (both sexes). Gnathopod 2 strongly subchelate (larger in ♂); segment 5 short, lobe short, deep.

Peraeopods 3 and 4 slender, dactyls short. Peraeopods 5–7, basis expanded, similar, 6 and 7 subequal, larger than 5.

Uropods 1 and 2, rami long, subequal. Uropod 3, outer ramus elongate, 2-segmented, margins spinose; inner ramus very short. Telson deeply cleft, lobes diverging, apices acute.

KEY TO NORTH ATLANTIC SPECIES OF *MELITA*

1. Pleon and urosome segments lacking dorsal teeth or carinations; urosome 2 spinose; pleon side plate 3, hind corner quadrate . . . . *M. nitida* (p. 65).
   Pleon segments usually with dorsal teeth, urosome 1 always with at least one dorsal tooth; pleon side plate 3 acute, produced . . . . 2.
2. Urosome segment 2 with one or more dorsal teeth or carinations; gnathopod 1 (♂), dactyl normal; telson lobes with weak inner marginal spines . . . . 3.

Urosome segment 2 with 2 short spines on either side of middorsal line; gnathopod 1 (♂), dactyl short, curled inward; telson lobes with 2 strong inner marginal
spines . . . . *M. palmata* (Montagu) European.
3. Some or all pleon segments dorsally toothed or mucronate; urosome with more
than 4 dorsal teeth . . . . 4.
Pleon segments without dorsal teeth or mucronations; urosome segments with
total of 4 dorsal teeth. . . . . *M. quadrispinosa* Vosseler.
4. Pleon segments 1–3 each with a posterodorsal central tooth and several smaller
teeth; urosome segments each with several teeth . . . . . *M. dentata* (p. 65).
Pleon segment 1 lacking dorsal tooth, 2 and 3 with single tooth only; urosome
segments 1 and 2 with 3–4 dorsal teeth only . . . . . . *M. formosa* Murdoch.

Melita nitida Smith 1873 Plate IX.2 (nĭt′ĭd ă)

L. 12 mm ♂, 9 mm ♀. Coxal plates large, rounded below. Urosome 2 with 4–5
short spines on each side of middorsal line. Antenna 1 strong, flagellum with
whorls of short setae; accessory flagellum 2–3 segmented. Antenna 2 (♂), peduncle 5 and flagellum with whorls of stiff, moderately long setae.

Mandibular palp slender, segment 3 equal to or longer than segment 2.
Maxilliped, outer plate short, not reaching tip of palp segment 2.

Gnathopod 1 (♂), dactyl short, arising below anterior margin, tip curled
inwards, closing on inner face of propod; in ♀, dactyl closing normally on
palm. Gnathopod 2 (♂), dactyl tip closing on inner side of palmar angle; palm
smooth, convex; in ♀, smaller less robust, similar.

Peraeopods 5–7, basis large, with posterior-distal lobe produced,
rounded.

Pleon segments 1–3 and urosome segments 1–3 lacking dorsal teeth or
mucronations. Pleon side plate 3, hind corner quadrate, not produced, posterior border smooth. Uropods 1 and 2 heavy, tips of rami extending beyond
peduncle of uropod 3. Uropod 3, terminal segment of outer ramus minute.
Telson lobes with apical and subapical spines.

Distribution: Southwestern Gulf of St. Lawrence (Northumberland Strait),
southwestern Nova Scotia (Pubnico Hbr.), central Maine to Cape Cod and
southward to Georgia and northern Florida, Gulf of Mexico to Yucatan. Also
recorded at Piscataqua estuary, N.H., Barnstable Harbor and Bass River estuary, Pocasset estuary, Mass., Narragansett Bay, R.I.
Ecology: Muddy bottom areas, mesohaline regions of estuaries, in salinities
of 3 to 20 o/oo, occ. to 30 o/oo. Often at base of clumps of hydroids and
ectoprocts.
Life cycle: Annual; ovigerous females from May–September, several broods.

Melita dentata (Krøyer) 1842 Plate IX.1 (děn tāt′ ă)

L. 22 mm ♂, 18 mm ♀. Body slender, elongate. Coxal plates 1–4 relatively shallow, hind corner notched. All pleosome and urosome segments with several

posterodorsal teeth or mucronations, middorsal tooth strongest. Antenna 1 elongate, accessory flagellum 4-segmented. Antenna 2, flagellum short, weakly setose.

Mandible, palp segment 1 with distal angle produced. Maxilliped, outer plate large, reaching end of palp segment 2.

Gnathopod 1 (both sexes), dactyl normal, approximating convex palm. Gnathopod 2 (both sexes), propod subquadrate; palm oblique, irregularly low-toothed; dactyl strongly curved, closing on acute posterodistal angle; small in ♀. Peraeopods 6 and 7 subequal, basis posterodistal lobe subacute, little produced.

Pleon side plate 3 acute, strongly produced. Uropods 1 and 2 short, tips not exceeding peduncle or uropod 3. Uropod 3 long and slender (esp. in ♂), terminal segment of outer ramus very short. Telson lobes, inner marginal spines very weak.

Distribution: A subarctic and boreal species: North Atlantic; in North America south to the Gulf of Maine and Cape Cod Bay.
Ecology: Moderately common along cold-water rocky and stony shores; from low water level to nearly 300 m. Epibenthic and under stones.
Life Cycle: Presumably annual; ovigerous females in March–May.

Genus *Maera* Leach 1813–14   (mēr' ă)

Body slender, elongate, dorsally smooth. Coxal plates shallow, smooth below, tending to separation, 4th little emarginate; 5th and 6th anterior lobe deeper. Head, rostrum short, anterior lobe rounded, eye small, rounded; inferior antennal sinus deep, rounded. Antenna 1 slender, longer than 2, peduncle elongate, 3 short; accessory flagellum long.

Mandible, palp slender, weakly setose; segment 1 with distal tooth. Maxilla 1, inner plate with 3 apical setae. Maxilla 2, inner plate margin bare.

Gnathopod 1, coxa produced anterodistally. Gnathopod 2 powerfully subchelate in ♂ and ♀, larger in ♂; segment 5, posterior lobe broad. Peraeopods 5–7 slender, basis little expanded; dactyls with short nail and subapical tooth.

Uropods 1 and 2 strong, rami subequal. Uropod 3 very large, rami broad, subequal, outer 2-segmented. Telson short, lobes slightly diverging, broadly acute.

KEY TO NEW ENGLAND SPECIES OF *MAERA*

1. Uropod 3 short, rami little exceeding uropod 1; peraeopods 6 and 7, basis relatively broad, with posterodistal lobe; dactyls bicuspate . . *M. inaequipes* (Costa).
   Uropod 3 elongate, rami much exceeding uropod 1; peraeopods 6 and 7, basis narrow, hind margin lacking distal lobe; dactyls simple . . . . 2.
2. Eyes unpigmented; peraeopods 5–7, basis very narrow, length about 3 times width; pleon side plate 3, hind corner acuminate . . . . . . *M. loveni* (Bruz).

Eyes round, black, small; peraeopods 5–7, basis length about twice width; pleon side plate 3, posterior tooth near midpoint of margin . . . . *M. danae* (p. 67).

Maera danae Stimpson 1853 Plate X.1 (dā′ nē)

L. 22 mm. Body slender, elongate. Coxal plates shallow, lower margin of 1 and 2 smooth, tending to slight separation. Eye small, roundish, black. Antenna 1, accessory flagellum 6-segmented. Antenna 2, peduncle 4 and 5 and flagellum setose esp. posteriorly; gland cone not reaching segment 4.

Gnathopod 2 larger in ♂, similar; segment 6 subquadrate, palm oblique, straight, evenly toothed or serrate, posterior angle tooth strong; dactyl strong, outer margin setose; segment 5 moderately long, posterior lobe with 6–8 groups of setae, some rastellate. Peraeopods 5–7, dactyls short, bidentate at tip; basis little expanded, posterodistal lobe not produced, oblique.

Pleon side plates 2 and 3, hind corner weakly toothed. Uropods 1 and 2 short, barely extending beyond peduncle of uropod 3. Uropod 3 very large, much exceeding uropods 1 and 2, rami subequal, about 3 times length of peduncle. Telson cleft nearly to base, lobes broad.

Distribution: American Atlantic; from the Gulf of St. Lawrence, outer coast of Nova Scotia and Gulf of Maine south to New Jersey. Common in Passamaquoddy Bay, Mt. Desert Island, Cape Cod Bay.
Ecology: A benthic and epibenthic species of muddy bottoms along rocky shores; from immediately subtidal to depths of more than 50 m.
Life Cycle: Annual or biennial; ovigerous females from March–May.

Family CRANGONYCIDAE n. fam. (krăng ŏn ĭs′ĭ dē)

Body smooth, urosome smooth, segments sometimes coalesced. Sexual dimorphism prominent in gnathopods, antenna 2, and uropods 2 and 3. Coxal plates deep, 4th largest.

Head without rostrum or inferior antennal sinus. Eye small, rounded. Antennae slender, antenna 1 much longer than 2; peduncle 1 large, accessory flagellum 2-segmented, terminal segment small. Antenna 2, peduncle strong, often calceolate in ♂. Mandibular molar small, palp stout. Maxilla 1, inner plate medium, with few marginal setae; outer plate with 7 apical spine teeth. Maxilliped, plates relatively small, palp large, powerful. Lower lip shallow, broad inner lobes partly fused to outer.

Gnathopods powerfully subchelate (both sexes), 2 usually the larger, palmar spines notched. Peraeopods slender, basis of 5–7 broadly expanded; peraeopod 6 usually longest. Uropod 1, peduncle smooth, lacking proximal anterior spine. Uropod 2, outer ramus sometimes modified in ♂. Uropod 3 short, not foliaceous, inner ramus always very reduced or lacking. Telson entire or slightly cleft, lobes apically spinose.

Coxal gills present on segments 2–6, occ. on 7, accessory gills usually on segment 6 and 7, occ. sternal gills present. Brood plates large, lamellar, with many marginal setae.

Members are exclusively fresh-water, N. America and Eurasia.

N. American Coastal Plain genera include: *Crangonyx, Stygonectes, Stygobromus, Synurella;* also *Bactrurus* and *Allocrangonyx.*

KEY TO NEW ENGLAND GENERA OF CRANGONYCIDAE

1. Uropod 3 biramous, ramus longer than peduncle; eyes present; mainly epigean .
    . . . . *Crangonyx* (p. 68).
    Uropod 3 uniramous, ramus shorter than peduncle; eyes lacking; subterranean
    (hypogean) . . . . *Stygonectes.*

Genus *Crangonyx* Bate 1862   (krăng ŏn' ĭks)

Body smooth, moderately compressed. Coxal plates 1–4 deep, lightly setose below, 4th largest, deeply excavate behind; coxae 5 and 6, posterior lobe deeper. Head, rostrum very short, anterior lobe large, rounded, inferior antennal sinus lacking, eye small, lateral. Antenna 1 longer than 2, peduncle 1 stout, subequal to 2. Antenna 2, peduncle longer than flagellum, both calceolate in ♂. Lower lip shallow, broad, inner lobes weak. Maxilla 1, inner plate with 4–5 setae. Female usually larger than male.

Gnathopod 1 and 2 strongly subchelate (esp. in ♂), 2 larger than 1. Peraeopod 6 longest. Uropods 1 and 2 strongly subequally biramous, uropod 2, outer ramus modified in ♂. Uropod 3, unequally biramous, outer ramus longer than peduncle, 1-segmented, inner ramus a small lobe. Telson short, broadly cleft, apices spinose.

KEY TO NEW ENGLAND SPECIES OF *CRANGONYX*

1. Gnathopods 1 and 2 (♀), palm of segment 2 lined regularly with cleft spines; segment 6, superior lateral setae in groups of 2–4. . . . *C. richmondensis* (p. 69).
    Gnathopods 1 and 2 (♀), palm lined with short weak spines; superior lateral setae singly inserted . . . . *C. pseudogracilis* (p. 68).

Crangonyx pseudogracilis Bousfield 1958
(= *Melita parvimana* Holmes 1905) Plate VIII.1   (sū dō grǎ' sǐl ĭs)

L. 10 mm ♀. Coxal plates moderately deep, 1 shallower than head. Eye relatively large, esp. in ♂, black. Antenna 2, peduncle segments 4 and 5 with 3–4 posterior groups of setae. Maxilla 2, inner plate with facial row of setae.

Gnathopod 1 (♀) relatively weak, not broadest distally, posterior marginal setae singly inserted, segment 5, posterior lobe narrow, dactyl inner margin smooth, palmar teeth weak, unevenly spaced. Gnathopod 2 (♀) slightly stronger, segment 6, superior lateral setae singly inserted, posterior palmar angle with one strong spine and two short spines posterior to it. Gnathopods (♂) strongly subchelate, palmar spines strong, regular.

Peraeopods 3–7, dactyls simple, not elongate, with 1 inner marginal seta. Peraeopod 7, basis posterior margin with about 10–20 serrations.

Pleon side plates 2 and 3, hind corners acute, that of 2 slightly produced. Uropod 1, peduncle strong, posterior margins strongly spinose. Uropod 2, outer ramus shorter, simply spinose in ♀, slightly recurved and distally armed with numerous comb spines in ♂. Uropod 3, outer ramus short, not more than 2 times length of peduncle. Telson cleft about half to base, lobes apically 3-spined.

Distribution: Eastern North America; Atlantic drainages from the St. Lawrence and Hudson east to larger rivers of New England. Introduced in British Isles.
Ecology: Fresh-water to slightly oligohaline; larger rivers, streams, lakes and ponds. In somewhat turbid, summer-warm, weedy locations.
Life Cycle: Annual; ovigerous females from April–September; several broods per female.

Crangonyx richmondensis richmondensis Ellis 1940
Plate VIII.2   (rĭch mŏnd ĕn'sĭs)

L. 14mm ♀, 9 mm ♂. Coxal plates 1–4 large and deep, coxa 1 deeper than head. Eye relatively small, black. Antenna 2 slender, short. Maxilla 2, inner plate lacking facial row of setae.

Gnathopod 1 powerfully subchelate (similar in ♂ and ♀), segment 6 broadening distally; posterior marginal setae doubly inserted; palmar teeth strong, evenly spaced; dactyl simple, smooth behind. Gnathopod 2 (♀) powerfully subchelate; segment 2, superior lateral setae in groups of 2–4; palm evenly convex, marginal teeth strong, regular; posterior lobe of segment 5 short, with 1 outer marginal seta.

Peraeopods 5–7 elongate, bases broadly expanded, posterior margins weakly serrate, dactyls short, each simple with 1 marginal seta.

Pleon side plates 1–3, hind corners acute, that of 2 moderately produced. Uropod 1, peduncle posteriorly weakly spinose. Uropod 2, outer ramus shorter, normally spinose (♀), distinctly recurved, with a few simple marginal spines in ♂. Uropod 3, outer ramus more than 2 times length of peduncle. Telson short, shallowly notched, apices with 3 spines.

Distribution: American Atlantic coastal areas from South Carolina north to New England, Nova Scotia (including Sable Island), and southern Newfoundland (Avalon). Specifically recorded in Maine and central and coastal Massachusetts.
Ecology: Bog ponds and outflows, low *p*H, humic, highly colored, nutrient-

poor fresh waters; hiding under debris and under grassy banks, from shoreline to depths of over 10 feet.

Life Cycle: Annual; ovigerous females April–June, dying out completely, some males persist to August; only one brood per female.

Family LILJEBORGIIDAE   (līlj börg ē′ ĭ dē)

Body smooth; pleosome and/or urosome dentate or spinose dorsally. Coxal plates large, deep, 1st expanded anterodistally, 4th excavate behind.

Sexual dimorphism strongly shown in gnathopods, uropods, and telson. Head, rostrum short, inferior antennal sinus very oblique; eye small. Antenna 1 shorter than 2, accessory flagellum distinct. Antenna 2 noncalceolate.

Upper lip slightly bilobed. Lower lip, inner lobes fused or lacking. Mandible, molar feeble, not triturative; palp geniculate at 2nd segment, segment 1 elongate. Maxilla 1, inner plate small; outer plate with 7 apical spine teeth. Maxilliped, plates small; palp large, dactylate.

Gnathopods large, subchelate (both sexes); carpus short, posterior lobe pronounced. Peraeopods 3–4 slender. Peraeopods 5–7, basis expanded; 7th distinctly longest and heaviest. Uropods 1 and 2 strong, spinose. Uropod 3 biramous, outer shorter, rami flattened, lanceolate. Telson large, deeply cleft, apices bifid.

Coxal gills on peraeon segments 2–6, saclike. Brood plates small, sublinear, with few marginal setae.

KEY TO AMERICAN ATLANTIC GENERA OF LILJEBORGIIDAE

1. Gnathopod 1 larger than 2 (both sexes) . . . . *Idunella* (p. 70).
   Gnathopod 1 smaller than or equal to 1 (both sexes) . . . . 2.
2. Gnathopod 2, segment 5, posterior lobe greatly prolonged behind propod; outer ramus of uropod 3 always 1-segmented . . . . *Liljeborgia.*
   Gnathopod 2, segment 5, posterior lobe not prolonged behind propod; outer ramus of uropod 3 usually 2-segmented . . . . *Listriella* (p. 71).

Genus *Idunella* Chevreux 1920 (formerly *Sextonia*)   (ī dūn ĕl′ ă)

Head, rostrum developed. Urosome 3 with pair of dorsolateral spines. Antenna 1 and 2, flagella multisetose; accessory flagellum well developed.

Mandible, cutting edge bidentate; left molar process strong. Maxilla 1, inner plate with 4 apical setae.

Gnathopod 1 much larger and stronger than 2 (both sexes), larger in ♂. Peraeopod 7 longer and stouter than 5 and 6.

Uropod 3 rami large, subequal, outer 2-segmented. Telson long, deeply cleft.

KEY TO AMERICAN ATLANTIC SPECIES OF *IDUNELLA*

1. Rostrum long; gnathopod 1 (♂), palm very oblique, hind margin short . . . . .
   . . . . *I. longirostris* Chevr.
   Rostrum short; gnathopod 1, palm oblique, posterior margin distinct . . . . . .
   . . . . *Idunella* species (Ches. Bay south)

Genus *Listriella* Barnard 1959 (lĭstrē ĕl' ă)

Body smooth, not mucronate. Urosome 3 with lateral cusps and/or short dorsal spines. Coxal plates smooth below, 5th and 6th with deeper hind lobe. Antennae short, flagella few-segmented; accessory flagellum 2–3 segmented.

Maxilla, inner plate narrow, with 2 apical setae. Maxilliped, inner plate small.

Gnathopod 2 larger and stronger than 1 (both sexes), larger in ♂ ; segment 5, posterior lobe narrow, short, not prolonged below segment 6.

Peraeopods 5–7, posterior margins of segments 5 and 6 with spines and/or short setae. Uropod 3, outer ramus shorter, 2-segmented, or fusing to 1 segment (adults). Telson medium cleft (variably), apices acute, spinose.

KEY TO AMERICAN ATLANTIC SPECIES OF *LISTRIELLA*

1. Peraeopod 7, segments 6 and 7 heavy, dactyl short; gnathopod 2, palmar margin with short tooth near hinge; epimeral plate 3, hind corner blunt. . . . .
   . . . . , *L. clymenellae* (p. 72).
   Peraeopod 7, distal segments slender, nearly unarmed; gnathopod 2, palmar margin convex, lacking strong hinge tooth; epimeral plate 3, hind corner with tooth . .
   . . . . *L. barnardi* (p. 71).

Listriella barnardi Wigley 1963 Plate XII.2 (bârn' ârd ī)

L. 5.5 mm. Head, anterior lobe acute. Eye round, lightly pigmented. Coxal plates 5 and 6 relatively deep, hind lobes deeper. Antenna 1, flagellar segment 1= peduncular segment 3. Maxilliped, inner plate very short.

Gnathopod 1, palm evenly convex, posterior margin longer. Gnathopod 2, palm evenly convex in ♀ ; in ♂ , margin excavate with small tooth adjacent to hinge; dactyl finely toothed on inner margin, and low swelling opposite palmar excavation.

Peraeopods 3–7, dactyls relatively long and slender, more than one-fourth length of segment 6. Peraeopods 6 and 7, basis broadly expanded, posterior margins evenly convex, evenly lined with setae. Peraeopod 7 longest and heaviest, segments 5 and 6 slender, with a few marginal spine groups.

Pleon side plate 3, hind corner forming a small upturned tooth. Uropods 1 and 2 strong, rami and peduncles subequal. Uropod 3 (♀), rami slender lanceolate, longer than peduncle, outer ramus 1-segmented in adult; in ♂, rami broad lanceolate, short, about equal to peduncle. Telson cleft nearly halfway to base, deeper in ♀.

Distribution: South side of Cape Cod and southern New England to Georgia. New England localities: Martha's Vineyard, Hadley Harbor, Buzzards Bay.
Ecology: Commensal in tubes of *Amphitrite ornata* and other marine polychaetes. From low intertidal to channels and gutters between islands.
Life Cycle: Annual; ovigerous females in June–September; several broods.

Listriella clymenellae Mills 1963 Plate XII.1   (klī měn ĕl′ ē)

L. 5–6 mm. Head, anterior lobe broadly rounded; eye relatively small, black. Antenna 1, flagellar segment 1 longer than peduncle segment 3. Coxal plates 5 and 6 relatively shallow, hind lobe little deeper than front lobe. Maxilliped, inner plate narrow, relatively long, with 2 apical setae.

Gnathopod 1 (both sexes) considerably smaller than gnathopod 2; palm convex, posterior margin long; inner face of segment 6 with posterodistal clusters of setae. Gnathopod 2 very strong (♂) with strong tooth near hinge (both sexes); inner face of segment 6 with numerous posterodistal clusters of setae.

Peraeopods 5–7, basis moderately expanded, posterior margin shallow-convex or nearly straight, margins nearly bare; dactyls short, less than one-fourth length of segment 6. Peraeopod 7 longest and heaviest, segments 5 and 6 each with 3–4 marginal clusters of spines and setae.

Pleon segment 3, hind corner quadrate. Uropods 1 and 2, rami distinctly shorter than peduncle. Uropod 3, rami short, broad lanceolate, outer ramus 1-segmented (adults, both sexes). Telson short, cleft about half to base.

Distribution: Cape Cod peninsula (Barnstable Harbor), Massachusetts to Chesapeake Bay to southeastern states and northern Florida. Also recorded at Barnstable Harbor (type locality), Vineyard Sound, Hadley Harbor, Buzzards Bay, Massachusetts.
Ecology: Commensal in tubes of *Clymenella torquata;* low intertidal flats to depths of 10 m or more.
Life Cycle: Annual; ovigerous females from May–September; several broods.

$\mathcal{P}$ONTOGENEIIDAE AND RELATED FAMILIES

Powerfully free-swimming deep-bodied amphipods with accessory flagellum markedly reduced or lacking; peduncle of antenna 2 short, not powerful; head with large eyes and moderate to strong rostrum; mouthparts basic, tending to modification of upper and lower lips; subchelate gnathopods, tending to reduction; telson simple cleft or entire; sexual dimorphism tending to reduction, especially in gnathopods.

Family PONTOGENEIIDAE (excl. Eusiridae)　　　　　(pŏnt ö jĕn ē'ī dē)

Body smooth, abdomen occ. carinate; peraeon short, pleosome powerful. Coxal plates moderately deep. Sexual dimorphism in antenna, gnathopods, and uropod 3. Head, rostrum strong. Eyes large, lateral. Inferior antennal sinus sharply incised. Antennae slender, elongate, subequal, calceolate in ♂. Accessory flagellum lacking or forming a minute scale. Mouthparts normal. Lower lip, inner lobes weak. Mandible molar strong; palp large, segment 2 longest.

　　Gnathopods 1 and 2 subchelate, usually small; carpal lobe shallow, attachment between segments 5 and 6 normal. Peraeopods 3–7 not excessively elongate or slender; basis of 5–7 expanded. Pleopods powerful, long. Uropod 1 and 2 sublanceolate, outer ramus shorter. Uropod 3, rami lanceolate, foliaceous. Telson intermediate in length, not attaining end of uropods 1 and 2; sharply cleft, not to base.

　　Coxal gills on segments 2–7, pedunculate. Brood plates large, lamellate, margins multiple setose.

　　The family is essentially endemic to Pacific and Antarctic regions.

Genus *Pontogeneia* Boeck 1870–71　　　　　(pŏnt ō jĕn ē'ă)

Pleon dorsally smooth or weakly toothed. Coxal plates 1–4 rather small, much overlapping, 4th largest, excavate behind; coxae 5–7 moderately deep behind. Pleosome and pleopods powerful. Head, rostrum moderate to

73

strong; anterior lobe blunt; inferior antennal sinus sharply incised. Antennae long, slender, subequal, calceolate in ♂, accessory flagellum lacking. Eye large subrectangular, lateral, larger in ♂.

Lower lip, inner lobes small or obsolescent; outer lobes with small medial gap. Maxilla 1, inner plate with fewer than 4 terminal setae; palp segment 2 longer than 1. Upper lip, epistome lacking anterior process. Mandibular palp, segment 3 shorter than 2.

Gnathopods 1 and 2 subchelate, subequal, similar; segment 5, posterior lobe shallow. Peraeopods 3–7 relatively slender, dactyls stout. Peraeopods 3 and 4, segments 4–6 with posterior plumose setae in ♂. Peraeopods 5–7, basis expanded posteriorly; segments 4–6 about equal to basis.

Uropods 1 and 2 outer ramus shorter. Uropod 3, rami lanceolate, subequal, foliaceous, outer ramus 1-segmented. Telson narrowly and deeply cleft.

KEY TO AMERICAN ATLANTIC SPECIES OF *PONTOGENEIA*

1. Coxal plates 1–3 deep or deeper than long; telson lobes tapering distally; uropod 2 exceeding telson; mandible, palp 2 nearly straight . . . . 2.
   Coxal plates shallow, broader than deep; telson cleft nearly to base, lobe margins subparallel; uropod 2 not reaching tip of telson; mandibular palp segment 2, inner margin strongly convex . . . . *P. bartschi* Shoem.
2. Gnathopods 1 and 2, carpus shallow, longer than propod; antenna 1, peduncular segments normal, combined length greater than head; mandible, palp segment 3 about three-fourths length of 2 . . . . *P. inermis* (p. 74).
   Gnathopods 1 and 2, carpus shorter than propod; lobe subacute below; antenna 1, peduncular segments very short, combined length shorter than head; mandibular palp segment 3 very short, about half segment 2 . . . . . *P. longleyi* Shoem.

Pontogeneia inermis (Krøyer) 1842 Plate XIII.1   (ĭn ĕrm′ ĭs)

L. 11 mm. Coxal plates 1–4 relatively small, deeper than broad. Pleon smooth dorsally. Eye large, reddish, broad reniform, larger in ♂. Rostrum moderately strong, acute. Antenna 1 (♀), peduncle segments 1 and 2 with 3–4 posterior marginal setal groups; accessory flagellum lacking. Antennal peduncles somewhat expanded, calceolate in ♂.

Lower lip, inner lobes vestigial. Mandibular palp, segment 2 nearly straight; segment 3 about three-fourths length of 2. Maxilliped, inner plate large, reaching palp segment 2.

Gnathopods 1 and 2 weakly subchelate (both sexes), subequal, segment 5 shallow, longer than 6. Peraeopod 7, basis much larger than that of peraeopods 5 and 6, posterior margin weakly crenulate.

Pleon side plates 2 and 3, hind corners acuminate; posterior margin of 3 sharply convex. Uropods 1 and 2, margins spinose; uropod 2 exceeding telson. Uropod 3 little exceeding 1 and 2, rami lanceolate, foliaceous in both sexes, more strongly so in ♂.

Telson longer than broad, cleft about two-thirds to base, lobes tapering distally.

Distribution: Arctic boreal, North Pacific and North Atlantic; in America south to Long Island Sound. Gulf of Maine commonly, Cape Cod Bay, Martha's Vineyard (to June).
Ecology: An essentially pelagic cold-water species; clings to submerged plants and algae, from lower intertidal levels to more than 10m.
Life Cycle: Annual; ovigerous females during winter months to about May.

Family BATEIDAE (bāt ē′ ī dē)

Body smooth or carinate dorsally. Coxal plates 3–4 very large and deep; coxa 1 rudimentary; coxa 5, hind lobe deeper than anterior lobe. Pleosome strong, urosome segment 3 longer than 2. Sexual dimorphism slight (antennae and uropod 3). Head, rostrum strong, eye large, lateral, rectangular. Antennae slender, long, subequal, calceolate in ♂. Accessory flagellum lacking. Antenna 2, peduncle short. Mouthparts normal. Mandibular molar strong, palp large, segment 2 longest; lower lip lobes weak or lacking. Maxilliped palp relatively short.

Gnathopod 1 vestigial, 2-segmented. Gnathopod 2 subchelate, small; carpus short, dactyl toothed behind. Peraeopods 3 and 4 strong, segment 6 and dactyl strong. Peraeopods 5–7 increasing posteriorly, basis variously expanded. Uropod 3, rami lanceolate, foliaceous, subequal. Telson short, sharply cleft, not to base.

Coxal gills large, pedunculate, simple or pleated, on segments 2–7. Brood plates large, spatulate, margins setose.

KEY TO AMERICAN ATLANTIC GENERA OF BATEIDAE
(see Shoemaker, 1925)

1. Peraeon and pleon dorsally smooth; gnathopod 2, segments 3 and 4 short, subequal; uropods 1 and 2 stout, lightly spinose; telson lobes regular . *Batea* (p. 75). Pleon dorsally carinate; gnathopod 2, segment 3 elongate, longer than 4; uropods 1 and 2 very slender, heavily spinose; telson lobes abruptly narrowing distally, apices acute . . . . *Carinobatea.*

Genus *Batea* Muller 1865 (bāt′ ē ă)

Body smooth, not dentate. Head, rostrum very strong; interantennal lobe scarcely produced, inferior antennal sinus shallow; lower head process blunt; eye large, especially in ♂. Coxal plate 4 very large and deeply excavate behind. Antenna 1 slightly shorter than antenna 2.

Mandible, palp segment 3 longer than half segment 2. Lower lip, inner lobes small or lacking. Maxilla 1, palp normally 2-segmented. Maxilliped, palp segment 2 not shortened.

Gnathopod 1, second segment linear, longer in ♀ than in ♂. Gnathopod 2, segment 3 about equal to 4, not elongate.

Peraeopods 3 and 4, segments 4 and 5, posteriorly plumose-setose in ♂. Peraeopods 5–7, basis slightly expanded posteriorly.

Pleon side plate 3, posterior margin serrate. Uropods 1 and 2 stout, lightly spinose posteriorly. Uropod 3, rami lanceolate, foliaceous, subequal; outer ramus 1-segmented. Telson deeply cleft, lobes regularly narrowing apically. Gills simple, not pleated. Sexual dimorphism strongly pronounced in antennae, eyes, gnathopod 1, peraeopods 3 and 4, and uropod 3.

Batea catharinensis Muller 1865
(= *Batea secunda* Holmes) Plate XIII.2   (kăth âr ĭn ĕn′sĭs)

L. 5–8 mm ♀. Head, rostrum strong, down-curved, acute, reaching at least to middle of antennal peduncle 1. Anterior angle obtuse. Coxal plates 2 and 3, lower margins convex, angles rounded; coxa 4, posterior lobe nearly as wide as coxa proper at base. Eyes large, broad-reniform, purple-brown. Antenna 1, peduncle 2 nearly equal to 1, both setose posteriorly.

Mandibular palp, segment 2 slightly longer than 3, inner margin setose but not sharply expanded distally. Maxilliped, outer plate not reaching palp segment 3. Lower lip, inner lobes small.

Gnathopod 2, carpus shorter than propod, posterior lobe relatively narrow and deep; propod deep, palm oblique, longer than posterior margin. Peraeopod 5, basis little expanded distally. Peraeopod 6, basis suborbicular, wider than long. Peraeopod 7, basis much deeper than broad, prolonged posterodistally beyond segment 3 as rounded lobe.

Pleon side plate 3, hind margin convex, lined with about 10 fine upward-pointing teeth, at the base of each of which is a setule. Uropod 3, rami broad-lanceolate, margins finely spinose and plumose-setose. Telson cleft beyond the middle, apices blunt and rounded, inner edges convex.

The male differs in stronger rostrum, larger eyes, longer calceolate antennae, less setose rudimentary gnathopod 1, presence of plumose setae on posterior margins of segments 5 and 6 of peraeopods 3 and 4, and more foliaceous uropod 3.

Distribution: American Atlantic coast, from the south side of Cape Cod and Long Island Sound, the Middle Atlantic states to Georgia, Florida, the Gulf of Mexico, and the Caribbean region to Brazil.

Ecology: Subtidal, on stony, gravelly bottom, to depths of over 20 m.

Life Cycle: Annual; ovigerous females from May–September, several broods per female.

Family EUSIRIDAE (non Barnard 1969)   (ū sĭr' ĭ dē)

Pleon dorsally carinate, mucronate, or both. Coxal plates small to medium deep, usually smooth below, 4th largest, excavate behind. Head, rostrum medium to strong. Eyes well developed. Thoracic segments short. Abdomen large, powerful. Antennae slender, peduncles elongate, accessory flagellum small, usually 2-segmented. Sexual dimorphism surpressed, except in antennae.

Mouthparts normal (primary). Mandible, molar ridged; palp large, falcate. Upper lip rounded. Lower lip, inner lobes partly developed. Maxilla 1, inner plate small, weakly setose; outer plate with 8–9 apical spine teeth. Maxilla 2, plates small. Maxilliped, inner plate small, 5–8 apical spines; palp large, raptorial.

Gnathopods subchelate, typically large, powerful, subequal; dactyls smooth; propod often attached to prolonged apex of carpus. Peraeopods 3–7 slender, often elongate, 5–7 increasing posteriorly.

Pleon side plate 3 usually serrate behind. Pleopods large, strong. Urosome segment 2 shorter than 1 and 3. Uropods lanceolate, outer ramus shorter. Telson large, tapering, narrowly cleft apically.

Coxal gills simple, on peraeon segments 2–7. Brood plates large, lamellate, margins multisetose.

KEY TO AMERICAN ATLANTIC GENERA OF EUSIRIDAE

1. Gnathopod propods deeper than long, attached to distal prolongation of carpus; antenna 1 longer than 2   .   .   .   . 2.
   Gnathopod propods longer than deep, attached normally to carpus; antenna 1 shorter than 2 .   .   .   . *Rhachotropis* (p. 77).
2. Carpal lobe of gnathopods narrow, deep; coxal plates with notched or serrate hind corner, 4th deepest; mandibular palp, segment 3 longer than 2.   .   .   *Eusirus.*
   Carpal lobe broad, shallow; coxal plates evenly rounded, 2 and 3 deepest; mandibular palp, segment 3 shorter than 2 .   .   .   .   *Eusirogenes.*

Genus *Rhachotropis* Smith 1883   (răk ŏ trōp' ĭs)

Pleon carinate-dentate middorsally and dorsolaterally. Rostrum strong, acute. Anterior head lobe subacute, inferior antennal sinus sharply incised; inferior head lobe rounded. Coxa 1 produced anteriorly. Antenna 1 shorter than 2; accessory flagellum very small, subequally 2-segmented.

Lower lip with small inner lobes. Mandibular palp segment 3 longer than 2. Maxilla 1, inner plate with 2–4 setae, outer plate with 10 spine-teeth. Maxilla 2, plates very small, inner not broader than long. Maxilliped, inner plates separated to base.

Gnathopods 1 and 2 strongly and regularly subchelate, similar, 2nd larger (slightly); propod longer than deep, attached normally to carpus; carpus (segment 5) short, lobe deep, narrow. Peraeopods 3–7 very elongate, dactyls elongate; basis of peraeopods 5 and 6 distinctly smaller than 7.

Pleon side plate 3 evenly serrate behind. Uropod 3, rami subequal. Telson cleft one-fourth to one-half, lobes *not* diverging at apex.

KEY TO AMERICAN ATLANTIC SPECIES OF *RHACHOTROPIS*

1. Each segment of pleon and urosome with double-toothed middorsal carination; rostrum very long, nearly equal to peduncle 1 of antenna 1; peraeopods 5 and 6, basis with acute posterior marginal process . . . . *R. aculeata* Lepechin.
Dorsal carinations with single tooth per segment; rostrum not reaching more than half peduncle 1; peraeopods 5 and 6, basis posteriorly convex or nearly straight.   2.
2. Urosome segment 1 with middorsal tooth; telson with small V-cleft at apex.   .   . 3.
Urosome segment 1 lacking dorsal tooth; telson narrowly cleft more than one-third . . . . 4.
3. Eyes present; peraeon 7 with middorsal tooth; pleon with dorsolateral as well as dorsal carination and teeth . . . . *R. lobata* Shoem.
Eyes lacking; peraeon 7 lacking dorsal tooth; pleon with middorsal carination and teeth only. . . . . *R. distincta* Holmes.
4. Peraeon 7 with posterior middorsal tooth; eyes very large, nearly meeting middorsally; pleon side plates 1 and 2, hind corners acute, produced   .   *R. oculata* (p. 78).
Peraeon 7 lacking dorsal tooth; eyes moderately large but not nearly meeting middorsally; pleon side plates 1 and 2 rounded behind . . . . . *R. inflata* Sars.

Rhachotropis oculata (Hansen) 1887 Plate XI.2   (ŏk ū lāt′ă)

L. 10–12 mm. Pleon with single dorsal and dorsolateral low carina and posterior tooth. Peraeon 7 with single dorsal tooth. Urosome 1 dorsally smooth. Eyes very large, covering most of side of head and nearly meeting middorsally. Rostrum moderately strong, deflexed. Antenna 1, peduncle 3 short; peduncle 1 stout, peduncle 3 less than one-third segment 2; accessory flagellum of one plus a minute terminal segment. Antenna 2 about equal to antenna 1, peduncle 5 slightly longer than 4.

Mandible, palp segment 3 elongate, more than 1½ times segment 2. Maxilla 1, inner plate with 2 apical setae. Maxilliped palp, dactyl short, about two-thirds segment 3. Gnathopods, posterior margin of propod short, less than one-fourth convex palmar margin.

Peraeopod 4 slightly longer than 3, dactyls of both elongate, nearly equal to segment 6. Peraeopods 5 and 6 subequal, distinctly smaller than 7. Peraeopod 7, basis strongly expanded, hind margin distally shallow-concave.

Pleon 1 and 2, hind corners acute, slightly produced. Pleon 3 with about 20 posterior marginal serrations. Uropod 3, rami equal in length, both 1-segmented, more than twice peduncle. Telson not reaching to tip of uropods, cleft about two-fifths, outer margins convex, sparsely spined.

Distribution: Pan-arctic and arctic-boreal; in North Atlantic from Labrador south to the Gulf of St. Lawrence and Cape Cod Bay.
Ecology: A pelagic and epibenthic carnivorous species of cold-water areas, in depths from 5 m to more than 100 m.

Family CALLIOPIIDAE (kăl ĭ ōp ē′ĭ dē)

Body compressed, smooth or dorsally carinate or dentate posteriorly. Coxal plates moderately deep, often serrate below, 4th excavate behind; pleosome strong. Sexual dimorphism in eyes, antennae, gnathopod (slight), and uropod 3. Head, rostrum small; inferior antennal sinus deep; eye large. Antennae slender, subequal, calceolate; peduncle 2 short. Accessory flagellum lacking.

Mouthparts about normal. Upper lip rounded. Lower lip, inner lobes small or lacking. Mandible, molar strong, palp strong, segment 3 falcate. Maxilla 1, outer plate with 7 apical spine teeth.

Gnathopods subchelate; dactyls often serrate behind.

Peraeopods 3–7 normal, 7th longest, bases expanded. Pleosome and pleopods strong.

Uropods 1 and 2, rami narrow, lanceolate, outer ramus shorter. Urosome 3 longer than 2. Uropod 3, rami lanceolate, foliaceous, inner larger. Telson medium, entire, or with apex notched. Coxal gills pedunculate, present on segments 2–7. Brood plates large, lamellate, margins setose.

KEY TO NEW ENGLAND GENERA OF CALLIOPIIDAE

1. Body dorsally smooth, low-carinate, not toothed; gnathopods powerfully subchelate, dactyls smooth behind; segment 5 (gnathopods) with narrow, deep, posterior lobe; antenna 1, peduncle 3 with distinct posterodistal process . *Calliopius* (p. 80).
   Body (esp. pleon) usually dorsally toothed or mucronate; gnathopods moderately or weakly subchelate, dactyls serrate or denticulate behind; posterior lobe broad, shallow or lacking; antenna 1, peduncle 3 not posterodistally produced . . . 2.
2. Gnathopod 2, segments 5 and 6 together very elongate, 5 greater than 60 percent of elongate segment 2; uropod 3, outer ramus distinctly shorter than inner. . . 3.
   Gnathopod 2, segments 5 and 6 not greatly elongate, 5 usually less than one-half segment 2; uropod 3, rami subequal, outer slightly shorter . . . . 4.
3. Accessory flagellum 1-segmented; mandible palp segment 3 short, less than two-thirds segment 2; uropod 3, outer ramus about two-thirds inner ramus . *Oradarea*.
   Accessory flagellum lacking; mandibular palp, segment 3 nearly equal to 2; uropod 3, outer ramus not greater than half inner ramus . . . . . *Leptamphopus*.
4. Maxilla 1, palp very reduced, not reaching end of outer plate; coxa 4 distinctly larger and deeper than coxa 1. . . . . *Laothoes*.
   Maxilla 1, palp exceeding outer plate; coxa 4 slightly deeper than coxa 1 . . . 5.
5. Coxa 1 produced anteriorly to front of eye; telson narrowly notched about one-fifth to base; peraeopod dactyls very elongate (two-thirds segment 6) . *Rozinante*.

Coxa 1 not strongly produced in front; telson entire, acute, or widely notched at apex; peraeopod dactyls usually less than half segment 6 . . . . 6.

6. Telson and uropods elongate; antenna calceolate (♂); mandibular palp segment 3 = segment 2 . . . . *Halirages.*

Telson and uropods not elongate; antenna not calceolate (♂); mandibular palp segment 3 shorter than 2 . . . . *Apherusa.*

Genus *Calliopius* Liljeborg 1865   (kăl ĭ ōp' ĭ ŭs)

Body with low middorsal carina, esp. on pleosome. Coxal plates medium deep, 4th largest, 5th and 6th posterior lobe deeper. Head, anterior lobe truncate; eye large, lateral (esp. in ♂). Antenna 1, peduncle segment 3, posterodistal angle produced; accessory flagellum lacking. Antennae 1 and 2 calceolate (both sexes).

Mandible palp segment 3 = 2, falcate, inner margin strongly setose. Maxilla 1, inner plate with 4 apical setae. Maxilla 2, inner plate with single strong submarginal seta. Lower lip, inner lobes weak.

Gnathopods 1 and 2 strongly subchelate, subequal (both sexes); segment 5 short, posterior lobe narrow, deep. Peraeopods 3 and 4 medium, segment 4 anterodistally produced. Peraeopods 5–7 regularly increasing in size; basis broadly expanded; segment 4 broadened, about equal in length to segment 5. Uropod 3, rami subequal, foliaceous, little exceeding uropods 1 and 2. Telson entire, rounded.

Calliopius laeviusculus (Krøyer) 1838 * Plate XIV.1   (lēv ĭ ŭs kūl'ŭs)

L. 11–14 mm. Eye very large, subquadrate, black, nearly meeting middorsally (♂). Antenna 1, peduncle 1 with posterior marginal setae; peduncle 3, posterior margin slightly convex, calceolate.

Mandible, palp segment 3 narrowing distally, subfalcate. Coxal plates 1–4, lower borders smooth, not crenulate. Peraeopods 5–7, basis strongly expanded, posterior margin very weakly and irregularly serrate, not closely crenulate.

Pleon side plates 1–3, hind corner very weakly mucronate, not produced. Uropod 3 about twice length of peduncle, all ramal margins foliaceous. Telson about twice as long as wide, apex smoothly rounded.

Body whitish, strongly mottled and banded with red, brown, and orange spots.

Distribution: A widely distributed and very common arctic-boreal species; in North America from Baffin Island and Labrador, south through eastern Canada and the New England states to Long Island Sound, and off New Jersey.
Ecology: Pelagic and epibenthic throughout coastal areas, from the lower intertidal (pools and swash zone) to more than 20 m in depth; in higher sal-

* This species is most easily confused with *Gammarellus angulosus* (p. 60).

inities, not penetrating below 17 o/oo in estuaries; mostly along rocky and sandy-rocky coasts, esp. in semiprotected locations.

Life Cycle: Annual; ovigerous females from March–June, one brood per year in northernmost areas.

Family PLEUSTIDAE   (plē ŭsťī dē)

Body deep, broad, smooth or low carinate. Urosome segment 2 short. Coxal plates deep, smooth below with notch at posterior angle; 4th largest, excavate behind. Head, rostrum short; eye broad-lenticular, lateral. Antennae slender, long. Antenna 1 longer; accessory flagellum lacking (or forming a minute scale). Antenna 2 calceolate, peduncle short.

Upper lip asymmetrically bilobed. Lower lip characteristically shallow, outer lobes widely spread, inner lobes partly fused, upper margin straight. Mandible, molar small, tending to reduction; right lacinia lacking, left fan-shaped; palp strong, 3rd segment large, falciform. Maxilla 1, inner plate small, 0–2 apical setae. Maxilla 2, plates short, inner broad. Maxilliped plates short, palp large, dactylate.

Gnathopods 1 and 2 subchelate, 2nd larger; carpus short. Peraeopods 5–7 subequal, not elongate; basis expanded; segment 4 with posterior distal process. Uropods biramous; rami slender, lanceolate, outer shorter. Telson entire or slightly emarginate. Coxal gills small, simple; on peraeon segments 2–6. Brood plates large, lamellate.

KEY TO AMERICAN ATLANTIC GENERA OF PLEUSTIDAE

1. Mandibular molar strong, surface ridged or triturating . . . . 2.
   Mandibular molar weak, knoblike, surface not ridged. . . . . 4.
2. Gnathopods 1 and 2 unlike, 2 always powerful, carpal lobe narrow, deep (both sexes); pleon middorsally carinate; accessory flagellum present, minute
   . . . . *Sympleustes.*
   Gnathopods 1 and 2 similar (both sexes); pleon smooth or middorsally mucronate; accessory flagellum lacking . . . . 3.
3. Gnathopods 1 and 2 weak, carpal lobe broad, shallow; maxilliped palp segment 3 prolonged beyond base of dactyl; molar process of mandible compressed
   . . . . *Stenopleustes* (p. 82).
   Gnathopods 1 and 2 moderately strong, carpal lobe deep; maxilliped palp normal mandibular molar cylindrical . . . . *Pleusymtes* (p. 83).
4. Rostrum long, extending length of antenna 1 peduncular segment 1; antennae relatively short; body dorsally and dorsolaterally ridged or processiferous
   . . . . *Pleustes.*
   Rostrum short to medium, extending not more than half length of peduncular segment 1; antennae elongate; segments may be middorsally but not dorsolaterally carinate or toothed . . . . 5.
5. Gnathopods moderately strong, carpal lobes broad, shallow; maxilliped, palp seg-

ment 3 produced beyond base of dactyl; body dorsally toothed . . *Neopleustes.*
Gnathopods powerful; carpal lobe short, deep; maxilliped palp normal; body dorsally smooth . . . . *Parapleustes.*

Genus *Stenopleustes* G. O. Sars 1893 *   (stĕn ŏ plē ŭst′ēz)

Body (pleon) smooth or mucronate, compressed. Coxal plates medium deep, with weak hind corner notch. Rostrum medium, not extending beyond half antenna 1, peduncle 1. Head, anterior lobe rounded. Antennae elongate, subequal; accessory flagellum lacking.

Mandible, molar well developed, compressed, surface ridged or triturating. Maxilla 1, inner plate with 1 apical seta. Maxilliped, palp segment 3 prolonged distally beyond base of dactyl; inner plate closely approximating opposite member, inner margin unarmed.

Gnathopods 1 and 2 relatively weakly subchelate, similar, 2nd larger (both sexes); carpal lobe broad, shallow. Peraeopods 3–7 slender, dactyls long. Pleon side plates, posterior margins serrate. Telson short, rounded.

KEY TO NEW ENGLAND SPECIES OF *STENOPLEUSTES*

1. Peraeon segment 7 and pleon segments 1 and 2 with posterodorsal tooth; peraeopods 3–7, distal segments (esp. dactyl) not exceptionally elongate
   . . . . *S. gracilis* (p. 83).
   Body dorsally smooth; peraeopods 3–7, segments 4–7 slender, elongate, segment 6 distinctly longer than segment 4 . . . . *S. inermis* (p. 82).

Stenopleustes inermis Shoemaker 1949 Plate XV.1   (ĭn ĕrm′ ĭs)

L. 6 mm. Pleosome dorsally smooth, not mucronate. Eye subrectangular, black. Antenna 2 slightly longer than 1. Coxal plate 5, lobes subequal in depth.

Mandibular palp segment 3 only slightly longer than 2, distal half setose. Peraeopods 3–7 very long and slender; posterodistal process on segment 4 short; dactyls long and slender, length about one-half respective propod. Peraeopods 5–7, basis large, posterior margin evenly convex, with 12–20 evenly spaced fine setae.

Pleon side plate 3, posterior margin with 10–12 serrations. Uropod 3, outer ramus about three-fourths inner. Telson, margins smooth.

---

* This genus (of which *S. malmgreni* is the type) cannot be submerged in *Sympleustes* (of which *S. latipes* is the type), since the latter differs in the following generic characters: accessory flagellum present; anterior head lobe blunt, emarginate; mandible molar cylindrical; gnathopods 1 and 2 dissimilar, weakly subchelate, 1 with shallow carpal lobe; pleon 1–3 carinate; telson suborbicular; pleon side plates smooth behind.

Distribution: Gulf of Maine; Bay of Fundy near mouth, to Cape Cod Bay.
Ecology: Epibenthic on finer sediments, cold-water areas in depths from 5 to
more than 50 m.
Life Cycle: Annual; ovigerous females from March–June.

Stenopleustes gracilis (Holmes) 1905 Plate XV.2   (grăˈsĭl ĭs)

L. 6 mm. Peraeon segment 7 and pleon segments 1 and 2 with middorsal
posterior mucronation. Eye nearly circular. Antenna 1 longer than 2, peduncle 1 stout.

Mandibular palp segment 3 longer than 2, inner margin of 2 and distal
two-thirds of 3 setose.

Gnathopod 2 very weakly subchelate, little larger than gnathopod 1.
Peraeopods 3–7 relatively stout, segment 5 of peraeopods 3–5 much shorter
than segment 4. Peraeopod 3, posterior lobe of coxa 5 deeper than anterior
lobe. Peraeopods 5–7, basis posterior margin with 10–12 short setae.

Pleon 3, posterior margin with 8–10 serrations. Uropod 2, outer ramus
distinctly shorter than inner. Uropod 3, outer ramus two-thirds inner. Telson
with 3–4 fine marginal setae on each side.

Distribution: South side of Cape Cod to Long Island Sound and off New Jer-
sey; also mouth of Chesapeake Bay.
Ecology: Epibenthic on sandy bottoms, mainly deeper than 5 m.
Life Cycle: Probably annual.

Genus *Sympleustes* Stebbing 1899 (based on *S. latipes* as type)
                    (sĭm plē ŭstˈēz)

Body, esp. pleon, dorsally carinate. Head, rostrum short; anterior lobe emarginate; accessory flagellum present. Antennae long, 1st much longer.

Mandible molar process strong, compressed laterally. Maxilla 1, inner
plate with 2 apical setae. Maxilliped palp segment 3 prolonged beyond base
of dactyl; inner plate closely approximating opposite member, inner margin
usually setose. Upper lip, lobes subequal.

Gnathopod 2 much more powerfully subchelate than 1 (both sexes);
carpus of 2 with narrow, deep posterior lobe; carpus of 1 shallow, broad.
Peraeopods 3–7 short, stout. Uropod 3, inner ramus much longer and
broader than outer. Telson short, suborbicular.

Genus *Pleusymtes* Barnard 1969 *   (plē ŭ sĭmˈtēz)

Body usually smooth, moderately compressed. Coxal plates moderately
deep, hind corner notch distinct. Head, rostrum short; anterior lobe acute.
Antennae elongate, accessory flagellum lacking.

---

* Barnard (1969) has applied the new name *Pleusymtes* to most North American species formerly ascribed to *Sympleustes*.

Mandible molar large, cylindrical, surface ridged or triturating. Maxilla 1, inner plate usually with 2 apical setae. Maxilliped, palp segment 3 not produced beyond base of dactyl; inner plate, inner margin setose.

Gnathopods moderately strongly subchelate; similar (both sexes); carpal lobe short, relatively deep. Peraeopods 5–7 subequal, not elongate.

Pleon side plate 3, hind corner with tooth and notch. Telson long, rounded.

Pleusymtes glaber (Boeck) 1861 Plate XIV.2   (glā′bĕr)

L. 6 mm. Pleon dorsum smooth, not toothed or carinate. Antenna 1, peduncle 1 with distal process. Head, anterior lobe sharply acute.

Mandible, palp segment 3 longest, inner margin of segments 2 and 3 setose throughout. Maxilla 1, inner plate with 2 apical setae.

Gnathopods 1 and 2 moderately powerful, similar, 2 slightly larger; propod palm convex, oblique, moderately deep. Peraeopods 5–7, basis strongly expanded, posterior margin weakly serrate; segment 4 about equal in length to 5, posterodistal lobe extending less than one-fourth posterior margin of 5.

Pleon side plate 3, hind corner acuminate, notched above. Uropod 3 not extending beyond 2; outer margin short, slightly longer than peduncle and one-half inner ramus. Telson subovate with 4–5 fine marginal setae.

Distribution: Amphi-Atlantic boreal; in North America from western Greenland and Labrador south to the Gulf of Maine and southern New England. A variant occurs from Cape Cod south to Chesapeake Bay, in summer-warm and more brackish waters.
Ecology: A moderately common species of rocky coasts, from the lowest intertidal and immediate subtidal levels to more than 50 m in depth.
Life Cycle: Annual; ovigerous females from March–June; one brood per year.

Body small, usually dorsally smooth. Sexual dimorphism reduced. Coxal plates 2–4 (esp. 4) usually large, deep. Antenna 1 slender, not elongate; accessory flagellum vestigial or lacking. Antenna 2, peduncle long; flagellum noncalceolate. Head, anterior lobe prominent. Rostrum medium, sometimes strongly developed. Eyes lateral, orbicular.

Upper lip asymmetrically bilobed. Lower lip, outer lobes with short rounded molar process; inner lobes fused medially or lacking. Mandibular palp weak or lacking; molar variable, usually weak or lacking. Maxilla 1, inner plate small; outer plate with 6 apical spine-teeth. Maxilla 2, plates small. Maxilliped, inner plate small, tending to fusion with opposite member; outer plate small or vestigial; palp strong, dactylate.

Gnathopod 1 subchelate, carpochelate, or simple. Gnathopod 2 larger, subchelate, carpal lobe usually deep. Peraepods 5–7 subequal, segment 4, posterodistal angle produced. Urosome segments 2 and 3 short. Uropods slender, rami lanceolate, spinose. Uropod 3 biramous or 2-segmented uniramous. Telson entire.

Coxal gills simple, saclike, on peraeon segments 2–6. Brood lamellae large, margins setose.

Component Families: Amphilochidae, Anamixidae, Cressidae, Leucothoidae, Stenothoidae, Thaumatelsonidae.

Family STENOTHOIDAE   (stĕn ō thō'ĭ dē)

Body small, smooth (occ. carinate). Urosome segments free. Coxal plates 2–4 very large, shieldlike, covering appendages; coxa 4 not posteriorly excavate; coxa 1 small, hidden. Head, rostrum very short; eyes round, lateral. Antennae medium, subequal; accessory flagellum lacking or vestigial; antenna 2, peduncle long.

Upper lip variously bilobed; lower lip, inner lobes coalesced, outer lobes mandibular processes short, blunt. Mandible, molar weak or lacking; palp variously reduced or lacking. Maxilla 1, outer plate with 6 spine-teeth. Maxilla 2, inner plate short. Maxilliped, inner plate partly or completely fused; outer plates vestigial, palp strong.

Gnathopod 1 normally subchelate or simple, smaller than 2; gnathopod 2 subchelate, segment 5 short, posterior lobe narrow. Gnathopods usually sexually dimorphic. Peraeopods 3–7 short, segment 4 prolonged distally. Peraeopods 5–7 subequal, 6 or 7 usually longest. Peraeopod 5, basis always narrow, 6 and 7 basis may be expanded. Uropod 3 uniramous, ramus 2-segmented, shorter than uropods 1 and 2. Telson entire, rounded.

85

KEY TO AMERICAN ATLANTIC GENERA OF STENOTHOIDAE

1. Peraeopods 6 and 7, segment 2 linear; coxal plate 4 extremely large, covering basis of peraeopod 7 . . . . 2.
   Peraeopods 6 and 7, segment 2 expanded; coxal plate 4 not covering basis of peraeopod 7 . . . . 4.
2. Coxal plate 3 small, lower margin hidden (Atlantic species); mandibular palp absent . . . . *Parametopella* (p. 91).
   Coxal plate 3 large, deep; mandibular palp 2–3 segmented . . . . 3.
3. Mandibular palp 2-segmented; peraeopods 3–7, segment 6 distally expanded inside dactyl; peraeopod 7, basis broader than peraeopod 6 . . . *Metopelloides*.
   Mandibular palp 3-segmented; peraeopods 3–7, segment 6 slender; peraeopods 6 and 7, basis similarly slender . . . . *Metopella* (p. 90).
4. Gnathopod 1 subchelate; mandibular palp lacking; maxilliped, inner plates entirely separated; peraeopods 3 and 4 similar . . . . *Stenothoe* (p. 86).
   Gnathopod 1 simple or subchelate; mandibular palp present; maxilliped, inner plates basally fused to opposite member; peraeopod 4 slightly heavier than 3 . 5.
5. Mandibular palp 1-segmented; peraeopod 7 not shorter than 6 . . . . *Stenula*.
   Mandibular palp 3-segmented; peraeopod 7 slightly shorter than 6 . . . . . 6.
6. Maxilla 1, palp 1-segmented; maxilliped inner plates fused, notched apically
   . . . . *Metopa* (p. 87).
   Maxilla 1, palp 2-segmented; maxilliped, inner plates separated distally
   . . . . *Proboloides* (p. 89).

Genus *Stenothoe* Dana 1852   (stĕn ö thö′ē)

Coxal plate 1 developed; coxa 2 and 3 large, 4 largest but not covering basis of peraeopod 6 and 7. Antennae relatively short and stout; accessory flagellum lacking.

Mandibular palp and molar lacking. Maxilla 1, palp strong, 2-segmented; inner plate with 1 apical seta. Maxilliped, inner plates separated; outer plate vestigial.

Gnathopod 1 more or less subchelate, much smaller than gnathopod 2. Gnathopod 2, propod large, sexually dimorphic; in both sexes posterior lobe of segment 4 prolonged beneath segment 5 (carpus). Peraeopods 3 and 4 similar. Peraeopods 6 and 7 subequal; basis expanded posteriorly. Peraeopod 5, basis narrow. Uropod 2, outer ramus shorter than inner.

KEY TO AMERICAN ATLANTIC SPECIES OF *STENOTHOE*

1. Peraeopod 5, basis slightly expanded; telson with 3 lateral marginal spines
   . . . . *S. gallensis.*
   Peraeopod 5, basis narrow; telson with 0–1 lateral marginal spines . . . . . 2.
2. Telson, apex rounded; peraeopod 5, anterior coxal lobe produced distally
   . . . . *S. minuta* (p. 87).
   Telson, apex acute; peraeopod 5, coxal plate lacking anterior lobe
   . . . . *S. brevicornis* var. *canadensis* Dunbar

Stenothoe minuta Holmes 1905 Plate XVI.1   (mĭn ūt′ ă)

L. 1.5–2.0 mm ♀. Coxal plate 5 with pronounced anterior lobe. Eye large, central. Antennae short. Antenna 1, flagellum 7–8 segmented; antenna 2, peduncle not elongate, flagellum 7-segmented.

Gnathopod 2 strongly subchelate, palm oblique; hind lobe of segment 4 not attaining setose hind lobe of 5. Gnathopod 2 (♀), segment 6 palm oblique, smooth, about equal to posterior margin.

Peraeopod 4 very slightly stronger than peraeopod 3. Peraeopods 5–7, segment 4, posterodistal lobe strongly developed; dactyls strong, nearly one-half length of segment 6. Peraeopods 6 and 7, basis broadly oval. Pleon side plates 2 and 3, hind corners subquadrate.

Uropod 2, outer ramus shorter than inner. Uropod 3, peduncle with single posteromedian spine. Telson rounded, with 1 lateral marginal spine.

Distribution: South side of Cape Cod to Chesapeake Bay and Georgia. Vineyard Sound; Martha's Vineyard, Hadley Harbor; Buzzards Bay; Long Island Sound.
Ecology: Abundant among hydroids and ectoprocts, in shallow bays and estuaries, from low-water level to more than 10 m, in polyhaline waters to about 10 o/oo; sometimes planktonic.
Life Cycle: Annual; ovigerous females from May–September and December–January; several broods per female.

Genus *Metopa* Boeck 1871   (mĕt ōp′ ă)

Coxal plate 1 developed; coxal plates 2 and 3 large; 4th largest, not covering peraeopod 7. Antennae slender; accessory flagellum lacking.

Mandible lacking molar process; palp weak, 3-segmented. Maxilla 1, palp 1-segmented. Maxilliped, inner plate fused throughout its length to opposite member.

Gnathopod 1 simple or weakly subchelate. Gnathopod 2 normally subchelate (sexually dimorphic).

Peraeopods 6 and 7, basis expanded posteriorly. Peraeopod 6 slightly the longer. Uropod 2, outer ramus slightly shorter than inner.

KEY TO AMERICAN ATLANTIC SPECIES OF *METOPA*

1. Antenna 1 and 2 markedly unequal in length (both sexes) . . . . 2.
   Antenna 1 and 2 subequal or slightly differing in length . . . . 3.
2. Antenna 1 shorter than antenna 2; gnathopod 1 subchelate (slender), dactyl long; telson rounded, margins bare . . . . *M. alderi* (p. 88).
   Antenna 1 longer than antenna 2; gnathopod 1, propod linear, dactyl very short; telson subacute, lateral margins with 3 spines . . . . *M. clypeata* (Kr).
3. Coxa 4, lower margin sinuous; peraeopods 6 and 7, segment 4 posterodistal lobe

large, curved, as long as segment 5. . . . . *M. propinqua* Sars.
Coxa 4, lower margin convex; peraeopods 6 and 7, segment 4 posterodistal lobe
moderate, not reaching to end of segment 5. . . . . 4.
4. Gnathopod 1 simple, propod sublinear; peraeopod 7, basis width equal to or
   greater than length; posterodistal lobe of segment 4 at least half segment 5 . . 5.
   Gnathopod 1 subchelate, propod slightly inflated; peraeopod 7, basis longer than
   wide; segment 4, posterodistal lobe short, about one-third segment 5 . . . . 7.
5. Gnathopod 2, palm irregularly serrate; telson margins bare . . *M. borealis* Sars.
   Gnathopod 2, palm smooth, oblique; telson lateral margins each 2-spined . . 6.
6. Gnathopod 2 very weak, slender; telson apex rounded; peraeopod 7, posterior
   lobe of segment 4 about half segment 5 . . . . *M. tenuimana* Sars.
   Gnathopod 2 normal, propod stout; telson apex acute . . . . *M. bruzelii* Goes.
7. Peraeopod 7, basis narrow, width about two-thirds length; telson apex rounded,
   lateral margins bare . . . . *M. solsbergi* Schneider.
   Peraeopod 7, basis width about four-fifths length; telson apex acute; lateral margins each with 3 stout spines . . . . *M. boecki* Sars.

Metopa alderi (Bate) 1871 (no illustration)   (ăl′dĕr ī)

L. 6–8 mm. Coxal plates 2 and 3 large, 2 sinuous behind. Eye large, round.
Antenna 1, flagellum 14-segmented, extending past peduncle of antenna 2
(♀); in ♂, antenna 2 much longer than 1, peduncular segments 3–5 elongate,
expanded.

Mandibular palp short, distinctly 3-segmented, terminal segment minute.
Maxilla 1, palp 1-segmented. Maxilliped, outer plates reduced to minute
rounded lobes; inner plates fused except for shallow median notch.

Gnathopod 1 (both sexes) weakly subchelate, palm very oblique; segments 5 and 6 shallow, subequal in length. Gnathopod 2 (♂) powerful, segment 6 broadening distally, palm serrate near hinge, with deep V-shaped excavation inside strong tooth at palmar angle; dactyl short, not closing on
palmar tooth; in ♀, segment 6 less powerful, palm convex, excavation shallow,
tooth at posterior angle short.

Peraeopod 5 longer but less robust than peraeopod 4; segment 4 in
each about equal to segment 6; dactyls short (less than one-third segment
6). Peraeopods 6 and 7 subequal; segment 4 in each, posterior angle produced strongly, more than halfway along posterior margin of segment 6;
dactyls about one-third length of segment 6.

Abdominal side plate 3, hind corner nearly quadrate, sharp. Uropod 2, .
outer ramus distinctly shorter than inner; peduncle with about 10 outer marginal spines. Uropod 3, peduncle with 4–5 short posterior marginal spines.
Telson unguiform, unarmed, apex rounded.

Distribution: Arctic-boreal; on American Atlantic coast south to Gulf of St.
Lawrence and northern Gulf of Maine.
Ecology: On rocky bottoms deeper than 10 m.

Genus *Proboloides* Della Valle 1893   (prō böl oi′ dēz)

Similar to *Metopa* in having:

Antenna 1, accessory flagellum lacking. Peraeopod 4 heavier than 3. Peraeopods 6 and 7, basis expanded. Mandibular palp slender, 3-segmented. Lower lip, inner lobes fused to distally truncate plate.

Except: Maxilla 1, palp 2-segmented. Maxilla 2, inner plate setose. Maxilliped inner plate separated distally from opposite member; outer plate small but distinct.

Gnathopod 1 simple or shallowly subchelate (both sexes). Peraeopods 3 and 4, 4th stronger and heavier. Peraeopods 4, 5, and 6, segment 4 longer than segment 6. Peraeopod 7 shorter than 6.

KEY TO NORTH ATLANTIC SPECIES OF *PROBOLOIDES*

1. Gnathopod 1 simple; gnathopod 2 (♀), palm smooth; peraeopods 5–7, segment 4 posterodistal lobe short, not extending past one-half segment 5 . *P. holmesi* n. sp.
   Gnathopod 1 shallow subchelate; gnathopod 2 (♀), palm toothed irregularly; peraeopods 5–7, posterodistal lobe of segment 4 produced length segment 5 . . . 2.
2. Gnathopod 2 (♀), dactyl inflated . . . . *P. nordmanni* (Steph).
   Gnathopod 2 (♀), dactyl normal . . . . *P. gregarius* (Sars).

Proboloides holmesi n. sp. Plate XVI.2   (hōmz′ī)

L. 2.5 mm ♀. Coxal plates 2 and 3 relatively large and deep, hind borders lightly spinose. Antenna 1, flagellum 8-segmented. Antenna 2, slightly longer, flagellum 7-segmented; peduncle not inflated.

Mandibular palp slender, distinctly 3-segmented, terminal segment about equal in length to 1st. Maxilla 1, palp broad, intersegmental line indistinct. Maxilliped, inner plates partially separated, apex blunt, weakly armed; outer plate a small acute lobe, apex with 2 setae; palp segment 2 with strong fan of setae at inner distal angle.

Gnathopod 1 (♀), anterior margin of basis evenly lined with setae: segment 5 slender, posterior margin longer than one-half anterior; segment 6 about equal to 5, posterodistal margin with 4–5 long setae; dactyl, inner margin setose. Gnathopod 2 (♀), segment 5, hind lobe small; segment 6 expanding distally, palm evenly convex, oblique, with 2 spines at posterior angle.

Peraeopod 4 stouter and more strongly spinose posteriorly than peraeopod 3; segment 4 in each longer than segment 6. Peraeopod 5, basis anterior and posterior margins short spinose; segment 4 longer than 6. Peraeopod 6, segment 4 longer than 6. Peraeopod 7 shorter than 6, basis more broadly expanded; segment 4 shorter than 6, posterodistal angle deeply prolonged. Pleon side plate corners subquadrate. Uropod 2, outer ramus slightly shorter than inner. Uropod 3, peduncle with posteromedial spine. Telson, short, broad, with 2 spines on either margin. Male unknown.

Distribution: South side of Cape Cod; Vineyard Sound, Elizabeth Islands, Buzzards Bay.

Ecology: Mainly on sandy and shelly sand bottoms, among hydroids and ectoprocts, in depths of 5–30 m.

Life Cycle: Presumably annual; ovigerous females from June–September.

Genus *Metopella* Sars 1892   (mĕt ōp ĕl′ ă)

Coxal plate 1 not developed; coxae 2 and 3 large; coxa 4 extremely large, covering basis of peraeopod 7. Antennae slender; access flagellum lacking.

Mandible, palp minute, 2–3 segmented. Maxilla 1, palp 1-segmented; inner plate 1-setose. Maxilliped, inner plate fused throughout length to opposite member.

Gnathopod 1 simple; segment 5, carpus long, lobe shallow. Gnathopod 2 strongly subchelate (sexually dimorphic).

Peraeopods slender, peraeopod 4 heavier than 3. Peraeopods 6 and 7, basis narrow; peraeopod 6 longest. Uropod 2, rami subequal.

KEY TO AMERICAN ATLANTIC SPECIES OF *METOPELLA*

1. Peraeon 4 with low middorsal carina; antenna 1, peduncle 2 short, about equal to 3 . . . . 2.
   Peraeon 4 dorsally smooth; antenna 1, peduncle 2 elongate, much longer than segment 3 . . . . 3.
2. Antenna 1, peduncle 1 with acute anterodistal process; urosome 1 carinate . . . . *M. nasuta* (Boeck).
   Antenna 1, peduncle 1 without process; urosome 1 smooth above . . . . *M. carinata* (Hansen).
3. Peraeopods 3–7, dactyls elongate, longer than one-half segment 6; telson without marginal spines . . . . *M. angusta* (p. 90).
   Peraeopods 3–7, dactyls short, less than one-half segment 6; telson with 2 marginal spines on each side . . . . *M. longimana* (Boeck).

Metopella angusta Shoemaker 1949 Plate XVII.1   (ăng ŭst′ ă)

L. 3.5 mm. Coxal plates 2 and 3, very deep, 3 narrow. Head shallow, eye small.

Antenna 1 and 2, peduncles strong, longer than flagellae, antenna 2 longer than 1. Mandibular palp a knoblike protuberance of three minute segments. Maxilla 1, inner plate with 1 terminal seta. Maxilla 2, inner plate with 5–6 marginal setae. Maxilliped palp, segment 3 slender, nearly twice length of segment 2.

Gnathopod 1 (both sexes), segments 5 and 6 subequal in length, carpus slender without posterior lobe. Gnathopod 2 (♂) strongly subchelate; palm evenly convex and very oblique, longer than posterior margin; carpal lobe narrow; in ♀, propod smaller, palm straighter, less oblique; carpal lobe broader.

Peraeopods 3–7 elongate, slender; dactyls elongate, nearly equal to respective propods. Peraeopod 4 longer than 3; peraeopod 5 longer than 6, peraeopod 7 shortest; in all, segment 4 distal process short.

Pleon side plate 3 much larger than plates 1 and 2, hind corner broadly acute. Uropod 2 much shorter than uropod 1, rami subequal, unarmed. Uropod 3, peduncle with 2 posterior marginal spines; terminal segment of ramus longer than proximal segment. Telson subovate, margins smooth.

Distribution: Gulf of Maine (Bay of Fundy), south to Long Island Sound and off New Jersey.

Ecology: On fine sediments in cold waters in depths of 5 to more than 40 m.

Life Cycle: Annual; ovigerous females April–June.

Genus *Parametopella* Gurjanova 1938   (păr ă mĕt öp ĕl′ ă)

Coxal plate 1 not developed; coxae 2 and 3 medium to small; coxa 4 large, extending anteriorly over coxa 3, posteriorly over coxa 7. Antennae slender; accessory flagellum lacking. Sexual dimorphism minimal.

Mandible, palp lacking. Maxilla 1, palp 1-segmented; inner plate lacking setae. Maxilla 2, inner plate small, lacking setae. Maxilliped, inner plates approximate, separated to base, outer plate very small.

Gnathopod 1 simple; carpal lobe shallow. Gnathopod 2 regularly subchelate; carpus short, posterior lobe developed, narrow.

Peraeopods 3 and 4 conspicuously larger than peraeopods 5–7. Peraeopods 5–7, basis narrow, not at all expanded. Peraeopod 6 longer than 7. Uropod 2, rami subequal. Telson lacking armature.

One American-Atlantic species is known, *P. cypris* Holmes.

KEY TO KNOWN SPECIES OF *PARAMETOPELLA*

1. Coxa 2 and 3, lower margins hidden by coxa 4 . . . . *P. cypris* (p. 91).
   Coxa 2, lower margin free (usually 3 also) . . . . 2.
2. Antenna 1, peduncle segment 1 with anterodistal process. . . . *P. stelleri* Gurj.
   Antenna 1, peduncle segment 1 lacking process . . . . . . *P. ninis* Barnard.

Parametopella cypris (Holmes) 1905 Plate XVII.2   (sĭp′ rĭs)

L. 2.0 mm. Coxa 4 very large, subovate, extending anteriorly to coxa 1, and posteriorly to coxa 7. Coxa 2 and 3, shallow, 3 smaller, both covered distally by 4. Head short, deep, eye large, black, central. Antenna short, subequal peduncles short.

Maxilla 1, outer plate with 4 spine-teeth, inner short. Gnathopod 1, dactyl with fine pectination and a few setae distally on inner margin; carpus, posterior margin with 2 groups of pectinate setae. Gnathopod 2, moderately powerful (♀), palm nearly straight, oblique, irregularly serrate near hinge, with 2 strong spines at posterior angle; carpal lobe rounded, contricted basally, extending below posterior margin or propod; in ♂, propod slightly larger, palmar serrations more pronounced.

Peraeopods 3 and 4, dactyl stout, longer than one-half segment 6. Peraeopods 5–7, margins weakly armed; segment 4, posterodistal lobe short.

Pleon side plate 3, hind corner nearly square. Uropod 2, nearly reaching tip of uropod 1, rami subequal, smooth. Uropod 3, peduncle and ramus unarmed, terminal segment of ramus shorter than proximal segment. Telson long-ovate, unarmed.

Distribution: South side of Cape Cod to the Middle Atlantic states (Chesapeake Bay), North Carolina, Georgia, and northern Florida. Common in Vineyard Sound, Woods Hole, and Buzzards Bay.
Ecology: A polyhaline species (salinities greater than 16 o/oo), on hydroids, ectoprocts, and sponges, from immediately subtidal to deeper channels; not common.
Life Cycle: Annual; ovigerous females from May–October; several broods per female.

Family LEUCOTHOIDAE   (lŏŏk ō thō' ĭ dē)

Body compact, moderately broad, smooth. Coxae variable, plates 2–4 usually moderately deep. Antennae short, flagellae usually much shorter than peduncles; accessory flagellum very small, vestigial.

Upper lip, epistome produced above, asymetrically incised below; lower lip, inner lobes rudimentary or lacking. Mandible palp slender, tending to reduction; molar lacking. Maxilla 1, inner plate small, lacking setae. Maxilla 2, plates small. Maxilliped, inner plate small, partly fused; outer plate very small; palp stout.

Gnathopod 1 complexly chelate (carpochelate); carpus posterior lobe very slender, acute, produced along entire margin of segment 6. Gnathopod 2 very powerfully subchelate (both sexes); carpal lobe variably produced along lower margin of propod.

Peraeopods 3–5, basis broadly expanded; peraeopod 5 longest. Uropod 2 much shorter than 1, rami of both styliform or slender. Uropod 3, peduncle elongate. Telson entire.

KEY TO AMERICAN GENERA OF LEUCOTHOIDAE

1. Coxa 1 much smaller than 2; mandible palp 1-segmented . . . . . *Leucothoides.*
   Coxa 1 slightly smaller than 2; mandible palp 3-segmented . . *Leucothoe* (p. 93).

Genus *Leucothoe* Leach 1814   (lōōk ō thō′ ē)

Antenna 1 equal to antenna 2, with small 1-segmented accessory flagellum. Coxal plate 1 normal, not concealed by coxa 2.

Mandibular palp 3-segmented, 3rd segment shorter than 2nd. Maxilla 1, palp 2-segmented. Maxillipeds, inner plates partly coalesced, outer plate rudimentary.

Gnathopod 1 chelate between 5 and 6 segments, 7th segment strong, dactylate. Gnathopod 2 powerfully subchelate, palm long and oblique, overhanging apex of produced carpal lobe of segment 5.

Peraeopods 5–7 subequal, not unusually reduced. Uropod 2, rami subequal. Uropod 3, rami lanceolate, subequal. Telson long, tapering, acute.

Leucothoe spinicarpa (Abildgaard) 1789 Plate XVIII.1   (spīn ĭ kărp′ă)

L. 8 mm. Eye moderately large, rounded, lateral, black. Antenna 1, accessory flagellum minute, 1-segmented; flagellum 6-segmented. Antenna 2 subequal to 1.

Mandibular palp, segment 2 twice length of 3, inner margin setose.

Gnathopod 1, dactyl long, about equal to half segment 6, which has lower margin bearing 6–10 evenly spaced fine spines. Gnathopod 2 larger in ♂; carpal lobe extends forward beyond tip of closed dactyl; propod large, inflated palm crenulate or toothed near hinge.

Pleon side plates 2 and 3, hind corners squarish or very slightly notched above. Uropod 3, rami subequal, shorter than peduncle. Telson long, narrow, tip sharply triangular, simple.

Distribution: Arctic-boreal, circum-Atlantic; in North America, south from Greenland to New England (Conn.).
Ecology: Commensal in sponges, tunicates, etc., subtidally to over 100 m in depth.

Free-swimming (non tube-building) smooth-, often broad-bodied amphipods with most paired appendages broadened, heavily spinose, or otherwise modified and armed for burrowing in unstable sandy or sandy mud substrata; antenna 1 shorter than 2, accessory flagellum present (except Oedicerotidae); eyes tending to reduction and/or loss of pigment; gnathopods often spadelike, powerful, similar, or reduced to function as accessory mouthparts (Haustoriidae); posterior peraeopods and/or uropod 3 often elongate, "rudder"-like; sexual dimorphism tending to reduction, especially in gnathopods (Haustoriinae, Oedicerotidae).

Family OEDICEROTIDAE (ē dĭ sĕr ŏt′ĭ dē)

Body smooth. Abdomen large, pleon powerful. Sexual dimorphism (antennae only) minimal. Rostrum strong, falcate. Eyes contiguous on dorsal border of head.

Antenna 1 usually shorter than 2; accessory flagellum lacking or rudimentary. Antenna 2, flagellum elongate ($\male$), noncalceolate. Coxal plates moderately deep, 4th excavate; coxa 5 and 6 deep.

Upper lip rounded or slightly emarginate. Lower lip rudimentary or lacking. Mandible, molar usually weak, of variable structure; palp 3-segmented. Maxilla 1 inner plate small, weakly setose; outer plate with 7 apical spine teeth. Maxilliped, inner plate small; palp large, dactylate.

Gnathopod 1 subchelate; gnathopod 2 subchelate or chelate; in both, carpal lobes elongate or prolonged behind propod.

Peraeopods 3 and 4, short, fossorial. Peraeopods 5 and 6 fossorial, subequal. Peraeopod 7, very elongate, dactyl styliform.

Uropods slender, elongate, lanceolate, biramous. Telson entire, small. Coxal gills on peraeon segments 2–6, saclike. Brood plates narrow, sublinear; relatively few marginal setae.

KEY TO AMERICAN ATLANTIC GENERA OF OEDICEROTIDAE

1. Gnathopod 2 chelate; segment 6 sublinear, length usually more than three times width. . . . . 2.
   Gnathopod 2 subchelate; segment 6, length not more than three times width . . 3.

2. Gnathopod 2, carpal lobe fused to propod throughout; mandible, molar small, conical; antenna 1 slightly longer than antenna 2 (♀) . . . *Synchelidium* (p. 98).
   Gnathopod 2, carpal lobe approximating but not fused to propod; mandible, molar cylindrical, ridged; antenna 1 shorter than 2 (♀) . . . . . *Pontocrates*.
3. Gnathopods 1 and 2 dissimilar; one or other carpal lobe not produced or not guarding propod like the other . . . . 4.
   Gnathopods 1 and 2 more or less similar; both carpal lobes similarly produced or not produced . . . . 6.
4. Gnathopod 1, carpal lobe vestigial or very short; mandible molar not ridged; rostrum short, subtruncate, eye terminal . . . . *Paroediceros*.
   Gnathopod 1, carpal lobe well developed; mandible molar ridged; rostrum usually strong, acute, eye on base of rostrum . . . . 5.
5. Antenna 1, peduncle segment 3 about equal to segment 1; antenna 1 longer than antenna 2 (both sexes) . . . . *Monoculopsis*.
   Antenna 1, peduncle segment 3, less than half segment 1; antenna 1 usually shorter than antenna 2 (♀). . . . . *Monoculodes* (p. 95).
6. Body dorsally strongly ridged or toothed (esp. pleon); rostrum very strongly developed . . . . *Acanthostephia*.
   Body dorsally smooth; rostrum moderately or weakly developed . . . . . . 7.
7. Rostrum moderately developed; gnathopods 1 and 2 strong; carpal lobes well developed; mandible, cutting edge (incisor) projecting, toothed . . . . 8.
   Rostrum usually short, truncate, or very weak; gnathopods relatively small, carpal lobes weak or broad at base; mandible, cutting edge short, untoothed . . . . 9.
8. Eyes completely fused across top of head; gnathopods 1 and 2, propods elongate, carpal lobes produced behind, antenna 1 longer than antenna 2 . . . . .
   . . . . *Perioculodes*.
   Eyes separated by middorsal line; gnathopods 1 and 2, propods relatively short, not guarded behind by carpal lobes; antenna 1 shorter than 2. . . . *Oediceros*.
9. Gnathopods strongly subchelate, carpal lobes moderately produced; rostrum very short or lacking; eyes very weak (lateral) or lacking. . . . . 10.
   Gnathopods relatively small; carpal lobes weakly produced, blunt; rostrum short or moderate; eyes usually present, well pigmented . . . . 11.
10. Antenna 1, peduncle segment 2 longer than 1; mandible, palp, segment 2 strongly curved . . . . *Arrhis*.
    Antenna 2, peduncle 2 shorter than 1; mandible, palp straight. . . . *Aceroides*.
11. Gnathopod 2, segment 5 (carpus) distinctly longer than in gnathopod 1; eyes lateral or lacking; mandibular palp, segment 2 straight . . . . . . *Bathymedon*.
    Gnathopod 2, segment 5 similar to gnathopod 1; eyes subterminally on rostrum; mandible, palp segment 2 curved . . . . *Westwoodilla*.

Genus Monoculodes Stimpson 1853   (mŏn ŏk u lōd′ ēz)

Rostrum acute, usually deflexed. Eye at base of rostrum, not completely fused middorsally. Coxal plates 4 and 5 largest. Antenna 1 usually distinctly shorter than antenna 2 (♀).

Mandibular molar strong, with triturating face; palp strong, segment 3 shorter than 2. Maxilla 1, inner plate with 2 apical setae. Maxilla 2, plates not reduced, normal. Maxilliped, outer plate moderately large, spinose. Lower lip, inner lobes distinct.

Gnathopod 1 and 2 subchelate. Gnathopod 1 short, carpal lobe broad, partly prolonged behind propod. Gnathopod 2 larger, carpal lobe slender, usually prolonged more than half length of propod.

Peraeopod 7 very elongate, segments 6 and 7 with posterior marginal plumose setae.

Uropod 3, rami not longer than peduncle.

KEY TO AMERICAN ATLANTIC SPECIES OF *MONOCULODES*

1. Rostrum blunt, eye at apex; peraeopods 3 and 4, dactyl thick . . . . 2.
   Rostrum acute, eye central or basally on rostrum; peraeopods 3 and 4, dactyl slender . . . . 3.
2. Coxal plates 1–4 small, shallow; gnathopod 2, carpus broad, extending to palmar angle of propod; pleon side plates 2 and 3, blunt quadrate . . *M. schneideri* Sars.
   Coxal plates deep, 4th deepest; gnathopod 2, carpal lobe slender, not reaching to palmar angle; pleon side plates rounded behind . . . *M. longirostris* (Goes).
3. Peraeopods 3 and 4, dactyls very short, less than one-fourth length of segment 6; antenna 1, peduncular segment 3 = segment 2; pleon side plates 2 and 3, blunt quadrate . . . . *M. edwardsi* (p. 97).
   Peraeopods 3 and 4, dactyls longer than one-half segment 6; antenna 1, segment 3 less than one-half segment 2; pleon side plates rounded behind . . . . . 4.
4. Antenna 1, peduncle segments 1 and 2 elongate, 2 distinctly longer than 1; gnathopod 1, carpal lobe broad, short . . . . *M. packardi* Boeck.
   Antenna 1, peduncle segment 2 not longer than 1; gnathopod 1, carpal lobe reaching or nearly reaching palmar angle of propod . . . . 5.
5. Antenna 1, peduncle segment 2 with anterodistal process . *M. tuberculatus* Boeck.
   Antenna 1, peduncular segments slender, without process . . . . 6.
6. Gnathopod 2, carpal lobe broad, short; telson apically rounded, apex with 2 stout spines . . . . *M. latimanus* (Goes).
   Gnathopod 2, carpal lobe slender, reaching to (or almost to) palmar angle of propod; telson truncate or notched at apex . . . . 7.
7. Gnathopod 2, carpal lobe not reaching palmar angle; eye located centrally on rostrum . . . . *M. borealis* Boeck.
   Gnathopod 2, carpal lobe reaching to palmar angle; eye located basally . . . 8.
8. Peraeopods 3 and 4, dactyls equal to or less than segment 6 . . . .
   . . . . *M. norvegicus* (Boeck).
   Peraeopods 3 and 4, dactyls equal to or exceeding length of segment 6 . . . 9.
9. Telson distinctly notched; eyes nearly spherical; peraeopod 7, segments 6 and 7 with short posterior marginal setae . . . . *M. tesselatus* Schneider.
   Telson truncate or very slightly emarginate; eyes ovate; peraeopod 7, segments 6 and 7, with long posterior plumose setae . . . . *M. intermedius* (p. 97).

Monoculodes edwardsi Holmes 1905 * Plate XIX.1   (ĕd′ wărdz ī)

L. 9 mm. Rostrum strong, moderately deflexed. Eye large, elliptical, remote from tip of rostrum. Coxal plates 1–4 as deep as body segments, lower margins lightly setose. Coxa 5 very deep, deeper than length of basis of peraeopod 5. Antenna 1, peduncle segments 2 and 3 subequal.

Mandibular palp slender, segment 3 about one-half segment 2, sparsely setose.

Gnathopod 1 slightly smaller than gnathopod 2; carpal lobe prolonged about two-thirds posterior margin of propod. Gnathopod 2, carpal lobe reaching to posterior palmar angle.

Peraeopods 3 and 4, segment 6 shorter than 5, densely setose; dactyl very short. Peraeopods 5 and 6, segments 5, 6, and 7, short. Peraeopod 7 extending little beyond uropods; basis posterodistally expanded into a rounded lobe.

Pleon side plates 2 and 3, hind corners bluntly quadrate. Uropod 3, outer ramus shorter than inner, tips not reaching tips of uropods 1 and 2. Telson rectangular, apex truncate, not emarginate.

Distribution: Gulf of St. Lawrence to Cape Cod, Middle Atlantic states south to Georgia and Northern Florida; Gulf of Mexico.
Ecology: Fine sand and silty sand; lower intertidal and immediately subtidal to less than 75 m. From fully marine to oligohaline brackish conditions.
Life Cycle: Annual; ovigerous females from May–September; several broods.

Monoculodes intermedius Shoemaker 1930 Plate XIX.2   (ĭn tĕr mĕd′ ī ŭs)

L. 6–8 mm. Head, rostrum strong, deflexed. Eye large, elliptical or ovate, on basal part of rostrum. Coxal plates 3–4 normal, 5 not enlarged, not deeper than basis of peraeopod 5, but slightly deeper than 6. Antennae normal, posterior peduncular margins thickly short-setose. Antenna 1, peduncle segment 3 very short (♂).

Mandible, palp strong; segment 3 longer than one-half segment 2; inner margin setose throughout.

Gnathopod 1 normally subchelate, posterior margin slightly convex, carpal lobe extending about two-thirds along posterior margin of propod. Gnathopod 2, propod longer than in gnathopod 1; posterior margin slightly concave; carpal lobe slender, barely attaining palmar posterior angle.

Peraeopods 3 and 4 slender, segment 6 = segment 5; dactyl elongate, exceeding segment 6, slender. Peraeopods 5 and 6, segments 5, 6, and 7

---

* Records from arctic regions are suspect; those from very low salinities in Chesapeake Bay and Florida (St. John's system) may be a different species.

slender. Peraeopod 7 very elongate, extending well beyond uropods; basis, posterior angle not produced distally as a lobe; segments 6 and 7 plumose behind.

Pleon side plates 2 and 3, posterior angles broadly rounded. Uropod 3 elongate, exceeding uropods 1 and 2; rami equal. Telson, apex very slightly emarginate, with pair of marginal spines.

Distribution: Amphi-Atlantic, boreal; in North America from the Gulf of St. Lawrence south to Cape Cod Bay.
Ecology: Moderately common on fine stable sand bottoms, from about 5 to more than 50 m.
Life Cycle: Annual; ovigerous females in winter.

Genus *Synchelidium* G. O.Sars 1892 (sĭn kēl ĭd' ĭ ŭm)

Rostrum short, deflexed. Eye not completely fused middorsally. Antenna 1 and 2 subequal, short (♀). Coxal plates 4 and 5 largest.

Mandible, molar very small, with apical seta; palp weak, slender. Maxilla 1, inner plate with 1 apical seta. Maxilla 2, plates very small. Maxilliped, outer plate small, a few heavy spines only. Lower lip, inner lobes partly fused, outer lobes widely separated.

Gnathopod 1 subchelate; border of palm denticulate. Carpal lobe prolonged beneath but not fused to propod. Gnathopod 2 chelate; propod long and slender; carpal lobe prolonged beneath and fused to propod along entire length.

Peraeopods 3 and 4, segment 6 and dactyl short. Peraeopod 7, segments 6 and 7 with slender spines along margins. Uropod 3, rami longer than peduncle.

KEY TO AMERICAN ATLANTIC SPECIES OF *SYNCHELIDIUM*

1. Carpal lobe of gnathopod 1, extending well beyond palmar angle
. . . . *S. americanum* (p. 98).
Carpal lobe of gnathopod 1, extending to palmar angle . *S. tenuimanum* Norman.

Synchelidium americanum n. sp. Plate XX.1 (ă měr ĭ kăn' ŭm)

L. 6 mm. Rostrum short, strongly deflexed. Eye touching anterior margin. Coxa 4 largest and deepest; coxa 5 large, posterior lobe deeper; coxa 6 much deeper than 7. Antenna 1, short, slightly larger than 2; peduncular segments 2 and 3 shorter than 1.

Mandibular palp slender; segment 3 nearly as long as segment 2. Maxilla 1, inner plate with 1 apical seta. Maxilla 2, outer plate not broadened. Maxilliped, palp short, palp segment 2 extending little beyond distal spines of outer plate.

Gnathopod 1, carpal lobe prolonged well beyond posterior angle of propod; segment 6, length about twice width, palm nearly vertical, with 12–15 fine denticles.

Gnathopod 2, carpal lobe fused throughout length to posterior margin of propod; dactyl short, about one-fifth length of propod; segment 6 (propod) about 1½ times length of segment 6 of gnathopod 1.

Peraeopods 3 and 4, segments 5 and 6 short, stout, very setose; dactyls very small. Peraeopods 5 and 6 similar, 6 distinctly larger than 5. Peraeopod 7, basis with posterodistal rounded lobe; segment 7 not longer than 6, posteriorly bare or with a few spines.

Pleon side plates 2 and 3 rounded posteriorly. Uropod 3, rami subequal, nearly attaining tips of uropods 1 and 2. Telson truncate, very slightly emarginate.

Distribution: From central Maine to Cape Cod, southward to Georgia.
Ecology: Fine sand beaches along semiprotected shores, from immediately subtidal to several meters in depth; a relatively rare sand-burrowing species.
Life Cycle: Annual; ovigerous females from May–September; several broods.

Family HAUSTORIIDAE   (hôst ōr ē′ĭ dē)

Body compressed or broad, smooth; urosome occasionally spinose or setose. Coxal plates deep, lower margins setose. Head, rostrum short or lacking. Eyes lateral, Pigmentation reduced or lacking. Sexual dimorphism striking or suppressed. Antenna 1 shorter than 2; accessory flagellum present.

Upper lip rounded; lower lip with inner lobes strongly developed. Mandible, molar strong, incisor weak, palp strong, geniculate. Maxilla 1, inner plate large, setose, outer plate with 8–10 spine teeth. Maxilla 2, outer plate variously enlarged, setose. Maxilliped, plates and palp large, setose.

Gnathopods 1 and 2 feebly subchelate, simple or minutely chelate, showing little sexual dimorphism. Peraeopods 3–7 modified for burrowing; segments variously expanded, spinose and/or setose. Peraeopods 5–7, basis expanded; peraeopod 6 longest. Pleopods powerful; peduncle short. Urosome and uropods variously reduced; uropods normally biramous. Uropod 3 biramous, differing in form from uropods 1 and 2. Telson small, usually cleft. Coxal gills on peraeon segments 2–6, occ. 2–7? Brood plates lamellate or narrow, with numerous marginal setae.

KEY TO SUBFAMILIES OF HAUSTORIIDAE

1. Body usually slender; gnathopod 1 subchelate; peraeopods 3–7 with dactyls; uropod 3 elongate, rami unequal; urosome not reduced; maxilliped palp 4-segmented; lower lip outer lobes normal . . . . . . PONTOPOREIINAE (p. 100).

Body broad; gnathopod 1 simple; peraeopods 3–7 lacking dactyls; uropod 3, rami short, subequal; urosome reduced; maxilliped palp 3-segmented; lower lip, outer lobes lacking mandibular process . . . . HAUSTORIINAE (p. 106).

## Subfamily PONTOPOREIINAE Bousfield 1965					(pŏnt ō pôr ī'ī nē)

Head and body segments of normal width, seldom tumid or produced laterally. Rostrum weak or lacking. Eyes usually present, pigmented. Pleon not abruptly narrowing beyond peraeon. Urosome normal, often dorsally toothed, spinose or pilose; posteroventral lappet lacking. Antenna 1 variously geniculate, peduncular segments 2 and 3 short. Antenna 2, peduncular segments 4 and 5 not lobate or excessively plumose behind.

Lower lip, mandibular processes subacute. Maxilla 2, plates subequal. Maxilliped, palp slender, 4-segmented.

Gnathopod 1 subchelate. Gnathopod 2, subchelate or simple, usually unlike gnathopod 1. Peraeopods 3–7 with dactyls. Peraeopods 3 and 4 similar, 2nd shorter. Peraeopods 6–7, segments 4 and 5 not expanded, dactyls very small. Pleopod peduncles not powerfully developed, rami subequal.

Uropods 1 and 2 normal, similar, equally biramous; uropod 2 smaller. Uropod 3 large, rami foliaceous (esp. ♂), very unequal. Telson deeply cleft.

Brood plates large, moderately developed on peraeon 5. Coxal gills usually on peraeon segments 2–7. Sexes dimorphic; female larger. In ♂, antennae are calceolate; uropod 3 natatory.

KEY TO NEW ENGLAND GENERA OF PONTOPOREIINAE

1. Body broad, (pleon) with middorsal prominences; urosome unarmed; coxal plates subacute below . . . . *Priscillina.*
   Body slender; pleosome dorsally smooth; urosome toothed, weakly spinose or setose; coxal plates square or rounded below . . . . 2.
2. Antenna 1, peduncle not geniculate; peraeopod 5 not doubly geniculate at segment 4; uropod 3, outer ramus not more than twice length of inner. . . . . . .
   . . . . *Pontoporeia* (p. 100).
   Antenna 1 geniculate at second joint; peraeopod 5 doubly geniculate at 4th segment; uropod 3, inner ramus much shorter than outer. . . . . 3.
3. Gnathopod 2 subchelate; maxilliped palp normal; urosome dorsally smooth; uropod 3, inner ramus one-half length of proximal outer segment . . . .
   . . . . Amphiporeia (p. 102).
   Gnathopod 2 simple, subfusiform; maxilliped palp, segment 2 produced apically, segment 3 arched; urosome dorsally with spinules; uropod 3, inner ramus very short, scalelike . . . . *Bathyporeia* (p. 104).

Genus *Pontoporeia* Krøyer 1842					(pŏnt ō pôr ī' ă)

Body dorsally smooth; urosome may be dorsally setose and/or with toothed process. Coxal plates deep, quadrate; 4th weakly excavate; coxa 5, anterior lobe deeper. Antenna 1, peduncle not geniculate at segment 2, directed anteriorly. Antenna 2 elongate in pelagic males.

Mandible, palp normal, segment 2 not geniculate. Maxilliped very small, palp segments 3 and 4 small, short.

Gnathopods 1 and 2 subchelate, unequal. Gnathopod 2 weakly subchelate in ♀. Peraeopods 5–7 normally flexed; segments 4–6 not expanded. Peraeopod 7, basis enormously expanded, much larger than in 5 and 6.

Uropod 3, rami short, inner slightly shorter; both foliaceous, esp in ♂; outer ramus 1-segmented (♀), biarticulate in ♂.

KEY TO AMERICAN ATLANTIC SPECIES OF *PONTOPOREIA*

1. Urosome 1, with dorsally directed bifid cusp; coxa 5, anterior lobe narrowing distally . . . . *P. femorata* (p. 101).
   Urosome segment 1, smooth above; coxa 5, anterior lobe rounded below . . . . .
   . . . . *P. affinis* Lindstrom.

Pontoporeia femorata Krøyer 1842 Plate XXI.1   (fĕm ŏ rāt′ă)

L. 13 mm. Coxal plates deep, 1st broad distally, 4th scarcely excavate, 5th anterior lobe deep, subacute. Pleosome, dorsal surface lightly setose. Urosome 1 with strong middorsal bifid recurved process. Head, anterior lobe produced, acute. Eye reniform. Antenna 1, accessory flagellum 2-segmented; segment 2 very short.

Mandibular palp strong, not geniculate, segment 2 not inflated. Maxilliped palp, segment 1 large; segment 2 truncate distally; dactyl very short.

Gnathopod 1 (♀) subchelate, small; segment 5 longer than 6, with deep rounded carpal lobe. Gnathopod 2 (♀) weakly subchelate, dissimilar to gnathopod 1, segment 5 longer than 6, posterior lobe shallow.

Peraeopods 5 and 6, basis not much enlarged, hind margins nearly straight, setose; dactyls very short. Peraeopod 7, basis very enlarged, posterior margin evenly rounded, richly long-setose.

Abdominal side plates 1–3, hind corners blunt, lower margins with short spines. Uropod 1 with strong interramal spine. Uropod 2, outer ramus distinctly shorter than inner. Uropod 3 (♀) short, spinose, not foliaceous, little exceeding 1 and 2; inner ramus shorter than 1-segmented outer ramus, both longer than broad peduncle. Telson short, apices subacute, cleft more than half.

In the male, antenna 2 elongate, calceolate. Peraeopod 7, segments 4–6 elongate, 4 produced posterodistally, posterior margin closely spinose. Uropod 3, rami large, foliaceous.
Distribution: Circumpolar and subartic; in North America from Baffin Island and Hudson Strait throughout eastern Canada and Nova Scotia to northern New England (Penobscot Bay).

Ecology: Burrowing in muddy and sandy mud bottoms, especially along channel banks, in shallow water, from just subtidal to more than 50 m.

Life Cycle: Annual; ovigerous females October–February; one brood per year. Mature males pelagic in fall and winter; common in February in Nova Scotian coastal waters.

Genus *Amphiporeia* Shoemaker 1929   (ăm fĭ pör ĭ'ă)

Body slender-fusiform. Head, all eye facets pigmented, lenticular. Like *Pontoporeia,* but antenna 1 geniculate at segment 2; segment 1 combining with opposite member to form prowlike pseudorostrum. Head, anterior lobe acute. Antenna 2 not elongate in ♂, peduncle 4 and 5 subequal. Coxa 1 not directed forward.

Mandible, palp geniculate, segment 1 broad. Maxilliped, segment 3 broad, not prolonged distally beyond base of dactyl.

Gnathopods 1 and 2 subchelate, small. Gnathopod 2 larger, both stronger in ♂. Peraeopod 5 doubly geniculate at 4th segment, which is broadened. Peraeopod 6, segment 4 somewhat expanded. Peraeopod 7, basis broadly expanded, with distal lobe produced, not excavate.

Urosome segments dorsally smooth. Uropod 3 unequally biramous; inner ramus slender, outer ramus prominently 2-segmented. Telson deeply cleft, lobes parallel, lacking inner marginal spines.

KEY TO AMERICAN ATLANTIC SPECIES OF *AMPHIPOREIA*

1. Gnathopods 1 and 2 similar in ♂ and ♀; coxa 5 deeper than one-half base of peraeopod 5; uropod 3, terminal segment of outer ramus elongate, about equal to inner ramus   .   .   .   . 2.

   Gnathopods 1 and 2 unlike in ♀, propod of gnathopod 2 sublinear; coxa 5 shallow, less than one-half length of basis; uropod 3, terminal segment of outer ramus short, less than one-half inner ramus.   .   .   .   .   *A. lawrenciana* (p. 102).

2. Antenna 1, peduncular segment 1 completely overhangs segment 2; abdominal side plate 3, hind margin deeply notched proximally; animal large (>10 mm) .   .   .

   .   .   .   .   *A. gigantea* n. sp. (p. 103).

   Antenna 1, peduncular segment 2 not completely overhung by segment 1; abdominal side plate 3 not sharply notched posteroproximally; animal small (3.5–6)   .   .

   .   .   .   .   *A. virginiana* (p. 103).

Amphiporeia lawrenciana Shoemaker 1929 Plate XXI.2   (lâr ĕnts ĭ ăn'ă)

L. 7–9 mm. Coxal plates 5 and 6 relatively shallow, less than half depth of corresponding basis.

Mandible, outer margin of palp segment 2 with 1–2 setae only; maxilliped palp, segment 3 not swollen.

Gnathopod 1 (♂) small, subchelate. Gnathopod 2 (♂) subchelate, larger than, but differing in form from gnathopod 1; segment 6, relatively narrow, palm short.

Peraeopod 5, segment 4 broad-ovate, posterior margin strongly convex. Peraeopod 7, segment 6 distinctly longer than 4 and 5.

Pleon side plate 3, hind corner acute, slightly produced. Uropods 1 and 2, rami with 2–3 longish marginal spines. Uropod 3 relatively short, outer ramus less than 3 times length of peduncle; terminal segment short, less than half the inner ramus. Telson short, extending slightly beyond peduncle of uropod 3; lobes each with 1 lateral marginal spine group.

Distribution: Southern Labrador and north shore of Gulf of St. Lawrence along the outer coast of Nova Scotia to the Bay of Fundy and northern Gulf of Maine. Recorded off Campobello Island, off Machias; inside Mt. Desert Island.
Ecology: Medium coarse sand, extreme low water level to nearly 200 m in depth; in summer-cold and cool waters.
Life Cycle: Probably annual; ovigerous females not taken in summer.

Amphiporeia virginiana Shoemaker 1933 Plate XXII.1   (vër jĭn ĭ ān′ ă)

L. 5.0 mm ♀; 3.5 mm ♂. Coxal plates 5 and 6 relatively deep, more than half depth of corresponding basis.

Mandible, outer margin of palp segment 2 with several setae. Maxilliped, palp segment 3 very large and inflated distally.

Gnathopods 1 and 2 (♂) small but strongly subchelate, palm oblique, similar. In ♀, gnathopods less strong, palm more vertical, segment 6 longer.

Peraeopod 5, segment 4 long-oval, posterior margin gently convex. Peraeopod 7, segments 4–6 subequal in length.

Pleon side plates 2 and 3, hind corner subquadrate, not produced. Uropods 1 and 2, marginal spines of rami few, short. Uropod 3 elongate; outer ramus slender, more than 3 times length of peduncle; terminal segment long, about equal to inner ramus; inner margin foliaceous. Telson long, extending well past peduncle of uropod 3; lobes divergent, each with 2 outer marginal groups of spines.

Distribution: From eastern Nova Scotia (Guysborough Co.) south along the Gulf of Maine to the Middle Atlantic states and North Carolina. Recorded from Nantucket and Martha's Vineyard, outer shores.
Ecology: Surf sand beaches, mid water to slightly subtidal levels; often concentrated at f.w. stream outflows over sand flats.
Life Cycle: Annual; ovigerous females from April –July; one brood per year.

Amphiporeia gigantea n. sp. Plate XXII.2   (jī gănt ē′ ă)

L. 11.0 mm ♀. Coxal plates 1–4, lower margins completely lined with stiff setae. Coxa 5 deeper than half coxa 4 and half basis of peraeopod 5. Antenna 1, peduncular segment 1 elongate distally, completely overhanging peduncular segment 2.

Mandible, palp segment 2 expanded medially, outer margin with 4 groups of slender spines. Maxilliped, outer plate reaching one-half palp segment 2; segment 3 not usually expanded.

Gnathopods 1 and 2, propod (segment 6) short, deep, palm very oblique, spinulose; gnathopod 2, segment 5 twice the length of segment 6.

Peraeopod 5, basis nearly as broad as deep; segment 4 broad-ovate, posterior margin strongly convex. Peraeopod 6, segment 4 little expanded, segment 6 distinctly longer than 5. Peraeopod 7, basis much broader than deep (segments 5–7 lacking in all three known specimens at hand).

Pleon side plate 2, hind corner subacute, somewhat produced; side plate 3, hind margin lightly setose, sharply and deeply incised proximally. Uropods 1 and 2, rami with 8–10 slender marginal spines. Uropod 3 elongate, outer ramus about three times length of peduncle, foliaceous on inner margin; terminal segment as long as inner ramus. Telson elongate, twice as long as wide; lobes narrow, 4–5 spines on outer margin.

Male unknown.

Distribution: Cape Cod Bay (Census station 1812N–type locality), near Cape Cod; off Sandy Hook, N.J., Pearce survey station 36.

Ecology: Medium coarse sand; subtidally in 10–15 m.

Life Cycle: Ovigerous females in August? Four brood young were noted in the brood lamellae of one female; each young is about one-fifth length of adult ♀.

Genus *Bathyporeia* Lindstrom 1855   (băth ĭ pör ī′ ă)

Body very slender-fusiform, small. Antenna 1 sharply geniculate at segment 2. Segment 1 very large, overhanging vertically directed antenna 1, forming (with opposite member) a pseudorostrum. Antenna 2 (♂), elongate in European species, short in N. American species. Urosome segments with dorsal spinules, and/or setae. Coxa 1 narrow, directed forwards.

Mandible, palp geniculate. Maxilliped, palp segment 2 prolonged distally beyond base of strongly curved segment 3.

Gnathopod 1 very weak, small, subchelate. Gnathopod 2 fossorial, subfusiform, dactyl lacking (?).

Peraeopod 6, segment 4 moderately expanded. Peraeopod 7, basis expanded behind, emarginate or excavate distally, lobe not developed.

Uropod 3, outer ramus elongate, cylindrical, 2-segmented; inner ramus a short lobe. Telson cleft to base, lobes diverging, apices heavily spinose.

KEY TO NORTH AMERICAN ATLANTIC SPECIES OF *BATHYPOREIA*

1. Telson lobes with inner marginal spines; eye with 4 pigmented marginal facets; peraeopod 7, posterodistal margin oblique, slender setose . . . .
. . . . *B. quoddyensis* (p. 105).
Telson lobes, inner margins bare; eye with 6 marginal pigmented facets and central group of 3; peraeopod 7, basis, posterior margin sharply incised, distal margin nearly horizontal, with 4–5 spines . . . . *B. parkeri* n. sp. (p. 105).

Bathyporeia parkeri n. sp. Plate XXIII.1   (pârk′ ĕr ī)

L. 4.0 mm ♀, 3.0 mm ♂. Head, anterior lobe subacute. Eye with 6 marginal pigmented facets, and an inner group of 3 pigmented facets. Antenna 1, peduncle segment 1, distal margin bulbous, strongly overhanging segment 2. Coxa 1 small, sharply directed forward; coxa 4 not exceptionally large; coxa 6 with small anterior lobe.

Maxilliped, palp segment 2 inner distal lobe rounded. Gnathopod 1, segment 6 narrower and shorter than segment 5; palm very oblique, dactyl stout. Gnathopod 2, segment 6 relatively short, obliquely truncate.

Peraeopod 5, segment 4 ovate, posterior margin smoothly convex. Peraeopod 6, segments 4 and 5 subequal in length. Peraeopod 7, posterior margin of basis serrate, sharply incised and excavate at junction with limb proper, lower (distal) margin with 5 stiff spines.

Pleon side plates 1 and 2, posterior margin proximally produced as sharp tooth. Side plate 3, posterior margin distally convex, overhanging small tooth at hind corner. Uropod 2, inner ramus shorter than outer, inner margin lacking spines. Urosome segment 1 with proximal and distal middorsal and dorsolateral clusters of small spines; urosome segments 2 and 3 dorsally bare. Uropod 3 relatively short, outer ramus about 3 times length of peduncle; terminal segment short. Telson lobes slightly diverging, inner margins lacking spines.

Distribution: South side of Cape Cod to northern Florida. New England localities: Monomoy Island, Vineyard Sound.
Ecology: Exposed sandy beaches, fine sand, from just below the breaker zone to 10 m. Seldom taken abundantly.
Life Cycle: Annual? Ovigerous females June–September.

Bathyporeia quoddyensis Shoemaker 1949 Plate XXIII.2   (kwôd ē ĕn′ sĭs)

L. 5.0 mm ♀, 4.0 mm ♂. Head, anterior lobe rounded. Eye with 4 weakly pigmented marginal facets. Coxal plate 1 relatively broad, hind margin straight; coxa 4 longer than 1–3, more than twice width of 3; coxa 5 large, width twice depth; coxa 6 with anterior lobe deeper. Antenna 1, distal margin of peduncle 1 nearly vertical, little overhanging segment 2.

Maxilliped, palp segment 2 inner distal lobe subacute; segment 3 form-ing a 90-degree curve. Gnathopod 1, segment 6 inflated, palm oblique, dac-tyl slender. Gnathopod 2, segment 6 nearly as long as 5, posterodistal (se-tose) margin gently rounding. Peraeopod 5, segment 4 subrhomboid, posterior margin angular. Peraeopod 7, posterior margin of basis nearly smooth, lower (distal) margin oblique, a few long setae at hind "corner." Peraeopod 6, segment 4 distinctly longer than 5.

Pleon segments 1–3, hind corners produced as small tooth; posterior margins gently convex. Urosome 1 with posteromiddorsal and dorsolateral clusters of short spines. Urosome 2 with 2 middorsal setae; urosome 3 with pair of small middorsal spines. Uropod 2, outer ramus longer, equal to pe-duncle. Uropod 3 elongate, outer ramus thick, heavy, more than 4 times length of peduncle; terminal segment much longer than inner ramus (♀), sub-equal to it (♂). Telson lobes diverging, inner margin with 1–2 spines.

Distribution: Outer coast of Nova Scotia and the Gulf of Maine (incl. Bay of Fundy) south to Chesapeake Bay. Recorded in Cape Cod Bay and off Long Island.
Ecology: Fine sand, just subtidal to more than 40 m, along semiprotected shores. Relatively low-density populations.
Life Cycle: Annual; ovigerous females from April–July.

Subfamily HAUSTORIINAE Bousfield 1965   (hôst ōr ē′ī nē)

Head and body segments broadly arched, truncate-fusiform in dorsal out-line; rostrum moderately strong, antennal sinuses deep; eyes small, weakly or not pigmented. Peraeon segments produced laterally in short lobes to which limbs are attached. Pleon abruptly narrowing beyond peraeon. Uro-some segments variously reduced, segments short, narrow, dorsally smooth; urosome 1 with posteroventral lappet. Antennal peduncular segments short, broad, lobate margins thickly plumose-setose; antenna 1, flagellum short; antenna 2, peduncle segment 4 lobate behind; flagellum short, noncalceo-late (both sexes).

Mouthparts modified for filter feeding. Lower lip, inner lobes large, elongate; outer lobes lacking mandibular processes. Mandible, palp elon-gate, segment 3 with proximal comb spines; incisor very weak or lacking. Maxilla 1 coxal segment with lateral baler lobe. Maxilla 2, outer plate tend-ing to large, lunate, strongly setose. Maxilliped, palp 3-segmented, 3rd seg-ment clavate or falcate.

Gnathopod 1 simple. Gnathopod 2 minutely chelate, otherwise similar to 1. Coxal plates 1–4 deep, arcuate, acute below. Peraeopods 3–7 without dactyls. Peraeopods 3 and 4 powerfully fossorial, usually similar, 2 usually larger than 1. Peraeopods 5–7, segments 4 and 5 broadly expanded, mar-gins and outer face richly spinose and plumose.

Pleopods powerful; peduncle short, broad; outer ramus longer than inner. Uropod 1 strong, spinose, unequally biramous (rarely uniramous). Uropod 2 smaller and unlike 1, thickly setose, occasionally uniramous. Uropod 3 short, rami subequal, setose at apices. Telson entire, variously cleft, or bilobed.

    Sexes similar, male smaller. Female brood plates small or vestigial on peraeon 5. Coxal gills on peraeon segments 2–6 only.

KEY TO NORTH AMERICAN GENERA OF HAUSTORIINAE

1. Peraeopod 4 distinctly smaller than and unlike peraeopod 3; coxal plates 1 and 2 much smaller than 3 and 4; telson of two widely separated lobes . . . .
. . . . *Eohaustorius* (Amer. Pacific).
Peraeopods 3 and 4 similar, subequal (4 occ. larger); coxal plates 1 and 2 deep and broad, telson entire or variously cleft, lobes approximate . . . . 2.
2. Posterodorsal border of pleon segment 3 free or slightly decurved, not recurved; urosome and uropods strong, rami spinose (except in *Lepidactylus*); pleon side plate 3 rounded . . . . 3.
Posterodorsal border of pleon segment 3 strongly recurved, forming a lobe overhanging urosome; pleon side plate 3 with posterior spinous process (except in *Haustorius*), urosome segments short, small. . . . . 4.
3. Body relatively slender, lateral lobes of peraeon weak; pleon gradually narrowing behind; head not very broad, rostrum weak; maxilliped, palp 3rd segment clavate; maxilla 2, outer plate little larger than inner . . . . . *Protohaustorius* (p. 108).
Body broadly fusiform; abdomen narrowing abruptly at pleon; head very broad, rostrum distinct; maxilliped, palp 3rd segment falcate or geniculate; maxilla 2, with large setose outer plate. . . . . 4.
4. Uropod 2 strong, biramous; telson broad, sharply or broadly cleft; antenna 2, peduncle 5 broad but not lobate behind; mandible with incisor; gnathopod 2 (♂) normally minutely chelate . . . . 5.
Uropod 2, small, uniramous, spatulate; telson small, entire; antenna 2, peduncular segment 5 lobate behind; mandible lacking incisor; gnathopod 2 (♂) carpochelate
. . . . *Neohaustorius* (p. 113).
5. Antenna 1, accessory flagellum 3–4 segmented; maxilla 2, outer plate lunate; uropod 1, rami subequal, setose and spinose . . . . *Lepidactylus* Say.
Antenna 1, accessory flagellum 2-segmented; maxilla 2, inner plate broad, outer margin convex; uropod 1, inner ramus shorter, both with spines only . . . . . .
. . . . *Parahaustorius* (p. 109).
6. Head broadest posteriorly, lateral margins subparallel, rostrum short; uropod 1, rami distally expanding, inner ramus distinctly the longer; uropod 3, terminal segment of outer ramus small or vestigial; maxilla 2, outer plate little larger than inner; telson cleft, shallow . . . . *Pseudohaustorius* (p. 119).
Head broadest medially, lateral margins convex; rostrum strong; uropod 1, rami distally tapering, outer ramus usually longer; uropod 3, terminal segment of outer ramus normal, distinct; maxilla 2, outer plate much enlarged; telson sharply cleft, short, spinose. . . . . 7.
7. Pleon side plate 3 rounded behind; uropod 1, rami subequal; urosome 2, dorsal margin short, nearly occluded; accessory flagellum 4–5 segmented . . . . . .
. . . . *Haustorius* (p. 112).

Pleon side plate 3 with posterior spinose process; uropod 1, inner ramus shorter, more slender than outer; urosome 2, dorsal margin broad, equal to urosome 3; accessory flagellum 2-segmented . . . . *Acanthohaustorius* (p. 115).

Genus *Protohaustorius* Bousfield 1965   (prōt ō hôst ör′ī ŭs)

Body relatively narrow; head narrow, rostrum short. Pleosome not overhanging urosome. Antenna 1, peduncular segment 1 longest; accessory flagellum 2-segmented. Antenna 2, peduncle 4 shallow-lobate behind; flagellar segment 1 elongate.

Mandible, incisor bidentate. Maxilla 2, outer plate little expanded. Maxilliped, segment 3 clavate.

Gnathopod 2, segment 6, much shorter than 5. Peraeopods 3 and 4 similar; peraeopod 4 longer and stouter (esp. segment 4) than in peraeopod 3.

Abdominal side plate 3 rounded behind. Urosome large; uropod 1, peduncle large, biramous; uropod 2 biramous, small; uropod 3 biramous, terminal segment of outer ramus relatively short. Telson entire, slightly cleft.

KEY TO AMERICAN ATLANTIC SPECIES OF *PROTOHAUSTORIUS*

1. Peraeopod 7, segment 4 narrowing toward posterior border, border with 2 spine groups; uropod 1, posterior margin of peduncle with distal spines only; mandible, palp segment 3 with about 4 comb spines . . . . . . *P. deichmannae* (p. 109). Peraeopod 7, segment 4, posterior margin as broad or broader than anterior margin, with 3–4 spine groups; uropod 1, posterior margin of peduncle spinose throughout; mandible palp segment 3 with about 10 comb spines . . . . *P. wigleyi* (p. 108).

Protohaustorius wigleyi Bousfield 1965 Plate XXIV.1   (wĭg′ lē ī)

L. 7.5 mm ♀. 6.5 mm ♂. Head as long as broad. Rostrum broad, obtuse. Eyes small, slitlike. Antenna 2, segment 4 posterior lobe, axis diverging distally from anterior margin. Flagellum 4-segmented, 1st not widening distally.

Mandibular palp, segment 3 proximally with about 10 short comb spines. Maxilliped, segment 3, arched, distally widest.

Coxal plates 1, 2, and 3, with 5, 7, and 7 plumose setae at posterior angle, respectively. Coxa 4, posterior lobe subacute; coxal width about equal depth. Peraeopod 5, segment 4, not broader than 5. Peraeopod 6, posterior coxal lobe large, deep; segment 5 subrectangular, distal margin with U-shaped incision; segment 6 as long as 5. Peraeopod 7, basis proximal posterior margin not sharply incised; segment 4, subrectangular, posterior margin subtruncate, with 3–4 stout marginal spines; segment 6 broadest proximally, length about equal to segment 5.

Pleosome 3, entire posterior and ventral margin continuous, smoothly rounded. Uropod 1, posterior margin of peduncle lined throughout with stout spines and minute serrations. Uropod 3, terminal segment of outer ramus long, nearly equal peduncle.

Distribution: Central Maine and Georges Bank to off North Carolina.
Ecology: Subtidal clean sands, from shoreline to over 80 fathoms.
Life Cycle: Annual; ovigerous females April–August.

Protohaustorius deichmannae Bousfield 1965 Plate XXIV.2   (dīk′ măn ē)

L. 6 mm ♀, 4.5 mm ♂. Head longer than wide; superior antennal sinus shallow; eye slit-shaped. Antenna 2, peduncle segment 4 medially broadest, long axis parallel to anterior margin. Flagellum 4-segmented, 1st broadening distally.

Mandible, palp segment 3 with 4 proximal marginal comb spines. Maxilliped, palp segment 3 short clavate.

Coxa 1, 2, and 3, posterior angle with 2, 3, and 2 plumose setae, respectively. Coxa 4, posterior lobe produced, acute, coxa much broader than deep. Peraeopod 5, segment 4 sharply expanding distally, wider than subquadrate segment 4. Peraeopod 6, posterior coxal lobe shallow; segment 5 narrowing distally; distal margin slightly concave; segment 6 distinctly shorter than 5. Peraeopod 7, basis sharply incised on proximal margin of posterior lobe; segment 4, posterior lobe narrowing, hind margin oblique, with 2 spine groups above distal angle; segment 6 relatively narrow.

Abdominal side plate 3, lower margin nearly straight. Uropod 1, posterior margin of peduncle with proximal minute serrations and distal row of spines. Uropod 3, terminal segment of outer ramus shorter than peduncle.

Distribution: Central Maine (Saco Bay) south to North Carolina and Georgia.
Ecology: Shallow, warm-water protected bays and estuaries, subtidal to about 20 m along edges of channels; fine sand with some admixture of silt.
Life Cycle: Annual; ovigerous females May–August; more than one brood per year.

Genus *Parahaustorius* Bousfield 1965   (păr ă hôst ör′ē ŭs)

Body broad. Pleon rounding normally, not overhanging urosome. Head widest posteriorly, rostrum short, acute. Antenna 1, peduncle segment 1 strong; accessory flagellum 2-segmented. Antenna 2, peduncle 4 broadly lobate posteriorly.

Mandibular incisor with 1 tooth. Maxilla 1, coxal baler lobe large. Maxilla 2, outer plate broadly expanded. Maxilliped, segment 3, stout, geniculate.

Peraeopods 3 and 4 similar; 3 longer. Peraeopod 7, segment 5, posterior margin long, more than one-half the anterior margin.

Abdominal side plate 3 rounded behind. Uropod 1 strong, normally biramous; interramal spines strong. Uropod 2 strongly biramous. Uropod 3 biramous, slender. Telson broad, moderately cleft.

KEY TO NEW ENGLAND SPECIES OF *PARAHAUSTORIUS*

1. Peraeopod 7, coxal plate broadly acute behind; segment 6 about equal to 5; uropod 1, posterior margin of peduncle spinose throughout; telson narrowly cleft, lobes rounded. . . . . 2.
   Peraeopod 7, coxal plate sharply elongated behind; segment 6 distinctly longer than 5; uropod 1, posterior margin of peduncle centrally unarmed; telson very broad, shallowly V-cleft. . . . . *P. attenuatus* (p. 110).
2. Peraeopod 7, segment 4 subrectangular, posterior margin with two (or three) prominent spines; peraeopod 6, segment 4 1½ times longer than segment 5, medially broadest; uropod 1, interramal spines less than one-half inner ramus. . . .
   . . . . *P. longimerus* (p. 111).
   Peraeopod 7, segment 4 narrowing behind, posterior margin oblique, with 1 stout spine; peraeopod 6, segment 4 slightly longer than 5, broadening distally; uropod 1, interramal spines more than one-half the length of inner ramus . . . .
   . . . . *P. holmesi* (p. 111).

Parahaustorius attenuatus Bousfield 1965 Plate XXVI.2   (ă tĕn ū āt′ ŭs)

L. to 14 mm ♀, 10.5 mm ♂. Head, rostrum and lateral lobes strong, sharply acute, superior antennal sinus deep. Eyes unpigmented in adults. Antenna 2, peduncular segment 4, deeper than long, posterior lobe short, strongly convex.

Maxilliped, terminal segment elongate, narrowing distally.

Peraeopod 5, segment 2 suborbicular, posterior margin finely setose proximally; segment 4 broader than deep, strongly rounding behind; segment 5 short, as broad as 4. Peraeopod 6, segment 4 long and broad, posterior margin shallow convex; segment 5 much narrower, distal margin short, concave, slightly oblique; segment 6 longer than 5, with 5–6 paired posterior spine groups. Peraeopod 7, coxa shallow, posterior lobe very attenuated, acute; segment 4 narrow behind, proximal and distal borders merging, "angle" with single spine; segment 5 squarish, deep, posterior margin nearly equal to anterior, with 2 spine groups; segment 6 broad, tapering, distinctly longer than 5.

Uropod 1, peduncle strongly arched, proximal hump bearing large spine and row of stiff setae; interramal spines slightly longer than one-half inner ramus. Uropod 3, terminal segment of outer ramus shorter than peduncle. Telson very broad, short, apex with shallow V-cleft.

Distribution: Northeast slope of Georges Bank and off Cape Cod, south to the mouth of the Chesapeake Bay. Recorded at Narragansett Beach near Pt. Judith, R.I.; Quick's Hole, Mass.

Ecology: In clean sand from just below low tide level to more than 50 m in depth; frequently cooccurring with *P. holmesi*.

Life Cycle: Annual? Ovigerous females in June–July, immatures in September samples.

Parahaustorius holmesi Bousfield 1965 Plate XXV.2   (hōmz'ī)

L. 13 mm ♀, 10 mm ♂. Head, rostrum strong, acute. Eyes unpigmented in adults. Antenna 2, peduncle segment 4 slightly deeper than long.

Mandibular palp, segment 2 very long, setose on inner margin. Maxilliped, segment 3 long, narrowing distally.

Peraeopod 5, segment 4 not wider than long, hind margin rounded; segment 5 squarish. Peraeopod 6, segment 4 relatively short, broadening distally; segment 5 almost square, distal margin truncate. Peraeopod 7, coxal plate broadly acute posteriorly; segment 4, posterior lobe long, narrowing, hind margin short, with 2 spine clusters above lower angle; segment 5 broad medially, posterior margin greater than 1½ anterior, with 2 spine clusters; segment 6 slender, about equal to 5 in length.

Uropod 1, peduncle strongly arched, proximal protuberance with very strong spine and row of slender spines; interramal spines stout, less than one-half inner ramus. Uropod 3, terminal segment of outer ramus equal to peduncle. Telson narrowly cleft one-half to base, inner margin of apices bare.

Distribution: Georges Bank and off Cape Cod, Mass., southward to the mouth of Chesapeake Bay. Recorded at Quick's Hole, Mass., "Albatross" stations on south side of Georges Bank.
Ecology: In fine clean sand, in depth of 20–50 m.
Life History: Ovigerous females, July–August. Immatures in August–September.

Parahaustorius longimerus Bousfield 1965 Plate XXV.1   (lŏnj ĭ měr' ŭs)

L. 10 mm ♀, 9.5 mm ♂. Head, rostrum broad, blunt conical. Eyes small, ellipsoid, very weakly pigmented, persisting in adult stage. Antenna 2, peduncle of segment 4 longer than deep, lobe relatively shallow.

Maxilliped, palp, segment 3 short, broad distally.

Peraeopod 5, segment 4 much broader than deep, hind margin truncate, segment 5 narrower. Peraeopod 6, segment 4 very long, more than 1½ times segment 5, margins subparallel, broadest in middle; segment 5 longer than wide, narrowing distally; segment 6 about as long as 5. Peraeopod 7, coxa squarish, posterior lobe broad; segment 4 subrectangular, posterior margin sharply truncate, with 2 heavy spines above lower angle; segment 5 broadest distally, hind margin short; segment 6 broad, not longer than 5.

Uropod 1, peduncle lined posteriorly with spines, with raised cluster of stout spines and stiff setae proximally; 2 interramal spines shorter than one-half inner ramus. Uropod 3, terminal segment of outer ramus shorter than peduncle. Telson cleft narrowly about one-half to base, inner side of apices bare.

Distribution: Cape Cod Bay (New Inlet, and Barnstable Harbor), Georges Bank and southern New England southward to northern Florida (Cape Kennedy); a close variety occurs in eastern Gulf of Mexico (Horn Island, Miss.). Common at Beach Pt., Barnstable Harbor, Nobska Pt., Nantucket and Martha's Vineyard; Narragansett Beach, R.I.

Ecology: Clean ripple sand, mainly surf and wave-exposed beaches, from the lower intertidal to depths of more than 25 ft. Burrows in the damp sand to depth of more than 4 inches when the tide is out.

Life Cycle: Annual; ovigerous females from May–December.

Genus *Haustorius* Müller 1775   (hôst ōr'ĭ ŭs)

Body broad-fusiform, medium to large. Head broadest medially, lateral margins strongly convex; rostrum strong; superior antennal sinus deep. Pleosome abruptly narrowing beyond peraeon 7; hind margin of 3 strongly decurved and recurved; side plate 3 rounded behind. Urosome greatly reduced, strongly overhung by pleosome; urosome 2 very short, dorsal margin nearly occluded. Antenna 1, accessory flagellum 3–5 segmented. Antenna 2, peduncular segment 5, deep, not lobate behind.

Mandibular palp long, segment 3 with numerous marginal comb spines. Maxilla 1, coxal baler lobe very large and well developed. Maxilla 2, outer plate extremely large and lunate. Maxilliped plates narrow; palp, segment 3 sharply geniculate, distally attenuated.

Peraeopods 3 and 4 similar, 3 slightly longer; posterior lobe of peraeopod 3 with both marginal and median spines. Peraeopod 6, segment 6 sublinear, not spatulate. Peraeopod 7, posterior border of segment 5 very short.

Uropods all biramous. Uropod 1, inner ramus not shorter than outer, rami closely subequal, with posterior marginal spines and slender setae. Uropod 2, peduncle long, rami short. Uropod 3, outer ramus with distinct terminal segment. Telson sharply incised; lobes apically spinose.

KEY TO NORTH ATLANTIC SPECIES OF *HAUSTORIUS*

1. Uropod 1, posterior margin of peduncle lined throughout with stout spines; peraeopod 7, segment 4, proximal margin rounding, continuous with posterior margin; European species . . . . 2.
   Uropod 1, posterior margin of peduncle with proximal and distal groups of spines; peraeopod 7, segment 4, proximal margin sharply set off from posterior margin; American species. 3
2. Accessory flagellum 4+-segmented; peraeopod 7, posterior margin of seg. 4 with 3–4 stiff spine groups . . . . *H. arenarius* Slabber.
   Accessory flagellum 3-segmented; peraeopod 7, prox. and posterior margin of segment 4 lined continuously with long stiff setae . . . . *H. algeriensis* Mulot.
3. Rostrum very long, attaining end of antenna 1, peduncular segment 1. . . . . *Haustorius* sp. (long rostrate form)
   Rostrum shorter, not exceeding midpoint of antenna peduncular segment 1 . . . . *H. canadensis* (p. 113)

Haustorius canadensis Bousfield 1962 Plate XXVI.1   (kăn ă děn′ sĭs)

L. to 18 mm ♀, 14 mm ♂. Head, rostrum long, somewhat variable, very acute, reaching to midpoint of peduncular segment 1 of antenna 1. Eyes present, but unpigmented in adults, very small, weakly pigmented in juveniles. Antenna 1, accessory flagellum 3–5 segmented. Antenna 2, peduncular segment 5 about as wide as long.

Mandibular palp, segment 3 with 12–15 lateral marginal comb spines. Maxilla 2, outer plate lunate, narrow, nearly twice length of inner plate. Maxilliped palp, segment 3, proximal stem narrower than distal lobe.

Peraeopod 5, basis posterior margin entirely setose; segment 4 wider than 5; segment 6 with 4–5 short anterior marginal spine groups. Peraeopod 6, all but distal margin of basis setose; segment 5 subquadrate, anterodistal angle broadly rounding; segment 6 with 5–6 short posterior marginal spine groups. Peraeopod 7, segment 4, upper distal angle sharply rounding, nearly 90 degrees sharply set off from posterior margin with 1 spine group, lower margin with spines only; segment 5, posterior margin with 1 spine group.

Uropod 1, peduncle, posterior margin with spines proximally and distally; posterior surface of both rami with spines and long setae. Uropod 3, terminal segment of outer ramus shorter than peduncle. Telson apex concave with sharp shallow median notch; lobes terminally with 7–10 slender spines.

Distribution: Southwestern Gulf of St. Lawrence (P.E.I. and Northumberland Strait) to eastern Cape Breton Island; Central Maine (Pine Pt., Ogunquit, Casco Bay) and southern New England to Chesapeake Bay. Common at Newburyport Beach and Brewster, Mass., Naushon Island, Falmouth and W. Falmouth, Mass., Seabrook Beach, N.H., Nobscussett Beach, Newport, R.I., and near Bridgehampton, N.Y.

Ecology: Intertidal, from low water level to about mean lower high water, on fine, clean, wave- and surf-exposed beaches; also in marsh creeks, in banks and bars exposed at low water.

Life Cycle: Annual or biennial; ovigerous females present from May–September, persisting through second winter. One brood per female per season.

Genus *Neohaustorius* Bousfield 1965   (nĕ ō hôst ōr′ĭ ŭs)

Body broad, pleosome without posterior shelf. Urosome stout, moderately strong. Head short, broad, rostrum short. Antenna 1, segment 1 not longer than 2; accessory flagellum 2-segmented. Antenna 2, peduncular segments 4 and 5 posteriorly lobate.

Mandible, incisor lacking; palp relatively short, stout. Maxilla 1, coxal baler lobe moderately developed. Maxilla 2, outer plate large, broadly lunate. Maxilliped, palp segment 3 broadly geniculate.

Gnathopod 2 carpochelate in mature male (segment 5 broadly expanded, with palm).

Peraeopods 3 and 4 similar, 3 larger. Peraeopods 5–7, segments 3–5 relatively weakly expanded, lightly setose. Abdominal side plate 3 rounded behind. Uropod 1, inner ramus short or lacking. Uropod 2 uniramous. Uropod 3, rami subequal, terminal segment of outer ramus distinct. Telson small, entire, apex slightly emarginate.

KEY TO AMERICAN ATLANTIC SPECIES OF *NEOHAUSTORIUS*

1. Antenna 2, flagellar segment 1 about equal to 2 and 3 combined; peraeopod 6, segment 6 with 2 groups of anterior marginal setae; uropod 1 uniramous   .   .   .   .
.   .   .   . *N. schmitzi* (p. 114).
Antenna 2, flagellar segment 1 not elongate; peraeopod 6, segment 6 with 1 group of anterior marginal spines; uropod 1 unequally biramous   .   .   .   .
.   .   .   . *N. biarticulatus* (p. 114).

Neohaustorius biarticulatus Bousfield 1965 Plate XXIX.1   (bī âr tĭk ū lāt′ ŭs)

L. 5.5 mm ♀. Head very broad, anterolateral lobes sharply rounding, flaring outwards. Eyes slitlike, weakly pigmented. Antenna 1, peduncular segment 1 deep, lateral margin with 6 heavy setae; accessory flagellum, outer segment shorter than inner. Antenna 2, posterior lobe of segment 4 deep, broadly arcuate; flagellum 4-segmented, 1st shorter than 2 and 3 combined.

Mandible with 5 accessory blades; palp segment 3, with 8–9 short comb spines. Maxilla 2, outer lobe broadly lunate. Maxilliped, inner plate large, narrowing distally; outer plate with convex outer margin; palp, segment 2, inner margin convex; segment 3 short, geniculate.

Peraeopod 6, segment 2 longer than broad; segment 6 short and cylindrical, with one group of anterior marginal spines. Peraeopod 7, segment 6 slightly longer than 5, basally broadest.

Uropod 1 biramous; peduncle stout, longer than rami; inner ramus shorter than outer, with posterior setae and terminal spines. Uropod 3, inner ramus shorter than outer. Telson with 3 posterior marginal spines on each side.

Distribution: South side of Cape Cod, from Nantucket Island, Martha's Vineyard, and Buzzards Bay to Long Island Sound; the Middle Atlantic states (Virginia) and Chesapeake Bay. Recorded at Sippewissett Marsh, Nantucket, and Martha's Vineyard.
Ecology: Burrows in sandy banks and bars, from low water to nearly mean high water level; in top 2–3 inches, occasionally down to 7 inches; often in company with *Haustorius canadensis;* migrates off shore in winter.
Life Cycle: Annual; ovigerous females from May–September. Several broods per female.

Neohaustorius schmitzi Bousfield Plate XXIX.2   (shmĭts′ ī)

L. 5.5 mm ♀, 4.0 mm ♂. Head short and wide; anterolateral angles acute. Eyes very small, unpigmented. Antenna 1, peduncular segment 1 sublinear sparsely setose; accessory flagellum segments subequal. Antenna 2, posterior lobe of peduncular segment 4 narrow; flagellum 4-segmented, segment 1 longer than 2 and 3 combined.

Mandible with 3 accessory blades; palp segment 3, with 6 comb spines. Maxilla 2, outer plate slender, lunate, double length of inner plate. Maxilliped, inner plate short, apex truncate; palp segment 2, inner margin slightly concave; terminal segment arcuate.

Peraeopod 6, segment 2 slightly broader than long; segment 6 cylindrical, with 2 groups of anterior marginal spines. Peraeopod 7, segment 6 sublinear, distinctly longer than segment 5.

Uropod 1 uniramous; ramus heavy, spinose, longer than peduncle. Uropod 3, rami subequal. Telson very small, with 2 short spines on each side of slightly emarginate apex.

Male: Flagellar segments of antenna 1 each with elongate aesthetascs.
Distribution: Cape Cod Bay (eastern shore); Vineyard Sound, Nantucket and Martha's Vineyard; Monomoy Island, outer part of Buzzards Bay; Pt. Judith Shore, and Long Island Sound, south to Georgia and northern Florida.
Ecology: From just below low water level to mean high water level; fine to medium sand, on both exposed and protected beaches; often in company with *H. canadensis*. In salinities from full sea salt diminishing to a summer average of about 6 o/oo (in Chesapeake Bay). Burrows in top 1–2 inches of substratum; females are higher up in tidal zone than males and immatures.
Life Cycle: Annual; ovigerous females from May–September; very few (4–6) large eggs per female; several broods per summer. Young may crawl back into brood pouch several times before finally leaving.

Genus *Acanthohaustorius* Bousfield 1965   (ă kănth ö hôst ör' ĭ ŭs)

Body broad, head broadest medially, rostrum broadly acute. Pleon narrowing abruptly behind peraeon; side plates acuminate posteriorly; plate variously prolonged into spinose process. Urosome somewhat reduced, longer than pleon 3, which overhangs it. Urosome 2 as long as 3, narrower than urosome 1. Antenna 1, accessory flagellum 2-segmented. Antenna 2, segment 4 deeply lobate; segment 5 expanded but not lobate behind.

Mandibular incisor mono- or bicuspate. Maxilla 1, coxal baler plate well developed. Maxilla 2, outer plate large, broad. Maxilliped, plates broad; palp segment 3 stout, geniculate.

Gnathopod 1 simple, segment 5 powerfully expanded, esp. in ♂. Gnathopod 2, segment 5 moderately expanded. Peraeopods 3 and 4 similar, 3 slightly larger. Peraeopod 5 long in relation to 7. Uropod 1, inner ramus shorter than outer, spinose and setose posteriorly. Uropod 2, rami and peduncle strong, subequal. Uropod 3, terminal segment of outer ramus long. Telson broad, sharply and deeply notched.

KEY TO NEW ENGLAND SPECIES OF *ACANTHOHAUSTORIUS*

1. Telson cleft nearly to base, posteromedial margins broadly convex; peraeopod 6, distal margin of segment 5 subtruncate; uropod 1, inner ramus slender, setae singly inserted; mature animals small (4–12 mm) . . . . 2.
   Telson sharply U-cleft one-half to base, lobes subtruncate behind; peraeopod 6, distal margin of segment 5 oblique; uropod 1, inner ramus strong, posterior marginal setae in clusters; animal medium-large (10–14 mm) . . *A. spinosus* (p. 117).
2. Pleosome 3, posterodorsal margin produced as a large subconical process, side plate 3 with short weak spinous process; peraeopod 7, coxal plate posteriorly quadrate, posterior lobe of segment 4 short, margins continuous, "angle" with 1 spine . . . . *A. intermedius* (p. 117).
   Pleosome 3, posterodorsal margin normally rounded behind, side plate with large spinous process; peraeopod 7, coxal plate posteriorly acute, hindlobe of segment 4 elongate, posterior margin distinct, with 2–3 spine groups . . . . 3.
3. Uropod 1, inner ramus very slender and short, about one-half length of outer ramus; peduncle strongly spinose posteriorly; telson lobes broad, lobes broader than long . . . . *A. shoemakeri* (p. 118).
   Uropod 1, inner ramus slightly shorter than outer, peduncle with a few posterior marginal spines; telson lobes as broad as long. . . . . . . *A. millsi* (p. 116).

Acanthohaustorius millsi Bousfield 1965 Plate XXVII.1   (mĭlz′ī)

L. 5.5–8.0 mm ♀, 4.5–7.0 mm ♂. Head, width about twice length, lateral margins evenly convex; rostrum short, blunt conical. Eyes very small, slitlike, weakly pigmented. Antenna 1, peduncle 1 as deep as long; flagellum 7-segmented. Antenna 2, peduncle 4 deeper than long; flagellum 7-segmented.

Mandible, palp segment 3 with 18 comb-row spines. Maxilliped, palp segment 3 with long distal margin.

Peraeopod 5, segment 5 narrower than 4; segment 6 as long as width of 5, terminal spines elongate. Peraeopod 6, segment 5 quadrate, distal margin at right angles to long axis. Peraeopod 7, coxal plate posteriorly acute; segment 4 with long posterior lobe, posterior margin oblique, with 3 spine groups; segment 5 without distinct posterior margin or spine group; segment 6 shorter than 5, broad.

Pleon side plates 1 and 2, hind corners acutely produced; pleon 3, posterior process long, strong. Uropod 1, peduncle posterior margin with a few spines only; rami slender, subequal in length, spinose posteriorly; inner ramus with 4 long, singly inserted, marginal setae and terminal spine clusters. Telson cleft nearly to base, "shoulders" of groove with slender spines.

Distribution: The most common and widely distributed essentially subtidal species of the American Atlantic coast, from Saco Bay, Me., south through New England to the Middle Atlantic states and (varietally) to Georgia and Florida (Cape Kennedy).
Ecology: In medium to fine sands, less highly oxygenated than those of in-

tertidal species; from mean low water level (sometimes in higher pools) to more than 50 m in depth. Occurs in shallows mainly in surf-protected areas, entrance to inlets, and outer sandy estuaries in salinities from fully marine to about 20 o/oo.

Life History: Annual; ovigerous females from April–August; several broods per year.

Acanthohaustorius intermedius Bousfield 1965
Plate XXVIII.1   (ĭn tĕr mēd′ ĭ ŭs)

L. 4.5 mm ♀. Head wider than long, outer margins smoothly convex; rostrum broad, blunt. Eyes unpigmented, not discernible. Body strongly arched. Pleosome segments broad; hind margin of 3 produced as a blunt-conical overhanging urosome process. Basal plate of peraeopod 5 relatively very large. Antenna 1, flagellum 5-segmented. Antenna 2, peduncle 4 relatively shallow; flagellum 5-segmented.

Mandible, incisor unicuspate; palp with comb row of 9 short spines. Maxilliped, palp segment 3 with short distal margin.

Peraeopods 3 and 4 relatively slender, segments relatively little expanded and weakly armed. Peraeopod 5, segment 5 much narrower than 4; segment 6 slightly longer than 5, terminal spines long. Peraeopod 6, segment 6 narrowly subquadrate, distal margin truncate, anterodistal corner slightly produced; anterior marginal spines long. Peraeopod 7, coxal plate truncate behind; segment 4, posterior lobes short, hind margin very oblique, with one slender spine group; segment 5 with distinct posterior margin bearing one spine group; segment 6 short, cylindrical, terminal spines long.

Pleosome side plates 1 and 2, hind corners not produced; side plate 3 with short, weak posterior process. Uropod 1, peduncle evenly spinose posteriorly with 3 moderately long interramal spines; inner ramus much shorter than outer, hind margin with 3 singly inserted setae. Telson cleft nearly to base, inner "shoulders" bare.

Distribution: East side of Cape Cod Bay and Georges Bank through southern New England, the Middle Atlantic states to Georgia and northern Florida (Cape Kennedy).

Ecology: Fine sands, from lower low water level and sandy channels of estuaries offshore to more than 40 m in depth.

Life Cycle: Annual; ovigerous females from May–September; several broods per female.

Acanthohaustorius spinosus Bousfield 1962 Plate XXVIII.2   (spīn ōs′ŭs)

L. 10–12 mm ♀. Head much broader than wide, broadest posteriorly; inferior angles blunt; rostrum short. Eyes apparently unpigmented; not discernible. Antenna 1, peduncle 1 longer than deep; flagellum 8–9 segmented. Antenna 2, segment 4 nearly as deep as long; flagellum 8-segmented.

Mandible, incisor unicuspate; palp segment 3 with more than 20 spines in comb row. Maxilla 1, palp elongate, outer margin heavily plumose. Maxilliped, segment 2 inner distal lobe narrow, subacute.

Peraeopod 5, segment 5 nearly as broad as 4, broader than length of segment 6. Peraeopod 6, segment 5 a parallelogram, distal margin oblique to long axis, anterior marginal spines not elongate, segment 6 relatively short, clavate, expanding distally. Peraeopod 7, coxal plate posteriorly broadly acute; segment 4, posterior lobe large, wider than segment 5, with truncate posterior margin bearing 3–4 stout spines; segment 6 short, broad.

Pleon side plate 2, hind corner acute, slightly produced; side plate 3 very deep and broad, hind corner process very strong. Uropod 1, peduncle very stout, hind margin lined with strong spines, with proximal cluster; rami subequal in length and thickness; posterior margin of inner ramus with 4–6 clusters of slender setae. Uropod 3, rami subequal. Telson very broad, cleft less than one-half to base, lobes apically truncate, lined with several slender spines.

Distribution: Outer coast of Nova Scotia (off Halifax), throughout the Bay of Fundy, Gulf of Maine, Georges Bank to south side of Cape Cod.
Ecology: Medium and coarse sand; in colder waters (less than 15°C summer) from the lower intertidal (Bay of Fundy) to depths of more than 200 m.
Life Cycle: Annual, possibly biennial; ovigerous females from May–July, probably only one brood per year.

Acanthohaustorius shoemakeri Bousfield Plate XXVII.2   (shoo māk'ĕr ĭ)

L. 6.5 mm ♀, 5.0 mm ♂. Head shape similar to that of *A. intermedius*. Eyes not visible in preserved material. Antenna 1, flagellum 9-segmented. Antenna 2, peduncle 4 as deep as long; flagellum 7-segmented, 1st longest.

Mandible, palp segment 3 with 20 spines in comb row. Maxilliped, palp segment 2, distal lobe sharply truncate.

Gnathopods 1 and 2, segment 6, very slender, not expanded. Peraeopod 5, segment 5 about as wide as 4; segment 6 large, heavy, longer than width of 5. Peraeopod 6, segment 6 broadly quadrate, distal margin at right angles to main axis, anterior marginal spines long. Peraeopod 7, coxal hind lobe acuminate; segment 4, posterior lobe short, oblique, posterior margin with 3–4 slender spines; segment 5 with single posterior seta; segment 6 slightly expanded, about as long as 5.

Pleon side plates 1 and 2 acute, produced. Pleon 3 relatively shallow, spinous process moderately strong. Uropod 1, peduncle as long as ramus, posterior margin spinose throughout, with 1 strong interramal spine; inner ramus weak; about two-thirds length of outer, tapering, with 3 groups of posterior setae. Uropod 3, rami slender; both segments of outer ramus longer than peduncle. Telson very broad, shallow; deeply cleft lobes apically rounded, inner shoulders bare.

Distribution: From south side of Cape Cod (off Martha's Vineyard) south to Middle Atlantic states (off Virginia); a variety off Georgia.
Ecology: Medium fine sand, 15–20 m.
Life Cycle: Probably annual; ovigerous female in August.

Genus *Pseudohaustorius* Bousfield 1965   (sū dö hôst ör'ĭ ŭs)

Body very broad-fusiform, short. Pleosome strongly overhanging urosome, abruptly narrowing beyond peraeon 7. Urosome very reduced; segment 2 occluded dorsally. Head very broad, short; rostrum short-acute. Antenna 1, segments 2 and 3 subequal; accessory flagellum 2-segmented. Antenna 2, segment 4 broadly lobate posteriorly.

Mandible, incisor unicuspate; 3 accessory blades in spine row. Maxilla 1 lacking accessory coxal baler lobe. Maxilla 2, outer plate not expanded. Maxilliped, palp segment 3 clavate.

Peraeopods 3 and 4 similar, segment 4 very powerful. Peraeopod 5, segments 4 and 5 extremely broad and heavily spinose. Peraeopod 6, segment 5 attached anterior to midpoint of segment 4. Abdominal side plate 3, posterior angle produced as stout spine.

Uropod 1, rami expanded distally, inner ramus longer. Uropod 2 biramous, strong. Uropod 3 short, outer segment of outer ramus very short or vestigial. Telson broad, slightly cleft, apex with long slender spines.

KEY TO KNOWN SPECIES OF *PSEUDOHAUSTORIUS*

1. Peraeopod 7, coxal lobe not exceptionally elongate; hind lobes of segments 4 and 5 lacking a distinct posterior border; hind margin of telson nearly straight. . . . . *P. borealis* (p. 120).
   Peraeopod 7, coxal lobe large and elongate; hind lobes of segments 4 and 5 with distinct posterior border, each with 2 spine groups; telson broadly V-cleft. . . 2.
2. Antenna 1, peduncular segment 3 longer than the first 3 flagellar segments; peraeopod 5, segment 6 shorter than width of segment 5; uropod 3, outer ramus distinctly 2-segmented . . . . *P. caroliniensis* (p. 119).
   Antenna 1, peduncular segment 3 shorter than first 2 flagellar segments; peraeopod 5, segment 6 longer than width of segment 5; uropod 3, outer ramus appearing 1-segmented (Gulf of Mexico). . . . . *P. americanus* Pearse.

Pseudohaustorius caroliniensis Bousfield 1965   (kăr ö lĭn ĭ ĕn' sĭs)
Plate XXX.1

L. 6.5–8.0 mm ♀. Head, width more than twice length, broadest posteriorly, rostrum short, broad, acute. Eyes unpigmented. Antenna 1, accessory flagellum markedly subterminal on peduncle segment 3; flagellum of 5–6 subequal segments. Antenna 2, posterior lobe of peduncle segment 4 very large, extending distally to one-half segment 5; flagellum of 8 unequal segments, subterminal segment longest.

Mandible, palp segment 3 distinctly shorter than 2, with 14 comb spines. Maxilla 1, inner plate with 2 distal setae. Maxilla 2, outer plate, distal margin oblique. Maxilliped, palp segment 2 much longer than wide.

Peraeopod 3, basis slender, narrower than segment 4. Peraeopod 4, posterodistal angle of coxa quadrate. Peraeopod 5, segment 5, width less than twice length; segment 6 short, length less than width of segment 5. Peraeopod 6, basis longer than wide; segment 4 not elongate; segment 5, anterior margin nearly straight. Peraeopod 7, segment 4, posterior lobe short, hind margin truncate; segment 6 broad, spatulate.

Pleon side plate 3 with 4 groups of facial plumose setae, 4–8 per cluster. Uropod 3, terminal segment of outer ramus short, not exceeding inner ramus. Telson shallowly V-cleft at apex.

Distribution: South side of Cape Cod (Buzzards Bay); off New Jersey; off Virginia; mouth of Chesapeake Bay to Georgia and northern Florida; North Carolina (Wrightsville Sound). Recorded at North Falmouth, Mass.
Ecology: Subtidal; in fine organic silty and more stable, less high-energy sandy substrata. From low intertidal, in beach seeps, to more than 10 m in depth.
Life Cycle: Annual; ovigerous females from May–August; several broods per female.

Pseudohaustorius borealis Bousfield 1965 Plate XXX.2   (bör ē ăl′ ĭs)

L. 6.5 mm ♂, 7 mm ♀. Head more than twice as wide as long; rostrum short, broadly acute. Eyes small, slitlike, weakly pigmented. Antenna 1, segment relatively short, accessory flagellum nearly terminal; flagellum of 6 segments, 1st longest. Antenna 2, peduncle segment 4 about as deep as long, lobe little extended distally; flagellum of 5 segments, 1st much the longest.

Mandible, palp segment 3 about equal to 2; comb row of 10 short spines merging into apical cluster of long pectinate spines. Maxilla 1, inner plate with 3 apical setae. Maxilla 2, outer plate, distal margin sharply truncate. Maxilliped, palp segment 2 nearly as broad as long, apical process large.

Peraeopod 3, basis strong, about as wide as segment 4. Peraeopod 4, coxal plate distal angle broadly rounded. Peraeopod 5, segment 5, width about twice length; segment 6 longer than width of segment 5. Peraeopod 6, basis wider than long; segment 4 elongate; segment 5, anterior margin convex, extending well forward of border of 4. Peraeopod 7, coxal plate short, narrowly rounding behind; segment 4, posterior lobe elongate, subacute, distal border indistinct, very oblique with one spine group; segment 6 shorter than 5.

Pleon side plate 3 with 5–6 groups of lateral plumose setae, up to 10 per cluster. Uropod 3, terminal segment of outer ramus about as long as proximal article and slightly exceeding tip of inner ramus. Telson very broad, apex truncate, or slightly convex, not emarginate.

Distribution: From south side of Georges Bank (lat. 42°N) and Cape Cod (Quick's Hole) to off Wachepreague, Va.

Ecology: Medium coarse to medium fine, more stable sands; in deeper water, from 10–60 m.

Life Cycle: Annual; ovigerous females in September; probably more than one brood per female.

Family ARGISSIDAE (ăr jĭs' ĭ dē)

Body smooth, slender. Peraeon segments short, deep; coxal plates 1–3 decreasing in size. Coxa 4 large, deep. Pleosome, powerful. Sexual dimorphism in antenna and urosome. Head, rostrum short, eye round, lateral. Antennae slender; antenna 1 shorter than 2, accessory flagellum present. Coxa 4 largest; coxa 5–7, posterior lobe larger than anterior.

Upper lip slightly emarginate; lower lip, inner lobes distinct. Mandible, molar strong, palp weak. Maxilla 1, outer plate with 8 apical spine-teeth. Maxilliped normal.

Gnathopods 1 and 2 simple or weakly subchelate, similar. Peraeopods 3 and 4 weak. Peraeopods 5–7 subequal, short, bases broadly expanded, dactyls short.

Uropods 1 and 2 biramous. Uropod 3 biramous; rami foliaceous. Telson deeply cleft.

Coxal gills saclike, on peraeon segments 2–7. Brood plates large, lamellate, margins setose.

Genus *Argissa* Boeck 1871 (ăr jĭs' ă)

Accessory flagellum 2-segmented. Coxa 4 large, rounded below.

Mandible with 7–8 spines in blade row; palp 3-segmented, short.

Peraeopod 7, segments 4–6, margins with plumose setae. Peraeopods 3–7, dactyls short.

Urosome segments 1–3 middorsally carinate or 2 produced posteriorly (♂). Uropod 3, outer ramus 2-segmented.

Argissa hamatipes (Norman) Plate XX.2 (hăm ăt'ĭ pās)

L. 6 mm. Urosome segment 2 (♂), posterior expansion large, overhanging urosome 3. Head, anterior angles obtuse. Eyes round, with 4 evenly spaced marginal pairs of pigmented lenticular facets. Rostrum short, rounded in front. Coxal plates 1 and 4 rounding in front, 2 and 3 narrowly rounding below.

Antenna 1 (♀), flagellum 7-segmented; in ♂, first two segments fused, setose behind, about equal to peduncular segments 2 and 3 combined. Antenna 2, flagellum 7-segmented; peduncle 4 longer than 5.

Mandible, palp segment 5 with 3–4 inner marginal setae. Gnathopods 1 and 2 simple; segments 5 and 6 slender, with simple and plumose setae behind. Peraeopods 3–7, basis with glandular bodies; dactyls short, set at right angles to segment 6. Peraeopod 6, basis smaller than that of 5; segment 4 (of both) with a few posterior marginal plumose setae. Peraeopod 7, basis very large, broadly lobed distally; margins smooth; segment 5 with anterior and posterior marginal plumose setae.

Abdominal side plate 3, hind corner broadly acute, not produced, margins smooth. Uropod 2 extending beyond uropod 1; rami narrowly lanceolate, with a few slender marginal spines. Uropod 3, rami with plumose setose inner margins; outer ramus with small terminal segment. Telson lobes subtriangular in ♀; oval in ♂.

Distribution: Amphi-Atlantic boreal; in North America from Labrador and the Gulf of St. Lawrence south to Cape Cod Bay.
Ecology: Epibenthic and pelagic, sandy and stony bottoms, from 5–100+ m in depth.
Life Cycle: Annual; ovigerous females in late winter and early spring.

Family PHOXOCEPHALIDAE *   (fŏks ō sĕf ăl′ ĭ dē)

Body fusiform (dorsal aspect), smooth. Coxal plates deep, 4th largest, lower margins setose. Head with rostrum overhanging base of antennae, usually in the form of a broad hood. Antennae short, vertically inserted, bases separated; accessory flagellum well developed. Sexual dimorphism pronounced except in gnathopods. Antenna 2 elongate, calceolate, and uropod 3 fully foliaceous in ♂.

Upper lip rounded; lower lip with small inner lobes. Mandible, incisor strong; palp strong, 3-segmented; molar process weak or vestigial. Maxillae 1 and 2, plates small, weakly setose. Maxilliped, plates small, weakly armed; palp large, dactylate.

Gnathopods 1 and 2 subchelate or chelate, subequal, not sexually dimorphic. Peraeopods 3–7, fossorial, spinose. Peraeopod 6 usually longest; Peraeopod 7 short (except *Platyischnopus*). Basis of peraeopods 6 and 7 always expanded. Pleopods often stronger in ♂ than in ♀. Uropods 1 and 2 biramous, falcate, spinose. Uropod 3 unequally biramous, more or less foliaceous. Telson entire or deeply cleft, lobes approximate.

Coxal gills simple, on peraeon segments 2–7. Brood plates slender, sublinear, with few marginal setae.

---

* Bousfield (1970) has indicated that *Platyischnopinae* should be transferred from Haustoriidae to Phoxocephalidae on the basis of the morphology of mouthparts, gnathopods, and uropods. The chelate gnathopods, entire telson, and elongate peraeopod 7 might justify separate subfamily recognition within the Phoxocephalidae.

KEY TO NEW ENGLAND GENERA OF PHOXOCEPHALIDAE

1. Peraeopod 5, segment 2, basis narrow; coxal plates 1–4, lower marginal setae plumose; abdominal side plate, hind corner usually produced, hooklike
   . . . . *Harpinia* (p. 127).
   Peraeopod 5, segment 2, basis broadly expanded; coxal plates 1–4, marginal setae simple; abdominal side plate 3, hind corner not produced as hook . . . . 2.
2. Rostral hood abruptly narrowed in front of eyes, apex sharply rounded; peraeopods 5 and 6 subequal (♀), segments 4 and 5 expanded, heavily spinose; maxilliped, palp segment 2, outer margin spinose. . . . . . . *Trichophoxus* (p. 125).
   Rostral hood regularly narrowing to apex; peraeopod 6 much longer than 5, segments 4 and 5 little or slightly broadened, moderately spinose; maxilliped, palp segment 2 lacking outer marginal spines. . . . . 3.
3. Eyes well developed and pigmented (esp. ♂); rostral apex rounded
   . . . . *Paraphoxus* (p. 124).
   Eyes very small and unpigmented; rostral apex acute . . *Phoxocephalus* (p. 123).

Genus **Phoxocephalus** Stebbing 1888   (fŏks ō sĕf′ ăl ŭs)

Body slender-fusiform. Coxal plates 1–4, posterodistal margin lined with long moderately simple setae. Coxa 4 deeper than broad. Rostral hood large, narrowing regularly to acute apex. Eyes small or vestigial. Antennae 1 and 2 subequal (♀); antenna 2, flagellum elongate, weakly calceolate in ♂.

   Mandible, molar cylindrical, ridged. Maxilla 1, inner plate lacking apical setae; palp 1-segmented. Maxilliped strongly dactylate; inner plate small; palp segment 2, outer margin smooth, 3 not produced. Epistome not produced in front.

   Gnathopods 1 and 2 subequal, similar; propod much longer than carpus, posterior carpal lobe of 2 vestigial. Peraeopods 3 and 4, segments 4 and 5 strong. Peraeopod 5, basis broadly expanded; segments 4 and 5 little expanded, longer than wide. Peraeopod 6, distinctly longer than 5, basis broadened in front and back; segments 4, 5 and 6 linear. Peraeopod 7 short, basis enormously expanded.

   Pleosome large, side plate 3 broad, hind corner blunt. Uropod 3, inner ramus short, bare (♀), foliaceous, nearly equal to outer ramus (♂). Telson deeply cleft.

Phoxocephalus holbolli Krøyer 1842 Plate XXXV.1   (hŏl′bŏl ī)

L. 5–7 mm. Coxa 4 deeper than wide, narrowing below; coxa 5, posterior lobe rounded. Urosome 3 (♂) with sharply rounded middorsal hump. Rostrum, lateral margin sinuous, apex slender, acute, slightly downturned. Eye small, oval, unpigmented. Antenna 1, peduncle segment 2 not lobate nor plumose posteriorly; accessory flagellum 3–4 segmented. Antenna 2 (♀), flagellum 6-segmented; peduncle segment 4 shallowly lobate posteriorly.

Mandible, palp segment 2 slender with 4–5 inner marginal setae. Maxilliped, inner margin of inner plate apically setose.

Gnathopod 1 slightly smaller than gnathopod 2; segment 5 shorter in gnathopod 2. Peraeopods 3 and 4 dactyls short, less than one-third segment 6, not basally flanked by pair of stout spines. Peraeopod 5, basis nearly as broad as deep; segments 4 and 5 not expanded. Peraeopod 6, basis broadest distally, posterior margin nearly straight; dactyl about two-thirds segment 6. Peraeopod 7, basis very broadly expanded, posterior margin with about 15 weak serrations; segment 3 smallest.

Abdominal side plate 3 very broad, lower margin straight, hind corner subquadrate. Uropod 1, rami subequal, with 1–2 proximal marginal spines. Uropod 2, peduncle with 3–4 tall slender spines; outer ramus with 1 marginal spine. Uropods (♀) short, inner ramus bare, about one-half length of outer ramus; in ♂, rami subequal, margins with plumose setae. Telson, length greater than width, lobes tapering, apices with terminal spine and seta.

Distribution: Arctic boreal; in North America, south to the Gulf of St. Lawrence, Gulf of Maine, and Long Island Sound.
Ecology: Common on fine sand and sandy mud bottoms, in "eelgrass" sand, in shallows and protected bays, from low intertidal to depths over 400 m.
Life Cycle: Annual; ovigerous females from February–June; one brood per female.

Genus *Paraphoxus* G. O. Sars 1891   (păr ă fŏks′ ŭs)

Body slender, small. Rostral hood broad, regularly tapering to rounded apex. Eyes present, black, larger in ♂. Antenna 1 and 2 subequal (♀); flagellum of 2 elongate and calceolate (♂).

Mandible, molar very weak, apex with seta(e). Maxilla 1, palp 2-segmented (occasionally indistinctly); inner plate with 2 apical setae. Maxilliped, inner plate broader than outer; palp segment 2, outer margin smooth. Epistome usually not produced anteriorly.

Coxal plates 1–4 deeper than broad, with posterodistal row of stiff setae. Gnathopods subequal, similar, propod larger than carpus. Peraeopod 5, basis expanded posteriorly; segments 4 and 5 moderately expanded. Peraeopod 6 elongate, distinctly longer than 3 or 5, segments sublinear. Peraeopod 7 short, distal segments linear.

Abdominal side plate 3, hind corner blunt. Uropod 3, inner ramus short, nearly bare (♀), foliaceous, nearly equal to outer ramus (♂). Telson deeply cleft.

KEY TO NORTH ATLANTIC SPECIES OF *PARAPHOXUS*

1. Peraeopod 3, segment 4 wider than long; uropod 3 (♀), outer ramus with some marginal plumose setae; abdominal side plate 3, posterior margin with 4–5 setae
   .   .   .   . *P. spinosus* (p. 125).

Peraeopod 3, segment 4 not wider than long; uropod 3 (♀), outer ramus lacking plumose setae; abdominal side plate 3, posterior margin bare   . *P. oculatus* Sars.

Paraphoxus spinosus Holmes 1905 Plate XXXIV.1   (spīn ōs' ŭs)

L. 5 mm ♀, 4 mm ♂. Coxal plate 4 subquadrate, narrowing below; coxa 5, posterior lobe subacute below. Head, rostrum blunt, rounded in front. Eye (♀) small, ovate, black; in ♂, very large, subrectangular, nearly meeting middorsally. Antenna 1, accessory flagellum 4–5 segmented. Antenna 2, flagellum 10–12 segmented (♀), more than one-half body length (♂), alternate segments calceolate.

Mandibular palp segment 1 short, apex obliquely truncate. Maxilla 1, palp indistinctly 2-segmented. Maxilliped, inner plate with 5 apical and marginal setae.

Gnathopod 1 slightly smaller than gnathopod 2; segment 6, palm oblique; segment 5 of gnathopod 1 longer than in gnathopod 2.

Peraeopods 3 and 4, dactyl short, about one-half length of distally spinose segment 6; segment 5 with stout posterodistal spine. Peraeopod 5, basis posterior margin straight, segments 4 and 5 expanded, not longer than wide; segment 6 much narrower. Peraeopod 6, segment 4 broadened slightly behind; segment 6 longer than 5. Peraeopod 7, posterior margin of basis not crenulate or serrate; segment 5 setose behind.

Abdominal side plate 3, hind corner sharply rounding, posterior margin with group of 4–5 longish setae. Uropod 1, rami marginally spinose. Uropod 2, rami with 1–2 marginal spines. Uropod 3 (♀), outer ramus twice length of inner ramus; outer ramus spinose, inner ramus with distal plumose setae; in ♂, rami subequal, all margins with plumose setae. Telson longer than wide; apices with spine and seta in lateral notch.

Distribution: American Atlantic; from Cape Cod to Chesapeake Bay, and northern Florida (eastern Pacific records are doubtfully of this species).
Ecology: A shallow-water, warmer-water species of fine, somewhat unstable sandy bottoms of protected bays and estuaries; males pelagic, especially at night.
Life Cycle: Annual; ovigerous females from May–September; several broods.

Genus *Trichophoxus* K. H. Barnard 1932   (trĭk ō fŏks' ŭs)

Body relatively broad-fusiform especially in ♀. Rostral hood abruptly narrowing in front of eyes. Eyes present, black, larger in ♂. Antennae subequal (♀); flagellum of 2 elongate and calceolate in ♂.

Mandible, molar weak; spine row short. Maxilla 1, palp 2-segmented, inner plate with apical setae. Maxilliped, inner plate relatively long; palp segment 2, outer margin with long finely pectinate spines.

Coxal plates 1–4 deeper than wide, posterior distal margin with simple setae. Gnathopods 1 and 2 slender, subequal, propod not longer than carpus. Peraeopod 5, basis posteriorly expanded, slightly shorter than peraeopod 5, segments 4 and 5 (in both) stoutly expanded, heavily spinose. Peraeopod 7 short, segments 4 and 5 broadened.

Abdominal side plate 3, posterior angle usually not produced. Uropod 3, inner ramus relatively long, bearing plumose seta(e) (♀); distally foliaceous, about equal to outer ramus (♂).

KEY TO AMERICAN ATLANTIC SPECIES OF *TRICHOPHOXUS*

1. Rostrum strongly constricted in front of eyes, distally spatulate; epistome not produced; peraeopod 5, basis posterior margin convex . . . . . . *T. floridanus*.
   Rostrum shallowly constricted basally, distally narrowing; epistome acutely produced; peraeopod 5, posterior margin of basis slightly concave
   . . . . *T. epistomus* (p. 126).

Trichophoxus epistomus (Shoemaker) 1938 new comb.
Plate XXXIV.2   (ĕp ĭ stōm′ŭs)

L. 7–8 mm ♀, 5 mm ♂. Coxal plate 5, lower hind margin rounded; coxae 1–3, hind corners sharply rounded. Head, rostrum shallowly constricted beyond eyes, distally tapering, apex rounded, tip reaching halfway along peduncle 2 of antenna 1. Inferior antennal sinus sharply incised below eye. Eyes (♀), small, black, ovate; in ♂, large, subrectangular, nearly meeting middorsally. Antenna 1, accessory flagellum 6–7 segmented; antenna 2 (♀) slightly longer than 1; in ♂ nearly as long as body.

Epistome acute, strongly produced forward. Mandible, palp segment 3 with proximal facial spine group; apex obliquely truncate. Maxilla 1, palp distinctly 2-segmented. Maxilliped, palp segment 2 with 4–6 long stiff finely plumose setae along outer margin.

Gnathopod 1 smaller than gnathopod 2; segment 5 more elongate and palm of segment 6 more oblique than in gnathopod 2. Peraeopods 3 and 4, segments 4 and 5 powerful; spinose posteriorly; segment 6 slender, posterior margin strongly spinose; dactyl short, less than one-half segment 6. Peraeopod 5, basis longer than broad; segment 4 much broader than deep (esp. in ♀), outer face of segments 5 and 6 with spine groups; dactyl short, scarcely one-half narrow segment 6. Peraeopod 6 longest, segment 4 broader than 5; segment 6 narrow. Peraeopod 7, basis broader than deep, posterior margin weakly serrate, segment 3 about equal to 4.

Abdominal side plate 3, hind corner obtuse; posterior margin with group of long setae, lower margin long-setose. Uropods 1 and 2, rami with a few proximal marginal spines. Uropod 3 (♀), length of outer ramus about twice inner, distally armed with a few plumose and pectinate setae; in ♂, rami subequal, margins distally plumose-setose. Telson, length about 1½ times width (esp. in ♂), 2 short spines at lateral apical angles.

Distribution: American Atlantic: from southern Maine and Cape Cod region to the Middle Atlantic states and North Carolina.
Ecology: Medium fine unstable sands, from immediately subtidal to more than 50 m. Males occasionally occurring in the plankton.
Life Cycle: Annual; ovigerous females from May–September.

Genus *Harpinia* Boeck 1876   (hâr pĭn'ĭ ă)

Body small, very slender-fusiform. Coxal plates deep, lower border with plumose setae; 4th plate usually wider than deep. Rostral hood broad, narrowing evenly to acute apex; inferior head angle acutely produced forward. Eyes lacking. Antenna 1 and 2 short, subequal (both sexes); basal articles of both pairs armed with brushes of setae (♂).

Mandible, molar lacking or vestigial; maxilla 1, palp 2-segmented; inner plate with apical seta. Maxilliped, inner plate small; palp segment 2, outer margin smooth, segment 4, nail elongate.

Gnathopods 1 and 2, propods similar, 2 distinctly large, carpus short. Peraeopods 3 and 4, segment 6 with long stout apical spines on either side of dactyl. Peraeopod 5, basis narrow. Peraeopod 6 very much longer than 5 and 7; dactyl elongate. Peraeopod 7 very short, basis moderately expanded.

Abdominal side plates 2 and 3 acute, 3 often produced as hook. Uropod 1, posterior margin weakly armed. Uropod 3, inner ramus about equal to inner segment of outer ramus. Telson short, deeply cleft.

KEY TO NORTH AMERICAN ATLANTIC SPECIES OF *HARPINIA*

1. Abdominal side plate 3, hind corner produced as distinct tooth; peraeopod 7, posterior margin of segment 6 bare, spinose, or simple-setose . . . . . . 3.
   Abdominal side plate 3, hind corner rounded, hind margin notched or serrate; peraeopod 7, hind margin of segment 6, with 4–5 plumose setae . . . . 2.
2. Peraeopod 7, hind margin of basis serrate, each notch with long seta
   . . . . *H. crenulata* Sars.
   Peraeopod 7, hind margin of basis weakly serrate, setae very short
   . . . . *H. truncata* Boeck.
3. Abdominal side plate 3, hind tooth strongly produced, length about double its basal width . . . . 4.
   Abdominal side plate 3, tooth short, length about equal its basal width . . . . . 5.
4. Peraeopod 7, hind margin of basis moderately serrate, notches uniform, each bearing a medium-length seta; peraeopod 6, hind margin of segment 6 lined with 2–3 spines . . . . *H. neglecta* Sars.
   Peraeopod 7, hind margin of basis weakly serrate; one distal notch much larger than others; peraeopod 6, hind margin of segment 6 bare . . . *H. plumosa* (Kr.)
5. Peraeopod 7, posterior margin of basis strongly and deeply serrate; peraeopod 6, posterior margin of segment 6 with 3–4 pairs of simple setae . *H. serrata* Sars.

Peraeopod 7, basis weakly or moderately serrate behind; peraeopod 6, posterior margin of segment 6 bare or with 2–3 short spines . . . . 6.
6. Coxa 1, lower margin with 10–12 plumose setae; uropod 2, outer ramus with about 6 slender marginal spines . . . . *H. cabotensis* Shoem.
Coxa 1, lower margin with 4–8 plumose setae; uropod 2, outer ramus with 1–2 short spines or unarmed . . . . *H. propinqua* (p. 128).

Harpinia propinqua Sars 1895 Plate XXXV.2   (prō pĭnk′ wă)

L. 4 mm ♀. Coxa 4 broader than deep, lower margin setose posteriorly; coxae 1–3, lower margins with 4–8 well-spaced plumose setae. Head, rostrum large, deep, completely overhanging peduncle of antenna 1. Antenna 1, peduncular segment 2 posteriorly lobate; accessory flagellum 4–5 segmented. Antenna 2 slightly longer than 1; peduncular segment 4 deeply lobate behind.

Mandibular palp, segment 2 slightly expanded, inner margin with a few long plumose setae. Maxilliped, palp strong, spinose, dactyl nail long, slender.

Gnathopod 1, segment 6 distinctly smaller and more slender, and palm more oblique than in gnathopod 2; segment 5 (in both) short, posterior lobe plumose. Peraeopods 3 and 4, segment 4 powerful; segment 6 with 1–2 spines posteriorly, and distal pair of long stout spines; dactyl stout, length nearly equal to segment 6. Peraeopod 6, anterior margin of basis heavily setose; posterior margin smooth, slightly concave medially; segment 6, longer than 5, with 2–3 short spines on posterior margin; dactyl slender, as long as segment 6. Peraeopod 7, posterior margin of basis moderately serrate with about 10 notches, each bearing medium length seta; segment 3 strong; dactyl longer than segment 6.

Abdominal side plate 3, hind corner produced as short sharp slightly upturned tooth; lower margin slightly concave, with 3 plumose setae; outer face with row of about 6 plumose setae. Pleopods small, weak (♀). Uropods 1 and 2, rami with 1–2 spines, set proximally. Uropod 3 short, slightly exceeding uropod 2; outer ramus, margins with 4–5 setae; inner ramus bare. Telson short, broader than long, apices subacute, each with 2 slender lateral apical spines.

Distribution: Amphi-Atlantic; in North America south to Cape Cod Bay, Mass.
Ecology: On fine sand and sandy mud bottoms, 10–50 m or more in depth.

Body more or less strongly compressed. Urosome dorsally carinate or humped, segments 5 and 6 fused. Antennae long, slender, often with filter setae. Antenna 2 usually longer, peduncle elongate; accessory flagellum minute or lacking.

Mandible, molar well developed. Palp, variable, tending to reduction. Maxilla 1, inner plate small; outer plate with 7–11 spine teeth. Maxilla 2, plates subequal. Maxilliped, outer plate large, palp short, tending to reduction. Upper lip rounded; lower lip, inner lobes present or vestigial.

Gnathopods weakly subchelate or simple. Peraeopods 3 and 4, dactyls with or without glandular ducts; segment 5 tending to reduction. Peraeopods 5–7, segment 6 and dactyl variously reversed, hooked; bases more or less expanded.

Pleopods strongly developed. Uropods 1 and 2 biramous, marginally spinose. Uropod 3 biramous, lanceolate, foliaceous (always in ♂). Telson deeply cleft. Coxal gills large, pedunculate, often pleated, on segments 2–7. Brood plates large, lamellate, setose.

Component families: Dexaminidae, Atylidae, Ampeliscidae.

Family DEXAMINIDAE

(dĕks ă mĭn' ĭ dē)

Body not extremely compressed. Pleon dorsally carinate and/or mucronate. Fused urosome segments often spinose dorsally. Head, rostrum strong. Eyes emergent, large, on well-developed anterior head lobes. Coxal plate 5 deep, 4th deepest.

Mandible without palp. Lower lip, inner lobes weak. Maxilla 1, inner plate small with 2 apical setae. Maxilliped, inner plate very small; palp variously reduced, short, 4th segment small or lacking.

Gnathopods 1 and 2 weakly subchelate; gnathopod 1 (♂), segment 6, with median anterior notch. Peraeopod 3 not smaller than 4; segment 4 tending to elongate. Peraeopods 5–7 subequal, basis not exceptionally expanded, that of 7 usually narrow; dactyls reversed. Uropod 2 smallest. Uropod 3, rami subequal. Telson large, deeply cleft. Coxal gills simple or pleated (esp. in ♂) on 2–6 (7). Brood plates relatively narrow, multisetose.

KEY TO AMERICAN ATLANTIC GENERA OF DEXAMINIDAE

1. Peraeopods 3–7, segment 4 elongate, greater than 5 and 6 combined . . . . 2.
   Peraeopods 3–7, segment 4 shorter than 5 and 6 combined . . . . 3.
2. Peraeopods 3–7, segment 6 expanded to form distal palm with dactyl . *Polycheria.*
   Peraeopods 3–7 not subchelate . . . . *Tritaeta.*
3. Coxa 4 largest; maxilliped, palp 3-segmented . . . . . . *Dexamine* (p. 130).
   Coxa 5 largest; maxilliped, palp 4-segmented. . . . . *Guernea.*

129

Genus *Dexamine* Leach 1813–14   (děks ă mĭn′ ě)

Body stout or moderately compressed. Pleosome and urosome dorsally carinate and mucronate. Head, anterior lobe acute; rostrum acute, moderate.
Maxilliped, palp weak, 3-segmented; outer plate very large. Lower lip, inner
lobes rudimentary.

Gnathopods 1 and 2 small, propod of 2 longer. Peraeopods 3 and 4,
segment 4 shorter than 5 and 6 combined. Peraeopods 5 and 6 subequal,
basis broad. Peraeopod 7, basis narrower than in peraeopods 5 and 6.

Uropods 1 and 2 narrow-lanceolate; outer ramus shorter in 2. Telson
long, deeply cleft.

Dexamine thea Boeck Plate XXXIX.1   (thē ′ ă)

L. 4 mm ♀. Pleosome and urosome 1 each with middorsal carination and single posterodorsal tooth. Coxal plates 1–4 moderately deep, lower margins
setose. Coxal plates 5 and 6 nearly as deep as broad. Urosome 2 and 3
fused, with pair of single dorsal spines. Eye, small, oval, black (♀), larger in ♂.
Antenna 1, peduncle 1 not prolonged posterodistally. Antenna 2 shorter (♀),
longer (♂) than 1; peduncular segments 4 and 5 subequal.

Maxilla 1, inner plate with 2 setae; palp, basal segment partly fused to
outer plate. Maxilliped, inner plate broad and obliquely truncate.

Gnathopod 1 (♀), carpus and propod shorter and deeper than in gnathopod 2. Gnathopod 1 (♂), propod with sharp deep midanterior marginal notch.
Peraeopods 3 and 4, dactyls nearly as long as slender segment 6. Peraeopod 5, hind margin of basis nearly straight. Peraeopod 6, basis very broadly
convex behind. Peraeopod 7, basis narrow, slightly convex behind.

Pleon side plates 1 and 2, hind corners acute; pleon 3 corner acute,
slightly produced. Uropod 1 not reaching tip of uropod 3, posterior margin
of peduncle spinose throughout. Uropod 3, rami spinose, very slightly setose
(♀), spinose and foliaceous in ♂. Telson cleft almost to base, lateral margins
spinose distally (♀), nearly ⅔ of length in ♂.

Distribution: Amphi-Atlantic; in North America from Labrador and outer
coast of Newfoundland and Nova Scotia to Cape Cod and Long Island
Sound.
Ecology: A moderately common "nestling" species among algae, eelgrass,
and under stones, from immediately subtidal to about 20 m, on semiprotected and protected shores.
Life Cycle: Annual; ovigerous females from March–June; one brood per female.

Family ATYLIDAE   (ă tīl′ ĭ dē)

Body very compressed, abdomen carinated always on urosome. Urosome segments 2 and 3 coalesced. Head, rostrum strong, eyes lateral. Antenna 1, accessory flagellum minute or lacking.

Mandible with slender 3-segmented palp. Lower lip, inner lobes not distinct. Maxilla 1, inner plate strongly setose; outer plate with 9 pectinate spine-teeth.

Gnathopods 1 and 2 subchelate, subequal; gnathopods sexually dimorphic. Peraeopod 4 distinctly smaller than 3; segment 5 (of each) smallest. Peraeopods 5–7, dactyls usually reversed. Peraeopod 7, basis more broadly expanded than in 5 and 6.

Uropod 2 distinctly shorter than 1, outer ramus shorter. Uropod 3, rami strong, subequal. Telson medium, deeply cleft.

Coxal gills on peraeonites 2–7 large, variously pleated or simple. Brood plates relatively large, lamellate.

American Atlantic genera: *Atylus* Leach 1815 (= *Nototropis* Costa 1853).

KEY TO AMERICAN ATLANTIC SPECIES OF *ATYLUS*

1. Pleon segments 1–3 and urosome segment 1 middorsally carinate; peraeopod 7, posterodistal lobe of basis not incised near base . . . . . *A. minikoi* (Walker). Pleon segments 1–3 dorsally smooth; urosome segment 1 with middorsal carina, notched medially; peraeopod 7, posterodistal lobe of basis sharply incised near junction with limb proper . . . . *A. swammerdami* (p. 131).

Atylus swammerdami (M.-E.) 1830 (no illustration)   (swäm ĕr dăm′ ī)

L. 4–9.5 mm. Pleon segments 1–3 dorsally smooth. Urosome segment 1 middorsally carinate, medially notched; urosome 3 with low spinulose hump. Coxal plates deeper than broad; coxa 1 not narrow. Head, rostrum short; anterior lobe shallowly incised in front; eye large (esp. in ♂), broadly reniform. Antenna 1 slightly shorter than antenna 2, lacking accessory flagellum.

Mandible, palp slender, inner margin weakly setose, outer margin bare. Maxilliped, palp short, slender, segment 2 not exceeding outer plate.

Gnathopods 1 and 2 weakly subchelate, somewhat unlike. Gnathopod 2, segments 5 and 6 slender, segment 6 shorter, palm oblique. Peraeopods 3 and 4 slender, dactyls short; peraeopod 4 distinctly smaller than 3. Peraeopods 5 and 7, posterodistal lobe of basis acutely produced. Peraeopod 7, posterodistal lobe of basis sharply incised near junction with segment proper. Coxal gills present on segments 2–7, pleated on segments 3–5.

Abdominal side plate 3, hind corner acuminate, not produced. Uropod 3 extending beyond uropod 1; peduncle more than one-third length of subequal lanceolate rami; margins spinose. Telson little longer than broad, lobes approximate, apices obliquely truncate, with small spine at lateral angle.

Distribution: Arctic boreal and North Atlantic; on American Atlantic coast south to Vineyard Sound and off Gay Head, Mass.
Ecology: On sandy bottoms, from low water level to more than 30 m. Rare.
Life Cycle: Unknown.

Family AMPELISCIDAE   (ăm pĕl ĭsk′ ĭ dē)

Body smooth, carinate only on urosome (esp. in ♂). Head truncate, longer than deep, lacking rostrum. Eyes 2–6 (usually 4), simple ocelli, lateral, anterior. Antenna 1 short, slender, often attached far ahead of antenna 2.
    Mandible, palp very strong, setose. Lower lip, inner lobes well developed. Maxilliped, palp strong, dactylate.
    Coxal plates very deep, lower margin very setose. Peraeopods 3 and 4, dactyls elongate, with gland ducts; segment 5 very short. Peraeopod 4 longer than peraeopod 3. Peraeopods 5 and 6, dactyls very short, hooklike, reversed. Peraeopods 5–7 bases expanded. Peraeopod 7 shortest, unlike peraeopods 5 and 6. Gnathopods 1 and 2 simple or very weakly subchelate; gnathopod 2 elongate.
    Uropod 3, rami broadly lanceolate, usually foliaceous. Uropod 1 little or not exceeding 2. Coxal gills usually pleated on peraeon 2–7. Brood plates sublinear.

KEY TO NEW ENGLAND GENERA OF AMPELISCIDAE

1. Head broadly truncate; antenna 2 arising directly below antenna 1; peraeopod 7, posterior margin of segment 2 vertical, medially concave; uropod 1, inner ramus distinctly shorter than outer ramus . . . . *Haploops* (p. 138).
    Head narrowing to short truncate apex; antenna 2 arising well posterior to antenna 1; peraeopod 7, basis broadly expanded, posterior margin oblique, convex; uropod 1, rami subequal . . . . 2.
2. Peraeopod 7 normally 7-segmented, segment 2, distal lobe anterior margin bare; peraeopods 3 and 4, segment 5 very short, partly overhung by 4 *Ampelisca* (p. 132).
    Peraeopod 7, 7th segment reduced to a spine; segment 2, anterior margin of distal lobe setose; peraeopods 3 and 4, segment 5 longer than one-half segment 6
    . . . . *Byblis* (p. 137).

Genus *Ampelisca* Krøyer 1842   (ăm pĕl ĭsk′ ă)

Body smooth; urosome with middorsal hump, more strongly developed in ♂. Coxal plates deep, lower margins setose; coxae 5–7 shallow. Head shallow, narrowing to short truncate margin. Eyes each of two separated ocelli, with single lens and cornea.

Mandibular palp strong, setose; molar powerful, surface ridged. Maxilliped, inner plate short, broad.

Peraeopods 3 and 4, segment 5 very short, partly overhung by prolongation of 4. Peraeopod 7 with 7 distinct segments; dactyl broad at base; basis (segment 2), anterior border of posterodistal lobe lacking setae.

Uropod 1, rami subequal. Uropod 3, rami broad-lanceolate, foliaceous, extending well beyond uropods 1 and 2.

KEY TO NEW ENGLAND SPECIES OF *AMPELISCA* (after Mills, 1967)

1. Segment 3 of peraeopod 7 shorter than segment 4 . . . . 2.
   Segment 3 of peraeopod 7 longer than segment 4 . . . . 6.
2. Lower corner of pleon side plate 3 strongly toothed . . . . 3.
   Lower corner of pleon side plate 3 rounded or quadrate . . . *A. agassizi* (p. 135).
3. Antenna 1 shorter than peduncle of antenna 2; uropod 1, outer ramus with inner marginal spines . . . . *A. verrilli* (p. 134).
   Antenna 1 equal to or longer than peduncle of antenna 2; uropod 1, outer ramus lacking inner marginal spines . . . . 5.
4. Peraeopod 7, anterior margin of segment 5 not strongly produced distally, margin entire . . . . *A. macrocephala* (p. 133).
   Peraeopod 7, anterior margin of segment 5 strongly produced distally, anterior margin notched . . . . *A. eschrichti* Kr.
5. Antenna 1 almost equal antenna 2 . . . . *A. aequicornis* Bruz.
   Antenna 1 shorter than peduncle of antenna 2 or less than half antenna 2 . . . 6.
6. Antenna 1, peduncular segments 1 and 2 equal; uropod 3, rami almost equal in size . . . . *A. declivitatis* Mills.
   Antenna 1, segment 2 longer than 1; uropod 3, inner ramus broader than outer   7.
7. Uropod 1 extending slightly beyond uropod 2; uropod 2, outer margin of outer ramus with 1–2 spines; posterolateral corners of last segment of urosome rounded
   . . . . *A. abdita* (p. 136).
   Uropod 1, not extending beyond uropod 2; uropod 2, outer margin of outer ramus with 3 spines; posterolateral corners of third segment of urosome sharply upturned . . . . *A. vadorum* (p. 135).

Ampelisca macrocephala Liljeborg 1852 Plate XXXVI.1   (mă krō sĕf' ăl ă)

L. 12–20 mm. Head with diagonal and slightly sinuous lower front margin; lower eye on a slight projection. Antenna 1 extending to end of segment 5 of peduncle of antenna 2. Antenna 2 about one-half length of body (♀); much longer in ♂.

Peraeopod 7, expansion of basis almost straight distally; segment 4 with a short setose projection posterodistally; segment 5, anterior and posterior margins about equal in length; the anterior margin smooth and entire, a few spines and long plumose setae arising from the posterior part of the distal margin; segment 6 long and narrow; dactyl only slightly curved distally.

Pleon side plate 2 quadrate at hind corner, with only one or a few setae. Plate 3 with a strongly sinuous posterior margin ending in large tooth at hind corner. Urosome 1, carina dorsally slightly concave, ending in small rounded process above succeeding segment. Carina more strongly developed in ♂.

Uropod 1 not reaching end of uropod 2; peduncular margins strongly spinose; approximating margins of rami lacking spines. Uropod 2, peduncular margins lightly spinose; rami, approximating margins each with 2 small spines, outer ramus subapically emarginate, bearing long stout grooved spine. Uropod 3, inner ramus, inner margin completely bare of plumose setae. Telson lobes, outer margin distally straight.

Distribution: Circum-Atlantic; in America from Greenland south to Cape Cod, Mass., Georges Bank, and off Rhode Island.
Ecology: Stable sandy bottoms, at depth of 10–280 m.
Life Cycle: Life span of 1½–2 years; ovigerous females in fall and winter.

Ampelisca verrilli Mills 1967 Plate XXXVI.2   (věr′ ĭl ī)

L. 10.5–13.5 adult ♀. Head, narrow anteriorly, lower front margin almost parallel to upper. Lower pair of eyes set at lower front corners of head and directed laterally and down. Antenna 1 reaching just beyond peduncle 4 of antenna 2. Antenna 2 about two-thirds body length. Maxilla 2, inner plate with row of plumose setae.

Peraeopod 7, posterodistal expansion of basis broadly rounded and slightly angular; segment 4 with short-setose posterior projection; segment 5 with blunt posterior spinose lobe; segment 6 somewhat ovoid; dactyl stout, nail long and strongly curved forward.

Pleon side plate 2 quadrate at hind corner. Plate 3 with sinuous posterior margin, hind corner prolonged into large tooth. Urosome 1 with pronounced dorsal carina, slightly concave, and round posterior margin.

Uropod 1 short, reaching only to about middle of uropod 2 rami (in ♀); in ♂ to tip of uropod 2; peduncle, outer margin with raised rounded process fitting over base of uropod 2; rami with some spines on approximating margins. Uropod 2, peduncle outer margin heavily set with short curved spines; rami, approximating margins variably spinose; outer ramus with subapical large grooved spine. Uropod 3, both rami, both margins with long plumose setae, more extensive in ♂. Telson lobes long, distally narrowing, with 6 spines on dorsal surface.

Distribution: South side of Cape Cod to North Carolina. Common in Buzzards Bay, entrance to Quick's Hole, Mass.; Mason's Point, Noank, Conn.

Ecology: Abundant in coarse sands, from low intertidal to depths of about 50 m, overlapping in finer sands with *A. vadorum*.
Life Cycle: Annual; ovigerous females in summer.

Ampelisca agassizi (Judd) 1896 (formerly *A. compressa* Holmes)
Plate XXXVIII.1   (ăg ă sēz′ ī)

L. to 7.5 mm. Head narrowly compressed, lower front margin about equal to length of lower margin. Antenna 1 short; peduncular segment 2 longer than 1. Antenna 2 very elongate, longer than body (in ♂). Maxilla 2, inner plate with row of plumose setae.

Peraeopod 7, basis very broad, lower margin transverse, gently rounding; segment 4, posterior lobe setose. Coxal gills simple narrow lobes, often with a few pleats or folds distally (♀); in ♂, gills highly pleated.

Pleon ridged dorsally. Pleon side plates with rounded hind corners. Urosome segment 1 with projecting carina with rounded oblique posterior angle (♀); in ♂, with deep transverse trough anteriorly, and a posteriorly projecting massive lobelike carina; fused urosome 5 and 6, with deep dorsal depression, small knob on either side.

Uropod 1 not reaching tip of uropod 2; outer ramus, margins spineless. Uropod 2, approximating margins of rami spineless. Uropod 3 relatively short, rami narrow; long plumose setae only on outer margin of outer ramus (♀); in ♂, all margins of both rami with plumose setae throughout. Telson broad distally, abruptly narrowing to apex.

Distribution: Southwestern Nova Scotia (St. Mary Bay); central Maine to the Caribbean region; also American Pacific coast. Recorded from Narragansett Bay, R.I., Buzzards Bay, Vineyard Sound.
Ecology: From shallow inshore waters to deep water (450m); mostly on stable coarse sandy sediments.
Life Cycle: Annual; ovigerous females in summer.

Ampelisca vadorum Mills 1963 Plate XXXVII.1   (vă dōr′ ŭm)

L. 7–15 mm ♀; 7–11 mm ♂. Urosome 1 dorsally oblique and high (♀), extending over next segment (♂); urosome 3, posterodorsal corners sharply upturned. Head about equal in length to first three peraeon segments. Antenna 1, peduncular segments 2 and 3 longer than peduncle of antenna 2 with several posterior marginal setae. Coxal plate 1, lower margin gently convex. Gill pleating well developed.

Mandible, palp segment 3, with apical setae only (♀). Maxilliped, inner plate medium, reaching well up to palp segment 2.

Peraeopod 6 (♀), propod with 8 anterior marginal spines. Peraeopod 7, segment 3 longer than 4; segments 3, 4, 5, 6 linear, not conspicuously swollen or lobate.

Uropod 1 barely or not reaching to tip of uropod 2; inner ramus, inner margin with 6–7 spines. Uropod 2, inner margin of inner ramus with 6–7 spines; outer margin of outer ramus with 3+ spines. Uropod 3, inner margin of inner ramus almost straight, setation sparse. Telson broadest medially, lobes virtually lacking dorsal spinules.

Distribution: Southwestern Gulf of St. Lawrence (Chaleur Bay, Magdalen Islands), sporadically along Nova Scotia and northern New England to Cape Cod; common southward to South Carolina and Georgia.
Ecology: Builds tubes in medium to coarse sands (often with broken shell admixture) in protected bays and estuaries, from lower intertidal to depths of over 70 m. Common in eelgrass (stable) sands, in higher salinities (down to about 20 o/oo). Tube is short, broad, up to 2½ cm long and 5 mm wide.
Life Cycle: Annual; breeds in April, May; after temperatures reach 8–10° C; ovigerous females in late May and early June, and again from July–September. Two generations per year; juveniles reach maturity the first summer, breed, and die, their young overwintering and breeding the following spring.

Ampelisca abdita Mills 1964 Plate XXXVII.2 (ăb dĭt′ ă)

L. 4–8.2 mm ♀, 4.0–8.0 ♂. Urosome 1 dorsally level, low (♀): not extending over next segment (♂). Urosome 3, hind corners rounded. Head about five-sixths length of first three thoracic segments combined. Antenna 1 extending somewhat beyond peduncle of antenna 2; peduncular segments 2 and 3 nearly bare behind. Coxal plate 1 strongly rounded below. Gill pleating diffuse, pleats few.

Mandibular palp, segment 3 with some marginal as well as apical setae. Maxilliped, inner plate small, barely reaching level of palp segment 2.

Peraeopod 6 (♀), propodus with up to 3 anterior marginal spines. Peraeopod 7, segment 3 longer than 4; segments 3, 4, 5, and 6 linear, not conspicuously lobed or expanded; segment 4 only slightly shorter than segment 5.

Uropod 1 slightly exceeding tip of uropod 2; inner ramus, inner margin with 2–3 spines. Uropod 2, inner ramus, inner margin with 2–3 spines; outer ramus, outer margin with 1–2 spines. Uropod 3, rami lanceolate, setae and spines well developed. Telson, margins proximally subparallel; dorsal surface of lobes with 4–5 small spinules.

Tube long, narrow, up to 3½ cm long and 2–3 mm wide.

Distribution: Central Maine (Sheepscot estuary) and southern New England south to northern Florida and the eastern Gulf of Mexico.
Ecology: In fine sand, silty sand, silt; from low intertidal to depths of 60 m, mainly in protected waters, and bays, from fully marine to brackish (10 o/oo) conditions.
Life Cycle: Annual; two main populations per year. Overwintering adult females are ovigerous in May–June, releasing young which breed and die that summer. Their young overwinter and reproduce the following spring.

Genus *Byblis* Boeck 1871   (bĭb′ lĭs)

Body smooth; urosome weakly carinate dorsally. Head narrowing apically to short truncate anterior margin. Antenna 2 originating posterior to antenna 1. Eyes 4. Coxal plates moderately deep, 1 deepest.

Peraeopods 3 and 4, segment 5 not greatly reduced, length greater than one-half segment 6. Peraeopod 7, basis very deeply expanded, posterior margin oblique, convex below, anterior margin of distal lobe with plumose setae; segments 3–6 normal; segment 7 reduced to a spine.

Uropod 1, rami subequal, tips reaching about to end of uropod 2. Uropod 3 short, little exceeding uropods 1 and 2; rami narrow, lanceolate, margins nonfoliaceous (in ♀). Telson short, partly cleft.

KEY TO AMERICAN ATLANTIC SPECIES OF *BYBLIS*

1. Coxal plate 1, lower margin finely serrate; uropod 3, approximating margins of rami serrate (more extensive in ♂); telson cleft one-half; corneal lenses absent
. . . . *B. serrata* (p. 137).
Coxal plate 1, lower margin even, smooth; uropod 3, approximating margins of rami smooth; telson cleft minute; corneal lenses large . . . *B. gaimardi* Krøyer.

Byblis serrata Smith 1873 Plate XXXVIII.2   (sĕr āt′ ă)

L. 14 mm ♀. Coxal plates 2–4 shallower than 1; lower margin of coxa 1 serrate; coxa 2 and 3 not twice as deep as wide. Urosome 1 raised posterodorsally, not acute; urosome 3 with low dorsal projection. Head, eyes distinct, lacking corneal lenses. Antenna 1 slightly longer than peduncle of antenna 2; peduncle 2 slightly longer than peduncle 1 (♀); antenna 2 about three-fourths length of body.

Mandibular palp segment 2 broadly expanded, both margins setose. Maxilliped, inner plate narrow, short. Lower lip, mandibular process distinct.

Gnathopod 1, propod nearly equal to carpus, dactyl about one-half length of propod. Gnathopod 2, carpus about three-fourths basis, propod about one-half carpus; dactyl pectinate behind. Peraeopods 3 and 4, segment 4 moderately stout, strongly setose anteriorly and posteriorly; dactyls longer than segment 6, and nearly straight. Peraeopod 5, basis convex posteriorly. Peraeopod 7, basis, distal margin of posterior lobe evenly convex, very deep, extending almost to segment 6; segment 6 relatively stout, dactyl very slender, straight.

Pleon side plate 3 sharply rounded behind. Uropods 1 and 2 reaching almost to tip of uropod 3. Uropod 3, rami about 1½ times length of peduncle, narrow lanceolate, each with 3 outer marginal spines, approximating margins serrate. Telson as long as wide, narrowly cleft nearly one-half to base, lobes apically rounded.

Distribution: South side of Cape Cod to North Carolina. Common around Naushon Island, Buzzards Bay, and Long Island Sound.

Ecology: Constructs tubes in medium to coarse sand; from immediately subtidal to over 40 m.

Life Cycle: Annual; ovigerous females from May–September.

Genus *Haploops* Liljeborg 1855   (hăp′ lö ŏps)

Body smooth or dorsally setose. Head broadly truncate in front; lateral lobes indistinct. Upper pair of eyes usually developed, lower eyes weak or lacking. Coxal plates medium deep; coxa 1 deepest, rounding below.

Mandibular palp slender; segment 2 not broadened; segment 3 elongate.

Gnathopods 1 and 2 simple or weakly subchelate. Peraeopods 3 and 4, segment 5 not greatly reduced, length about one-half segment 6. Peraeopod 7 abnormally 7-segmented; segments 6 and 7 small, linear; basis moderately or narrowly expanded behind; posterior margin about parallel to anterior margin; anterior margin of distal free lobe setose.

Uropod 1, inner ramus shorter than outer. Uropod 3 extending well beyond uropod 2; rami subequal, foliaceous, apices blunt. Telson deeply cleft.

KEY TO AMERICAN ATLANTIC SPECIES OF *HAPLOOPS*

1. Coxa 4 distinctly shallower than coxa 3; peraeopod 7, basis distal lobe moderately broad, segment 6 as long as 5; pleosome and urosome segments 1 and 2 with posteromiddorsal tufts of long setae . . . . *H. setosa* Boeck.

   Coxa 4 about as deep as coxa 3; peraeopod 7, basis distal lobe narrow, segment 6 short, less than one-half length of segment 5; abdomen dorsally lacking setal tufts . . . . *H. tubicola Liljeborg = H. spinosa* Shoem.

The deep-bodied Lysianassidae are very distinctive in antennal, mouthpart, and gnathopod morphology, and sufficiently unlike other gammaridean families as to obscure their phylogenetic relationships. The sponge-dwelling Colomastigidae are apparently derived from stenothoid-like ancestors; the Lafystiidae are believed to have evolved from pleustids via acanthonotozomatid-like ancestors.

Family COLOMASTIGIDAE   (kōl ō măst ĭg' ĭ dē)

Body subcylindrical, depressed. Coxal plates shallow, overlapping. Head short; eye medium, lateral. Antennae short, stout; accessory flagellum lacking. Peduncular segments stout, tapering distally; flagella very short. Upper lip bilobed, epistome conical, projecting.

Mandible lacking blades or palp; incisor broad, long dentate; molar visible. Maxilla 1, palp 1-segmented, larger than outer plate. Maxilla 2 small, plates partly coalesced, outer plate smaller. Maxilliped, inner plate vestigial, coalesced medially; outer plate large, lacking armature; palp strong, 4-segmented.

Gnathopod 1 elongate, slender, simple or setae-tipped. Gnathopod 2 subchelate.

Peraeopods 3–7 slender, subequal, basis narrow. Pleopods slender. Urosome 2, short. Uropods 1 and 2 biramous. Uropod 3 biramous. Telson entire, short.

Coxal gills on peraeon segments 2–6. Brood plates moderately large, lamellate, marginal setae with hooked tips.

Genus *Colomastix* Grube 1861   (kōl ō măst' ĭks)

With the characters of the family. Head, rostrum short. Antenna 2, peduncle as stout as antenna 1. Antenna 1, flagellum, 3-segmented.

Mandibular incisor with 5 terminal blades, molar vestigial. Maxilla 2, rami proximally fused, apically spinose. Maxilliped inner plates small.   139

Gnathopod 1 shorter in mature ♂. Gnathopod 2 ♀, subchelate palm short. Brood plates large, setae curl-tipped. Uropod 3, both rami longer than peduncle.

KEY TO NORTH ATLANTIC SPECIES OF *COLOMASTIX*

1. Antennae 1 and 2 heavy, tubular, subequal; gnathopod 2 (♂), segment 2 not expanded distally, in ♀, propod extended beyond base of dactyl; peraeopod segment 5 much longer than 4; gnathopod 1 very elongate, segments 5 and 6 more than twice length of segments 3 and 4; uropod 3, rami unequal; telson subtruncate . . . . *C. halichondriae* (p. 140).
   Antenna 2, more slender than antenna 1, shorter; gnathopod 2 (♂), basis expanded distally; in ♂ propod not overhanging base of dactyl; peraeopod segment 4 longer than 5; gnathopod 1 relatively short, segments 5 and 6 less than twice segments 2 and 3 combined; uropod 3, rami subequal; telson subacute . . *C. pusilla* Grube.

Colomastix halichondriae n. sp. Plate XVIII.2   (hăl ĭ kŏn' drĭ ē)

L. 2–3 mm. Coxal plates 2–5 progressively shallower; coxae 1 and 2 prolonged anteriorly, distal angle acute. Head, rostrum short, narrow, acute. Eye round, reddish. Antenna 1 and 2, subequal (♂). Antenna 1, peduncle segments 1–3, posterodistal angles acutely produced. Antenna 2, peduncle segments 4 and 5 with posterior marginal spines and setae.

Gnathopod 1 (both sexes), very slender, elongate, terminated by 4–6 slender setae; segments 5 and 6 more than twice length of segments 3 and 4. Gnathopod 2 (♂), basis not greatly expanded distally; segment 5 short and deep; segment 6 moderately powerful, with single toothed process at palmar angle and large tooth near hinge; dactyl heavy at base, posterior margin evenly curved; in ♀, segments 5 and 6 subequal in length; segment 6 distally overhanging base of dactyl; palm very oblique and slightly convex.

Peraeopods 3–7 slender, subequal, margins bare; segment 5 (in all) subequal to or longer than segment 4. Abdominal side plate 3, hind margin smoothly rounding.

Uropods 1 and 2 not reaching tip of uropod 3. Uropod 1 (♂), inner ramus proximally broad, with long shallow setose subapical indentation; in ♀, rami very narrow, subequal, inner margins finely serrate. Uropod 3, inner ramus longer than outer, inner margin very finely serrate. Telson subrectangular; apex subtruncate, finely setulose, with pair of dorsal setae.

Distribution: American Atlantic coast, from south side of Cape Cod to Chesapeake Bay and Georgia. Gulf states?
Ecology: Commensal in sponges (*Halichondria bowerbanki, Haliclona permollis*) from below tide lines to several m; mainly in polyhaline waters (20–30 o/oo).
Life Cycle: Annual; ovigerous females in summer (June–August); several broods.

Family LAFYSTIIDAE   (lă fĭst ē′ ĭ dē)

Body depressed, smooth. Coxal plates 1–3 shallow, subquadrate, 4th acuminate below, emarginate behind. Head with large, flat, truncated rostrum. Eyes lateral. Antennae short, segments tapering. Accessory flagellum lacking. Antenna 1 longer, stronger.

Upper lip rounded; lower lip deep, without inner lobes. Mandible, incisor strong; molar and spine row lacking; palp 3-segmented, strong. Maxilla 1, palp vestigial. Maxilla 2, outer plate longer. Maxilliped, inner plate narrow; outer plate broad, inner margin crenulate; palp reduced, 2-segmented.

Gnathopod 1 simple, dactyl strong. Gnathopod 2 weakly subchelate. Peraeopods 3–7 strong, powerful, dactyls hooked. Peraeopods 5–7 subequal, bases expanded.

Uropods 1 and 2 biramous. Uropod 3 biramous. Telson small, entire. Coxal gills simple, saclike, on peraeon segments 2–6. Brood plates large.

Genus *Lafystius* Krøyer 1842   (lă fĭst′ ĭ ŭs)

With the characters of the family. Rostrum strongly developed. Coxal plates medium, touching serially. Antennae short, stout.

Mandible lacking molar; palp inserted near base. Maxilla 1, palp 1-segmented. Maxilliped, palp 2-segmented. Lower lip, lacking inner lobes.

Gnathopod 2 weakly subchelate. Peraeopods 3–7, distal segments and dactyls very stout, subequal. Uropod 3, rami unarmed.

Lafystius sturionis Krøyer 1842 Plate XXXIX.2   (stër ĭ ōn′ĭs)

L. 9.0 mm. Body smooth, urosome dorsally even. Rostrum as long as rest of head, narrowing, subtruncate. Eyes lateral, large, rounded, bulging. Coxal plates 1–3 subequal, rounded below; coxa 4 deepest, anterior angle acute; posterior excavation shallow. Coxae 5 and 7, hind lobe rounded below. Coxa 6 acutely produced below. Antenna 1, flagellum 7-segmented, about equal in length to stout peduncle. Antenna 2, flagellum 5-segmented, about equal to short, slender peduncle. Sexes similar.

Gnathopod 1 longer than gnathopod 2; segment 6 longer than 5; dactyl slender, nearly as long as 6. Gnathopod 2, segment 5 longer and deeper than 6; dactyl distinctly exceeding palmar angle, with subapical inner tooth.

Peraeopods 3 and 4, anterior margin of segment 4 smooth, not setose; dactyls strongly hooked, enlarged at base. Peraeopods 5–7 subequal; bases evenly rounded behind, dactyls not strongly hooked.

Abdominal side plate 2, hind corner acuminate, slightly produced. Hind corner of side plate 3 obtusely truncate. Uropods 1 and 2, rami subequal, sublanceolate, neither extending beyond uropod 3. Uropod 3, inner ramus slightly shorter, outer almost reaching tips of uropods 1 and 2. Telson subovate, short, apical margin with a few short spines.

Distribution: American Atlantic coast, Labrador to off the Carolinas.
Ecology: Externally parasitic on larger fishes, especially dogfish, skates, *Lophius,* cod, haddock, and other groundfish.

Family LYSIANASSIDAE   (lī sǐ ăn ăs'ī dē)

Body deep, smooth, subcylindrical or subfusiform, integument hard. Coxal plates 1–4 large, deep, smooth below, 4th excavate behind; plates 5–7, deep, 5th larger than basis. Head, rostrum weak; eyes large, lateral. Antenna 1, peduncle 1 inflated, robust, segments 2 and 3 short; accessory flagellum well developed. Antenna 2 slender, short (♀), elongate, calceolate (♂).

Upper lip rounded, epistome pronounced. Lower lip deep, lacking inner lobes. Mandible, incisor large, simple; molar variously modified or reduced; palp slender, 3-segmented. Maxilla 1, inner plate elongate, with 2 (1) strong apical setae; outer plate with 9 apical spine teeth; palp 2-segmented. Maxilla 2, outer plate larger; inner plate, proximal seta(e) stronger. Maxilliped, outer plate large; palp 4-segmented.

Gnathopod 1 simple, subchelate or chelate. Gnathopod 2 slender, minutely chelate or subchelate; segment 3 elongate (both sexes). Peraeopods 5–7 subequal or successively longer. Basis strongly expanded. Uropods 1 and 2 biramous, lanceolate. Uropod 3, rami lanceolate, usually foliaceous. Telson usually deeply cleft, occasionally notched or entire.

Coxal gills on peraeon segments 2–7 pedunculate, simple, laminar, or pleated. Brood plates sublinear, setae few, simple, apical.

KEY TO COMMON NEW ENGLAND GENERA OF LYSIANASSIDAE

1. Pleon side plate 3, hind corner produced as stout tooth or hook . . . . . . 2.
   Pleon side plate 3, hind corner acuminate, square, or rounded . . . . . . 4.
2. Upper lip or epistome (or both) produced in front; gnathopod 1, dactyl usually with posterior marginal tooth; mandible, molar setulose . . . . *Anonyx* (p. 149).
   Upper lip not produced in front; molar ridged; gnathopod 1, dactyl simple . . . 3.
3. Head small, lower part hidden behind coxa 1; uropod 2, rami sickle-shaped, inner ramus not medially constricted; telson medium to large, deeply cleft
   . . . . *Hippomedon* (p. 143).
   Head, normal, only slightly hidden by coxa 1; uropod 2, inner ramus medially contricted; telson short, apically notched . . . . *Onesimus.*
4. Mandibular palp arising opposite molar; uropod 3 (♀), both rami foliaceous, at least on one margin . . . . 5.
   Mandibular palp arising proximal to level of molar; uropod 3 (♀), rami spinose or with very few plumose setae (except *Opisa*) . . . . 7.
5. Antenna 1, peduncle segment 1 strongly inflated, segments 2 and 3, deeper than long; peraeopods 6 and 7 subequal; gnathopod 1, dactyl with posterior marginal tooth . . . . 6.
   Antenna 1, peduncle 1, little inflated, segment 2 and 3 longer than wide; peraeopod 7 distinctly longest; gnathopod 1, dactyl smooth (or seta only) behind
   . . . . *Psammonyx* (p. 144).

6. Coxa 1 narrowing distally; gnathopod 1, palm of segment 6 usually concave
. . . . *Tryphosella* Bonnier.
Coxa 1 not narrowing distally; gnathopod 1, palm of segment 6 convex
. . . . *Tmetonyx* Stebbing.
7. Gnathopod 1 strongly chelate; telson narrow, deeply cleft; animal externally parasitic on fishes (esp. cod) · · · · · · · · · · · · · · · · · *Opisa.*
Gnathopod 1 weakly subchelate or simple; telson moderately broad, variably cleft or entire; animal not parasitic on fish . . . . 8.
8. Gnathopod 1 simple; peraeopod 7, distinctly longest . . . . 9.
Gnathopod 1 weakly subchelate; peraeopods 6 and 7 subequal . . . . . . 10.
9. Telson entire; uropod 3, outer ramus 1-segmented; gills simple *Lysianopsis* (p. 146).
Telson cleft; uropod 3, outer ramus 2-segmented; gills pleated . . . *Socarnes.*
10. Telson longer than broad, deeply cleft; peraeopod 5, segment 4 more or less broadly expanded . . . . *Orchomenella* (p. 147).
Telson short, cleft not more than one-half; peraeopod 5, segment 4 only slightly broader than segment 5 . . . . *Orchomene.*

Genus *Hippomedon* Boeck 1870 (incl. *Paratryphosites* Stebbing 1899)

(hǐp ō mēd'ŏn)

Head, eye subrectangular. Antenna 1, peduncle 1 with anterodistal process. Coxal plates strongly overlapping, smooth below, partly hiding head.

Mandible, molar strong; spine row lacking; palp inserted opposite molar. Maxilliped, outer plate, inner margin lined with stout spine-teeth; palp strong.

Gnathopod 1 simple; segment 5 elongate (longer than 6). Peraeopods 3 and 4, segments 5 and 6 strongly setose posteriorly. Peraeopods 5–7 subequal, slender distally. Branchiae on segments 2–7.

Pleon side plate 3, hind corner produced as strong tooth. Uropods 1 and 2 long, rami slender, sicklelike. Uropod 3, rami margins spinose, outer ramus 2-segmented. Telson, apices acute, diverging.

KEY TO NEW ENGLAND SPECIES OF *HIPPOMEDON*

1. Telson short, squarish, shallowly cleft (not to middle); gnathopod 1 distinctly subchelate, palm well defined . . . . *H. abyssi* (Goes).
Telson distinctly elongate, deeply cleft; gnathopod 1, palm lacking or ill-defined 2.
2. Gnathopod 1 essentially simple, palm not definable; telson lobes deeply V-cleft
. . . . *H. serratus* (p. 144).
Gnathopod 1 weakly subchelate; telson narrowly cleft . . . . 3.
3. Antenna 1, peduncle 1 with anterodistal process overhanging segments 2 and 3; pleon side plate 3, with sharp notch above base of sharply upturned tooth
. . . . *H. denticulatus* (Bate).
Antenna 1, peduncle 1 not overhanging 2 and 3 distally; pleon side plate 3, hind tooth without basal notch . . . . *H. propinquus* Sars.

Hippomedon serratus Holmes 1905 Plate XL.1   (sĕr ăt′ ŭs)

L. 14 mm ♂. Urosome 1 middorsally elevated and rounded. Eye, subrectangular, normally or fairly pigmented. Antenna 1, peduncle 1 with anterodistal process overhanging segments 2 and 3; flagellum short (♂).

Mandibular molar large, ridged; palp long, slender, segment 2 nearly bare distally; segment 3, comb setae long. Maxilliped, inner plate relatively short, broad; palp heavy, relatively short.

Gnathopod 1 simple, propod slender, about one-half length of elongated carpus. Peraeopods 3 and 4, segments 5 and 6 moderately powerful and lined posteriorly with long setae (singly or paired). Peraeopods 5–7 subequal (5 smaller), bases moderately expanded, hind margins shallow-convex, distal lobes sharply rounded. Peraeopod 7, hind margin of basis with numerous fine serrations; dactyl shorter than in 5 and 6. Pleon side plate 2, hind corner quadrate, not produced. Side plate 3, hind tooth moderately strong, not notched above at base.

Uropods 1 and 2, rami very slender, a few spines proximoposteriorly; tips of 1 reaching tips of uropod 3. Telson lobes deeply V-cleft, apices somewhat diverging, subacute.

Distribution: From the Gulf of St. Lawrence and New England to the Middle Atlantic states and North Carolina.
Ecology: Sandy and sandy silt bottoms, from about 5 to more than 75 m.
Life Cycle: Unknown.

Genus *Psammonyx* n. gen.   (săm ŏn′ ĭks)

Head, anterior lobe subacute. Eye lenticular. Coxal plates 1–4 deep, fringed below. Antennae 1 and 2 elongate in both sexes, calceolate. Antenna 1, peduncle 1 little inflated, segments 2 and 3 not broader than long; flagellar segment 1 not longer than peduncle 3.

Mandible, molar strong, compressed, ridged; spine row weak; palp inserted opposite molar; toothed lacinia in left mandible only. Maxilla 1, inner plate with 1 terminal seta. Maxilla 2, plates not elongated or narrow. Maxilliped, plates normal; outer plate, inner margin with numerous short spine teeth; palp stout. Upper lip produced forward, blunt.

Gnathopod 1 subchelate; carpus and propod short; dactyl not toothed behind. Gnathopod 2 minutely subchelate, propod short. Peraeopods 3 and 4 strong, posterior margins of segments 5 and 6 setose; dactyls strong. Peraeopods 5–7 strong, unequal, 5 shortest, 7 distinctly longest. Peraeopod 5, segments 4 and 5 broadly expanded.

Pleon side plate 3, hind corner not a pronounced tooth. Uropods 1 and 2 strong, rami sicklelike; peduncle spinose. Uropod 3, rami broad-lanceolate, foliaceous, outer ramus 2-segmented. Telson deeply cleft; lobes apically truncate, spinose. Coxal gills simple, not pleated on segments 2–6; 5–6 with accessory lobes.

Type species: *Anonyx nobilis* Stimpson 1853.
Relationships: The species bears a superficial resemblance to *Pseudalibrotus litoralis,* but is probably more closely related to *Kyska* (Shoemaker) in its sand-burrowing morphology, and to *Anonyx* (sens. str.). It is quite different from N. Atlantic members of the genus *Tmetonyx* (sens. str.). See Barnard (1969) for definition of *Tmetonyx.*

KEY TO NEW ENGLAND SPECIES OF *TMETONYX* (SENS. STR.)

1. Coxa 1 narrowing distally; gnathopod 1, palm of segment 6 concave; urosome 1 with middorsal carina; side plate 3, sharply rounded .   .   .   .   *T. compressa* Sars.
   Coxa 1 not narrowing distally; gnathopod 1, palm convex; urosome 1 dorsally smooth; abdominal side plate 3, hind corner acute, slightly produced
   .   .   .   .   *T. cicada* (Fabr.).

Psammonyx nobilis (Stimpson) 1853   (nōb′il ĭs)
[formerly *Tmetonyx nobilis* (*Stimpson*)] Plate XL.2

L. 16–20 mm. Head, anterior lobe slightly produced. Eye medium, black. Antenna 1, accessory flagellum 5-segmented.

Mandibular palp segment 1 with distal tuft of long setae; 3 blades in spine row. Maxilliped, inner plate tapering distally, apex truncate.

Gnathopod 1, propod and carpus short, subequal in length; palm oblique, minutely and irregularly serrate, posterior angle with 2 stout spines.

Peraeopods 3 and 4, segment 4 anterodistally produced nearly one-half the length of segment 5. Peraeopod 5, basis wider than long, distal lobe sharply rounded; segment 4 wider than long, rounded; segment 6 about equal to length of 5. Peraeopods 6 and 7, segments 4 and 5 heavier but shorter than 6; dactyls slender, about one-half the length of segment 6. Pleon side plates 1 and 2, hind corners acuminate; side plate 3, corner almost quadrate.

Uropods 1 and 2, peduncles longer than rami, outer posterior margin with 5–10 long slender spines; rami basally short-spinose. Uropod 3 distinctly exceeding 1; rami subequal, inner margins foliaceous, outer spinose. Telson lobes subparallel, obliquely truncate apices each with 3–4 short spines.

Distribution: From Belle Isle Strait and throughout the Gulf of St. Lawrence to southern New England and Long Island Sound. Common along sandy shores of northern New England.
Ecology: A common "cold-water" species, burrowing in medium to medium-coarse sand, occasionally with organic admixture; in proximity to stones,

along exposed and semiprotected shores, from about mid-tide level to 6 m in depth.

Life Cycle: Annual or biennial; ovigerous females from February–April; young breed next winter and spring; some persisting into second year. One brood per ♀.

Genus *Lysianopsis* Holmes 1905   (līs ĭ ăn ŏp′ sĭs)

Body surface smooth, not ciliated. Urosome 3 with posterior distal process. Coxal plates deep, 1 expanded anteriorly. Coxa 5 large, anterior lobe deeper. Antenna 1 and 2 short, subequal (both sexes). Antenna 1, peduncle 1 little inflated, not with posterior distal process; peduncle 2 much longer than 3; accessory flagellum long.

Upper lip prolonged into tonguelike lobe separated from epistome by narrow incision. Mandible, palp attached behind molar, segment 3 with apical setae only. Maxilla 1, inner plate with 2 apical setae. Maxilla 2, plates long, subequal in width. Maxilliped, inner plate tapering; outer plate, inner margin crenulate, lacking teeth.

Gnathopod 1 simple, carpus short; dactyl short, simple. Peraeopods 5–7 increasing in length, 7th distinctly the longest.

Pleon side plate 3, hind corner broadly rounded. Uropod 2, inner ramus constricted distally. Uropod 3, peduncle short, with posterodistal lobe; rami bare (♀), outer 1-segmented. Telson entire, small.

Taxonomic notes: Barnard (1969) has synonomized this genus under *Lysianassa* M. E. 1830. Until the component species have been carefully re-analyzed, however, *Lysianopsis alba* is sufficiently distinctive from all other species of *Lysianassa* to warrant retention of *Lysianopsis*.

Lysianopsis alba Holmes 1905 Plate XLIII.1   (al′ bă)

L. 8–10 mm. Body smooth, not dorsally ciliated. Urosome 3, posterolateral corner slightly produced. Head, anterior lobe subacute, moderately produced. Eye reniform, relatively small. Coxal plates squarish below. Antenna 1 not longer than 2 (♀); peduncular segment 1 not produced posterodistally.

Mandible, palp segment 3 subfalcate, with several stout apical setae; segment 2 with apical fan of slender spines. Lower lip, outer lobes closely approximated. Maxilliped, inner plate, much shorter than outer plate.

Peraeopods 3 and 4, segment 4 stout, prolonged distally in front of 5; segment 6, not much longer than 5, hind margin with short stiff setae (♀). Peraeopod 5, basis wider than deep, hind margin straight; segments 4 and 6, subequal in length. Peraeopod 6, basis shallow concave behind; segment 4 nearly as broad as long. Peraeopod 7, basis shallow concave behind; segment 4 stout, segment 6 longest.

Pleon side plate 3 evenly rounded behind. Uropods 1 and 2, outer ramus longer than inner, tips well exceeding uropod 3. Uropod 3, peduncle short, little longer than subequal, nearly unarmed, acute rami, with hind margin flangelike. Telson nearly as wide as long, evenly rounding.

Distribution: South side of Cape Cod to the Middle Atlantic states, Carolinas, to northern Florida; also in eastern Gulf of Mexico.
Ecology: A common species of coarse shelly sand, in roots of eelgrass, from low tide level to depths of more than 20 m, along semiprotected and protected shores.
Life Cycle: Annual; ovigerous females from May–September; several broods. Pelagic males very rare; taken only in May?

Genus *Orchomenella* G. O. Sars 1890* (ôrk ö měn ěl'ă)

Coxal plates very deep; coxa 1 distally expanded; coxa 5 extremely deep, greater than basis of per. 5. Head, anterior lobe prominent. Antenna 1 short, peduncle 1 very inflated. Antenna 2 short.

Epistome prominent, overlapping upper lip. Mandible, palp arising behind molar process. Maxilla 1, inner plate apically narrowing. Maxilla 2, plates narrow. Maxilliped, outer plate, inner margin with distal spine-teeth, proximally crenulated; inner plate tall.

Gnathopod 1 subchelate, propod longer than carpus. Peraeopods 5–7 subequal; segment 4 expanded, posterodistal angle overhanging segment 5.

Pleon side plate 3, hind corner not produced. Uropod 3 weakly foliaceous, inner ramus shorter. Telson deeply cleft.

KEY TO NEW ENGLAND SPECIES OF *ORCHOMENELLA*
(after Stebbing 1906)

1. Coxal plate 5 broader than deep; pleon side plate 4, hind angle upturned; eye facets imperfectly developed . . . . *O. groenlandica* Hansen.
   Coxal plate 5 deeper than broad, posterior lobe extending downward; pleon 3, hind corner not upturned; eye fully and normally pigmented . . . . 2.
2. Gnathopod 1, carpal lobe very narrow behind; peraeopods 3–4, posterior margin with double row of spines, spines stout; peraeopod 5, segment 4, anterior margin spinose; epistome not prominent . . . . *O. pinguis* (p. 148).
   Gnathopod 1, carpus, length of posterior margin about one-half anterior margin; peraeopods 3 and 4, spines weak, singly inserted; peraeopod 5, segment 4, anterior margin nearly bare; epistome prominent . . . . . . *O. minuta* (p. 147).

Orchomenella minuta Krøyer 1846 Plate XLII.2 (mĭn ūt' ă)

L. 5mm. Head, anterior lobe acute, strongly produced. Eye reniform, black, medium. Antenna 1 slightly shorter than 2 (♀), flagellum 8–10 segmented; ac-

---

* Barnard (1969) has outlined an involved synonymy of *Orchomenella* with *Orchomene* Boeck 1871, excluding *O. groenlandica* (Hansen) which is treated as a distinct generic taxon.

cessory flagellum 3-segmented. Coxal plate 1 narrow, not expanding anter-odistally.

Maxilliped, inner plate short, barely reaching midpoint of outer plate. Gnathopod 1, carpal lobe relatively broad, about one-half length of upper margin of carpus; propod margins subparallel. Peraeopods 3 and 4, segment 6 slender, posterior margin with 4–5 short weak single spines. Peraeopod 5, basis broad-oval; segment 4, posterior lobe small, anterior margin with 1 short spine only. Peraeopods 6 and 7, basis evenly convex and finely serrate behind; segment 5 shorter than 4; segment 6 slender; dactyls slender.

Pleon side plate 3 broadly acute, hind margin gently convex. Uropod 3 short, barely exceeding tips of uropods 1 and 2; inner margin of outer ramus with 1 or 2 plumose setae as well as short spines. Telson broader than long, cleft four-fifths.

Distribution: Pan-arctic and arctic-boreal; in the western North Atlantic from Hudson Strait and Baffin Island south to the Gulf of Maine, Vineyard Sound, and off New Jersey.

Ecology: Very common in fine stable sands, or sandy mud, along channels, roots of eelgrass, etc., in cold-water bays and semiprotected shores; from lowest intertidal levels to depths of more than 100 m.

Life Cycle: Annual; ovigerous females from February–April; one brood per female. Males present from late August into winter.

Orchomenella pinguis (Boeck) 1861 Plate XLII.1   (pǐng′ wǐs)

L. 6–7 mm. Head, anterior lobe subacute, pronounced. Eye large, deep-reni-form, reddish. Coxal plate 1 broad, anterodistally expanded. Antenna 1 dis-tinctly shorter than antenna 2 (♀); flagellum 12–14 segmented; accessory fla-gellum 4–5 segmented.

Mandibular palp, segment 2 with 2 rows of apical comb spines. Maxilliped, inner plate long and narrow, reaching beyond midpoint of outer plate, which has only a few marginal spine-teeth apically. Upper lip project-ing as far as epistome.

Gnathopod 1, carpal lobe short, narrow; propod narrowing distally to vertical, slightly convex palm. Peraeopods 3 and 4, segment 6 stout, with 6–8 pairs of short posterior marginal spines. Peraeopod 5, basis broadly ex-panded, widest proximally; segment 4, posterodistal lobe extending one-half segment 5, anterior margin with 4–6 slender spines. Peraeopods 6 and 7, segments 4 and 5 subequal in length; segment 6 slightly longer and less heavy; dactyls relatively short and heavy.

Pleon 3, hind corner subquadrate, posterior margin finely crenulate. Uropod 3 extending beyond uropod 1; inner margin of outer ramus with 3–4 plumose setae as well as short spines. Telson cleft nearly to base, lobes each with 4 dorsal spinules.

Distribution: Arctic-boreal and pan-arctic; in western North Atlantic south-ward through the Gulf of St. Lawrence, outer Nova Scotia and Gulf of Maine to Cape Cod, Buzzards Bay, and North Carolina.

Ecology: In fine to medium coarse sands, in cold-water areas from low tide level to more than 100 m. An abundant and voracious scavenger, often strip-ping baited fishing lines and lobster pots.

Life Cycle: Annual; ovigerous females February–April; one brood per fe-male. Pelagic males in winter (February).

Genus *Anonyx* Krøyer 1842   (ăn ŏn' ĭks)

Head, interantennal lobe very pronounced. Eye, large, widening below. An-tenna 1, flagellar segment 1 longer than peduncular segments 2 or 3; peduncle 1 strongly inflated. Coxal plates moderately deep; coxa 1, expand-ing anterodistally.

    Mandible, molar weak; palp inserted opposite molar. Maxilliped, inner plate short; outer plate, inner margin finely crenulate, lacking spine-teeth. Upper lip, epistome produced anteriorly, incised above.

    Gnathopod 1 weakly subchelate, hand narrowing distally to short palm; segments 5 and 6 subequal. Peraeopods 5–7 subequal, distal segments slender.

    Pleon side plate 3, hind corner produced strongly as an upturned tooth. Uropod 2, inner ramus with distal or subapical notch. Uropod 3 extending beyond uropods 1 and 2, rami variably foliaceous. Telson deeply cleft. Coxal gills 2–6 usually pleated, 7 simple; 5 and 6 with accessory lobes.

KEY TO NEW ENGLAND SPECIES OF *ANONYX*
(after Steele and Brunel 1968)

  1. Gnathopod 2 distinctly subchelate, both palm and dactyl larger than those of gnathopod 1; coxa 1, anterior margin straight or convex . . . . . . *A. debruyni.*
     Gnathopod 2 minutely chelate or subchelate, dactyl minute; coxa 1, anterior mar-gin proximally shallow-concave . . . . 2.
  2. Posterior margin of segment 6 of peraeopods 3 and 4 with a row of pairs of long setae and a few short setae and/or spines; epistome produced *A. liljeborgi* (p. 150).
     Posterior margin of segment 6 of peraeopods 3 and 4 with single row of long setae and corresponding row of spines; epistome not produced beyond anterior margin . . . . 3.
  3. Uropod 2, inner ramus completely constricted, spine long; pleon plate 3, margin below tooth straight or concave . . . . *A. compactus.*
     Uropod 2, inner ramus partly constricted, spine short; pleon side plate 3, margin below tooth convex . . . . *A. sarsi* (p. 149).

Anonyx sarsi Steele and Brunel 1968 Plate XLI.2   (sărs' ī)

L. 20–31 mm. Body medium-sized. Urosome 1 slightly carinate dorsally and projecting slightly posteriorly. Head, anterior lobe rounded, lower margin straight. Eye very large (esp. ♂), expanded at bottom, black. Antenna 2, peduncle extending beyond peduncle of antenna 1.

Upper lip low, slightly produced in front of epistome. Mandible, palp segment 1 with apical setae. Maxilliped, inner plate, margins subparallel, apex obliquely truncate.

Gnathopod 1, segment 6 narrowed distally; dactyl with posterior accessory tooth, palm width greater than base of dactyl. Gnathopod 2 minutely subchelate. Peraeopods 3 and 4, segment 6, posterior margin with single row of long setae and paired row of long spines. Peraeopod 7, basis distinctly shorter than rest of limb, posterior margin with more than 20 serrations that are distally more crowded and prominent.

Pleon side plate 1, lower front corner produced, acute. Pleon plate 2, lower margin setose, hind corner produced as acute tooth. Pleon plate 3 setose below; posterior tooth with deep base, hind margin convex.

Uropod 1, both rami with inner and outer marginal spines. Uropod 2, inner ramus slightly constricted just beyond middle, with medium-sized constricted spine. Uropod 3, rami tapering distally, inner edges only (both sexes) with plumose setae. Telson, apices broad, apical spines very short.

Antennae of male long (antenna 2 with calceoli); eyes usually larger; uropod 2, inner edge of outer ramus serrated, short blunt spine in each notch; uropod 3 more foliaceous than in female.

Distribution: Probably circumpolar; in the western North Atlantic south to the Gulf of Maine, Vineyard Sound, and Rhode Island.
Ecology: Relatively common in subtidal sandy substrata in cold-water areas, from extreme low water level to about 50 m. Recently numerous near shore and a voracious scavenger of trap bait, penned lobsters, etc.
Life Cycle: Annual or biennial; ovigerous females from midwinter to June; one brood per female.

Anonyx liljeborgi Boeck 1871 Plate XLI.1   (lĭlj′ bôrg ī)

L. 9–18 mm. Body relatively small. Urosome not carinate. Head, anterior lobe produced, lower margin concave. Eye medium, expanded slightly below, reddish brown. Antenna 2, peduncle short, extended little beyond peduncle of antenna 1.

Upper lip very large, projecting as a high rounded lobe in front of the epistome. Mandible, palp segment 1 and proximal part of 2 lacking setae. Maxilliped, inner plate tapering to oblique apex.

Gnathopod 1, segment 6 regularly narrowing distally, palm shorter than width of base of dactyl; dactyl with small posterior marginal accessory tooth. Gnathopod 2 minutely subchelate, dactyl about one-half length of distal end of propod.

Peraeopods 3 and 4, segment 6, posterior margin with row of pairs of long setae. Peraeopod 7, basis about as long as rest of limb, posterior margin with less than 20 regularly spaced, very weak serrations.

Pleon side plate 1, lower front corner produced, acute. Pleon plate 2 not setose below, hind corner weakly acuminate. Pleon plate 3, hind tooth slightly upturned, moderately long, hind margin convex.

Uropod 1, outer ramus lacking spines on inner edge. Uropod 2, inner ramus distinctly constricted at distal third, with a long constriction spine; outer ramus, end spine long. Uropod 3, rami narrow-lanceolate; inner edges only with setae in immatures, all four edges with some setae in mature females as well as males. Telson lobes narrowing toward apex, apical spines relatively long and prominent.

Antenna 2 of male longer, calceolate; uropod 3 more foliaceous than in female.

Distribution: Subarctic and amphiboreal; from the Canadian subarctic south to Nova Scotia, the Gulf of Maine; rare in southern New England to Delaware.

Ecology: On sandy substrates in cold-water areas, from the lower intertidal to depths of several hundred meters. Occurs in the very cold year-round negative-temperature bottom water.

Life Cycle: Annual or biennial; ovigerous females from midwinter to April–May; one brood only.

$\mathcal{T}$ALITRID AMPHIPODS (SUPERFAMILY TALITROIDEA)

Amphipods of very distinctive and primitive morphology, strongly sexually dimorphic (esp. in gnathopods) with males almost invariably larger, and more or less showing all stages of evolution from purely aquatic to strictly terrestrial existence. Primarily endemic to the southern hemisphere and to Indo-Pacific regions, the superfamily Talitroidea encompasses a complex of families, not yet fully defined, characterized by a relatively smooth, hard, weakly setose integument, and various reduction of antenna 1, mouthpart plates and palps, brood plates, pleopods, and uropod 3. Superficially similar to the Lysianassidae, especially in mouthpart morphology, the group may be related to Phliantidae and possibly Eophliantidae through the Kuriidae.

Superfamily TALITROIDEA Bulycheva 1957

Body smooth, occasionally spinose, seldom setose. Urosome shortened. Sexual dimorphism marked in gnathopods and antennae. Head lacking rostrum; eyes present. Antenna 1, short, lacking accessory flagellum. Coxal plates 2–4 usually with posterior process, 4th excavate behind.

Mandible lacking palp, molar strong. Lower lip lacking inner lobes. Upper lip rounded. Maxilla 1, palp reduced or vestigial; inner plate narrow, with 2 stout apical setae; outer plate with 9 terminal spine teeth. Maxilla 2, inner plate with strong proximal plumose seta. Maxilliped, inner plate with 3 apical teeth.

Gnathopods primarily subchelate. Gnathopod 2 usually larger, often very powerful in ♂. Uropod 3 small, uniramous (occ. minutely biramous). Telson short, cleft, or lobes fused. Coxal gills on peraeon segments 2–6. Brood plates primarily lamellate, margins primitively with hooked setae.

KEY TO NORTH AMERICAN FAMILIES OF TALITROIDEA

1. Antenna 1 longer than peduncle of antenna 2; coxal plate 1 similar to 2–4; maxilliped, palp strongly dactylate, 4-segmented; primarily aquatic; gnathopod 2 (♀ and young) normally subchelate . . . . 2.
   Antenna 1 shorter than peduncle of antenna 2; coxal plate 1 smaller and shorter than 2–4; maxilliped palp not dactylate, 4th segment vestigial or lacking; animals terrestrial or semiterrestrial; gnathopod 2 (♀ and young), propod mitten-shaped, minutely chelate . . . . Talitridae (p. 157).

2. Body appendages strongly fossorial, segments broad, very setose; maxilliped, palp clavate; coxal plates, lower margins setose . . . . . . Dogielinotidae.*
   Body appendages typically slender, nonsetose; maxilliped, palp normally dactylate; coxal plates, lower margins bare or weakly spinose . . . . 3.
3. Telson entire; accessory gills present on some peraeon segments (always on 7); coxal plates deep, lacking posterior process; maxilla 1, palp minute, vestigial. . .
   . . . . Hyalellidae (p. 153).
   Telson variably cleft; accessory gills lacking; maxilla 1, palp short, slender, 2-segmented; coxal plates with posterior process . . . . . . Hyalidae (p. 154).

Family HYALELLIDAE Bulycheva 1957   (hī ăl ĕl′ ĭ dē)

Coxal plates 1–4 deep, lacking posterior process. Antenna 1 longer than peduncle of antenna 2. Mandible palp vestigial or lacking. Maxilliped, palp 4-segmented, dactyl strong. Gnathopod 2 strongly subchelate (δ).

Peraeopods 5–7 slender, not fossorial, increasing in size posteriorly. Uropod 3 uniramous or lacking ramus. Telson entire or slightly cleft. Accessory gills present on some peraeon segments, always on 7. Brood plates large, triangular, margins lined with short hooked setae.

KEY TO NORTH AMERICAN GENERA OF HYALELLIDAE

1. Telson entire; accessory gills on 6 and 7; antenna 2, peduncle slender; fresh-water only. . . . . *Hyalella* (p. 153).
   Telson slightly cleft; accessory gills small or lacking; antenna 2, peduncle segments powerful (δ); marine . . . . *Parhyalella* Bate.

Genus *Hyalella* Smith 1873   (hī ăl ĕl′ ă)

Body smooth or dorsally and laterally mucronate or with spinose processes. Antenna 2, peduncle slender (δ), not expanded or incrassate; gland cone prominent. Coxal plates very deep, deeper than long.

Maxilla 1, palp vestigial, 1–2 segmented. Maxilliped palp, dactyl medium short. .

Gnathopods 1 and 2 ($\female$) subchelate, 2 not stronger than 1; carpal lobe short. Gnathopod 2 (δ) powerfully subchelate, posterior margin of propod slightly concave; carpal lobe present, always produced between segments 4 and 6. Accessory gills on segments 6 and 7, double on 6, saclike.

Uropod 1, rami marginally spinose, interramal spine lacking or very weak. Uropod 3, ramus present, narrowing distally. Telson entire, with pair of apical spines.

---

* N. Pacific endemic.

Hyalella azteca Saussure 1857 Plate XLIII.2   (ăz′ tĕk′ ă)

L. 8 mm ♂, 6 mm ♀. Body small, dorsally mucronate on pleon segments 1 and 2, occasionally on 3, or smooth (form *inermis*). Coxal plates very deep, 4th largest; lower margins lightly and evenly spinose. Head, eye subovate, black, slightly larger in ♂.

Antenna 1, peduncular segments 1 and 2 subequal, flagellum 8–10 segmented. Antenna 2, peduncle slender, segment 5 longer than 4; flagellum 9–10 segmented. Maxilliped, palp segment 2 wider than long, exceeding outer plate.

Gnathopod 1 (♂), propod shorter and less deep than carpus, expanding distally; palm oblique, convex. In ♀, propod narrow, short; palm vertical, convex. Gnathopod 2 (♂), propod large, distally broadest; posterior margin slightly concave; palm convex, with large low tooth near hinge; carpal lobe deep. In ♀, propod slender, elongate, expanding distally; palm short, convex, vertical.

Peraeopods 3 and 4, posterior margins of segments 5 and 6 with 3–5 short, stout spines. Peraeopods 5–7, basis broadly expanded, posterior margin with 4–10 very weak serrations; segments 5 and 6 lacking posterior marginal spines or setae. Abdominal side plates 2 and 3, hind corners sharply subquadrate, not produced.

Uropods 1 and 2, both rami with 2 slender marginal spines. Uropod 3, ramus and peduncle subequal in length, apex with long spine(s). Telson, apex rounded, with 2 slender wide-set spines.

Distribution: Permanent fresh waters of North and Central America and Caribbean islands north to the tree line. Coastally in larger rivers seaward into tidal fresh waters, and in fresh-water barrier beach lagoons.
Ecology: Nearly ubiquitous in permanent fresh waters; in algae and bottom aquatic vegetation; in sand, under stones; in shallow water, down to 20–30 ft.
Life Cycle: Annual; ovigerous females in April–October, after water has attained 10°C or more; several broods per female.

Family HYALIDAE   (hī ăl′ĭ dē)

Antenna 1 longer than peduncle of antenna 2 (usually). Coxal plates moderately deep, posterior process variously developed. Maxilla 1, palp reduced, 2-segmented. Maxilliped, palp dactylate, 4-segmented. Gnathopods 1 and 2 moderately to strongly subchelate, both sexes. Peraeopods slender, nonfossorial; 5–7 regularly increasing in size. Pleopods normal. Uropod 3 uniramous or unequally biramous. Telson usually cleft deeply. Accessory gills absent. Brood plates large, triangular, margins with numerous short hooked setae.

KEY TO NORTH AMERICAN GENERA OF HYALIDAE

1. Antenna 1 originating below level of eyes; telson entire; mandible lacking molar; gnathopods 1 and 2 (δ) similar . . . . *Najna* (Pacific).
   Antenna 1 originating at or above eye level; telson cleft; mandible with strong triturating molar; gnathopod 2 (δ) much more powerful than gnathopod 1 . . . 2.
2. Uropod 3 biramous, inner ramus minute . . . . 3.
   Uropod 3 uniramous. . . . . 4.
3. Maxilla 1, palp 1-segmented, vestigial; gnathopod 2 (δ), carpal lobe masked by segments 4 and 6; uropod 1 with interramal spine . . . . . *Parhyale* (tropical).
   Maxilla 1, palp large, 2-segmented; gnathopod 2 (δ), carpal lobe produced between articles 4 and 6; uropod 1 lacking interramal spine . . . .
   . . . . *Parallorchestes* (N. Pacific).
4. Gnathopod 2 (δ), carpal lobe produced between segments 4 and 6; telson usually apically notched, incompletely cleft . . . . *Allorchestes*.
   Gnathopod 2 (δ), carpal lobe masked by segments 4 and 6; telson cleft to base, bilobed . . . . *Hyale* (p. 155).

Genus *Hyale* Rathke 1837   (hī' ăl ě)

Coxal plates, medium deep, 4 much deeper than 5, posterior process present in plates 1–4. Maxilla 1, palp 1–2 segmented, reduced, not vestigial, attaining base of outer plate spine-teeth. Maxilliped, palp segments broad, segment 4 dactylate.

Gnathopod 2 powerfully subchelate in δ; posterior carpal lobe is vestigial or completely masked between merus and propod (in mature δ). Uropod 3 uniramous. Telson cleft to base, lobes separate.

Marine intertidal or coastal fresh water.

KEY TO NEW ENGLAND SPECIES OF *HYALE*

1. Coxal plates wider than deep, rounded below; abdominal side plates 2 and 3, hind corners blunt; antenna 2 (δ), peduncle 5 with clusters of fine setae on posterior margin; uropod 1 with strong inner distal spine . . . *H. plumulosa* (p. 155).
   Coxal plates subquadrate, lower margin straight; abdominal side plate corners sharply subquadrate; antenna 2 (δ) lacking posterior clusters of setae; segment 6 with single posterior marginal seta; uropod 1 with very short inner distal spine . . .
   . . . . *H. nilssoni* (p. 156).

Hyale plumulosa (Stimpson) 1853 * Plate XLIV.2   (plo͞om ūl ōs' ă)

L. 12 mm δ, 10 mm $\female$. Body smooth, slender. Coxal plates 1–4 moderately large, broad, rounded below, posterior process prominent on 2–4; coxa 5 al-

---

* Formerly *Allorchestes littoralis* of Holmes, Kunkel and previous authors.

most as deep as 4; coxa 1, margin evenly rounded. Head, eye deep rectangular, black, outer ring of facets unpigmented. Antenna 1 slender, peduncular segments 1–3 subequal; flagellum 10-segmented. Antenna 2 of medium length; peduncular segment 5, posterior margin with dense clusters of fine setae (in ♂); flagellum 12–14 segmented; gland cone prominent. Maxilliped, palp segment 2 largest, broad distally.

Gnathopod 1 (both sexes), basis not expanded; propod expanded distally, palm vertical, dactyl closely fitting. Gnathopod 2 (♂), propod very large, deep-ovate, palm oblique, evenly convex, longer than posterior margin; carpus, posterior lobe very small, not completely masked in subadult specimens. In ♀, propod considerably larger than in gnathopod 1, expanded distally, dactyl fitting palm; carpal lobe narrow.

Peraeopod dactyls short. Peraeopods 5–7, slender, basis broadly expanded, posterior margins weakly crenulate. Abdominal side plates 2 and 3, hind corners blunt.

Uropod 1, outer ramus with marginal spines, interramal spine strong, long, inserted at inner distal angle of peduncle. Uropod 2, rami longer than peduncle, margins spinose. Uropod 3, ramus longer than peduncle, apex strongly spinose. Telson cleft to base, lobes slightly separate.

Distribution: American Pacific coast, from Southern Alaska to Southern California; American Atlantic; from southern Maine (Casco Bay) to North Carolina.
Ecology: Intertidal on protected rocky and stony shores and in salt marshes at base of *Spartina* roots; under fucoids, under small stones and in crevices; occasionally in upper tide pools; mainly in the lower midlittoral, but occasionally up to the drift line. Animals may hop about in air when disturbed but tend to run and crawl primarily.
Life Cycle: Annual; ovigerous females May–September; several broods.

Hyale nilssoni (Rathke) 1843 Plate XLIV.1   (nĭls′ ŏn ī)

L. 9 mm ♂, 7 mm ♀. Body smooth, compact. Coxal plates 1–4 medium deep, lower margins straight, finely spinose, posterior marginal process well developed; coxa 5, lobes shallow. Eye small, oval, black. Antenna 1, peduncular segments very short; flagellum 6–8 segments. Maxilla 1, palp exceeding base of terminal spine-teeth of outer plate. Maxilliped, palp large, segment 3 larger than 2.

Gnathopod 1 (♂) small; propod narrow, sides parallel; palm oblique, convex, slightly exceeded by dactyl; in ♀, similar but smaller. Gnathopod 2 (♂), basis expanded distally into large rounded free lobe; propod large, ovate, narrowing distally; palm evenly convex, oblique, shorter than posterior margin. In ♀, basis with square distal corner; propod short, slightly narrowing distally, palm short.

Peraeopods 5–7 increasing slightly in size posteriorly, dactyls normal;

segment 4 somewhat expanded, posterior margin lined with short spines. Peraeopods 6–7, segment 6 with single posterior marginal seta. Peraeopod 7, posterior margin of basis unevenly low crenulate. Abdominal side plates 2 and 3, corners subacute. Uropod 1, outer ramus with marginal spine(s); interramal spine lacking. Uropod 2, outer ramus lacking marginal spine(s). Uropod 3 very short; apex blunt, spines short. Telson broad, cleft near to base.

Distribution: Amphi-Atlantic; northwestern Europe, Norway to western France. American Atlantic coast, from southern Labrador and St. Lawrence estuary to Connecticut.
Ecology: Intertidal on rocky and stony shores; under *Fucus* and *Ascophyllum* and in crevices, throughout the midlittoral zone. On exposed or protected shores, in higher salinities. Animals may hop actively in air when disturbed.
Life Cycle: Annual; ovigerous females April–August; more than one brood.

Family TALITRIDAE   (tăl ĭ' trĭ dē)

Body smooth, occasionally with dorsal abdominal processes. Head, buccal mass extending prominently below. Sexual dimorphism strongly expressed in gnathopods and antenna 2 in semiterrestrial, reduced or nearly absent in strictly terrestrial species. Coxal plates 2–4 larger and deeper than coxa 1, posterior process of 2–4 usually present. Antenna 1 shorter than peduncle of antenna 2. Antenna 2, peduncular segments (esp. 5th) long, often enlarged (δ); gland cone absent. Maxilla 1, palp vestigial, minutely 2-segmented. Maxilliped, palp segment 4 very small, fused to 3rd, or lacking.

Gnathopod 1 small, subchelate or simple; gnathopod 2 (δ) powerfully subchelate (occ. chelate) in semiterrestrial species, tending to neotenic reduction or loss in terrestrial species. In $\female$, and young, propod mitten-shaped, dactyl arising subapically, forming a small chela.

Peraeopods 6 and 7 generally much larger than 5. Pleopods tending to reduction, modification, or loss. Uropod 1 arising from prepeduncular extension of urosome. Uropod 3 uniramous. Telson entire, apically notched.

Coxal gills present on peraeon segments 2–6, saclike in semiterrestrial, highly modified or convoluted in terrestrial species. Brood plates oval or linear, marginal setae hooked or simple.

KEY TO AMERICAN ATLANTIC GENERA OF TALITRIDAE

1. Gnathopod 2 alike in δ and $\female$; pleopods very reduced . . . . . . *Talitroides.*
   Gnathopod 2 powerful in δ, mittenlike in $\female$ and juveniles; pleopods slightly reduced
   . . . . 2.

2. Peraeopod 6 longer than 7; gnathopod 1 simple in ♀, fossorial; uropod 3, ramus longer than peduncle, laterally compressed . . . . . . *Talorchestia* (p. 162). Peraeopod 6 shorter than 7; gnathopod 1 primarily subchelate, not fossorial; uropod 3, ramus cylindrical, narrowing to apex . . . . 3.
3. Maxilliped, palp segment 4 distinct; pleopod 3 markedly shorter than 1 and 2 . . . . *Parorchestia*. Maxilliped, palp segment 4 minute, masked by 3; pleopods subequal . . . . *Orchestia* (p. 158).

Genus *Orchestia* Leach 1813   (ŏr kĕst′ĭ ă)

Coxal plates 2–4 moderately deep, posterior process developed. Coxa 5 not as deep as 4. Body generally smooth, appendages not excessively spinose. Animals semiterrestrial. Antenna 2 (♂), peduncle variously enlarged. Antenna 1 very short. Maxilliped, palp 3-segmented, or 4th very minute and masked by 3rd, proximal segments broad, inner margins heavily spinose. Maxilla 1, palp present.

Gnathopod 1 (♂) small, subchelate, dactyl not exceeding palm; in ♀, subchelate dactyl usually exceeding palm. Gnathopod 2 (♂) powerfully subchelate; minutely chelate, mittenlike in ♀, basis expanded anteriorly; segments 5 and 6 not elongate, subequal; segment 3 short.

Peraeopods 3 and 4, dactyls short, similar. Peraeopods 5–7 successively longer. Pleopods about normal or slightly reduced, subequal.

Uropods 1 and 2 normal, not elongate or heavily spinose. Uropod 3, ramus cylindrical, shorter than peduncle. Telson not shorter than wide, apex notched. Gills on all segments short, saclike, slightly longer on 2 and 6. Brood plates large, marginal setae simple.

KEY TO NEW ENGLAND SPECIES OF *ORCHESTIA*

1. Uropod 1 extending beyond uropod 2; peraeopod 4, segment 5 long, about equal to segment 6; peraeopod 7 (♂), segments 4 and 5 slender, elongate . . . . . 2. Uropods 1 and 2 extending about equally posteriorly; peraeopod 4, segment 5 short, distinctly smaller than segment 6; peraeopod 7 (♂), segments 4 and 5 swollen or expanded . . . . 3.
2. Uropod 1 with interramal spine; gnathopod 1 (♀ and im.), dactyl not exceeding palm, gnathopod 2 (♂), dactyl extending back to segment 5 . . *O. uhleri* (p. 161). Uropod 1 lacking interramal spine; gnathopod 1 (♀ and im.), dactyl distinctly exceeding palm; gnathopod 2 (♂), dactyl closing on palmar angle . . . . *O. grillus* (p. 159).
3. Uropod 1, outer ramus with marginal spines; antenna 2 (♂), peduncular segments 4 and 5 not enlarged . . . . *O. gammarella* (p. 159). Uropod 1, outer ramus marginally smooth; antenna 1 (♂), peduncular segments enlarged, incrassate . . . . *O. platensis* (p. 160).

Orchestia gammarella (Pallas) 1766 Plate XLV.1   (găm âr ĕl′ă)

L. 16 mm ♂, 14 mm ♀. Body large, heavy. Coxal plates 2–4 relatively small, subequal. Coxa 5, anterior lobe deeper, rounded below; coxa 6 hind lobe much deeper than anterior lobe. Eye medium, squarish. Antenna 1 not exceeding antenna 2, peduncle 4; flagellum 4–5 segmented. Antenna 2 (♂), peduncle segments 4 and 5 stout, not incrassate; flagellum 16-segmented. Maxilliped, palp segment 3 large; 4th minute, masked by spines of 3rd.

Gnathopod 1 (♂), propod and carpus subequal in length; merus not tumescent posteriorly. In ♀, palm short, oblique, slightly exceeded by dactyl. Gnathopod 2 (♂), propod very large, broadest distally, palm evenly convex, slightly shorter than posterior margin; segment 4 as deep as long. In ♀, basis broadly expanded anteriorly, margin evenly convex; propod shorter than carpus; merus not tumescent posteriorly.

Peraeopods 3 and 4, segments relatively short, slightly inflated, segment 5 shorter than 6. Peraeopod 7 (♂), segments 4 and 5 broadly expanded. Peraeopod dactyls short.

Abdominal side plates 2 and 3, hind corners acuminate, not produced. Pleopods slender, weak, peduncles shorter than uropod 1 peduncle. Uropod 1 not exceeding uropod 2, both rami of uropods with marginal spines; interramal spine lacking. Uropod 2, rami marginally spinose. Uropod 3, ramus short, apex acute. Telson margins and apex spinose.

Distribution: Amphi-Atlantic; northwestern Europe, Norway to the Mediterranean; Atlantic coast, from eastern and southern Newfoundland to Cape Breton Island; from western end of Nova Scotia and throughout Bay of Fundy along coast of Maine to Casco Bay.
Ecology: At the drift line of rocky and gravelly shores, on exposed and protected beaches; occasionally in estuarine or salt-marsh locations.
Life Cycle: Annual; ovigerous females May–August; more than one brood per female.

Orchestia (grillus) Bosc 1802 (= *O. palustris* Smith)
Plate XLV.2   (grĭl′ŭs)

L. 18 mm ♂, 16 mm ♀. Body large, slender, with longitudinal mid-dorsal striping. Coxal plates 2–4 broader than deep; coxa 5 very large and deep, hind lobe equally deep. Coxa 6 very deep. Head, eye large, rounded, black. Antenna 1, flagellum 4–5 segmented, tip exceeding peduncle 4 of antenna 2. Antenna 2 slender, moderately long, peduncle not expanded (both sexes), flagellum of 20 or more segments. Maxilliped, outer plate short, palp segment 3 large, 4th segment minute, masked by 3rd.

Gnathopod 1 (♂), propod shorter than carpus; in ♀, palm very short, very oblique, greatly exceeded by dactyl. Gnathopod 2 (♂), propod large, ovate;

palm oblique, evenly convex, shorter than posterior margin; merus longer than deep; in ♀, basis anteriorly expanded, margin evenly convex; propod about equal to carpus; merus slightly tumescent posteriorly. Peraeopods 3 and 4, segments 4–6 relatively long and slender. Peraeopods 5–7 long, basis oval, segments 4–6, slender, not broadened; dactyls moderately long.

Abdominal side plates 2 and 3, hind corners square; posterior margin weakly spinose. Pleopods normal, long. Uropod 1 extending well beyond uropod 2; rami with spines on inner and outer margins; interramal spine lacking. Uropod 2, rami marginally spinose, longer than peduncle. Uropod 3, ramus large, apex blunt, strongly spinose. Telson apically spinose.

Distribution: American Atlantic coast, from Chaleur Bay and western Newfoundland south to Florida and the Gulf of Mexico coast.
Ecology: Salt marshes, intertidally, from slightly below mean high water level to the supralittoral drift. Under debris, among roots of *Spartina* and marsh grasses where it builds nests and runways. Occasionally under gravel on protected shores.
Life Cycle: Annual; ovigerous females May–August; several broods per female.

Orchestia platensis Krøyer 1845
(= *O. agilis* Smith) Plate XLVI.2   (plăt ĕn′ sĭs)

L. 12 mm ♂, 11 mm ♀. Body small to medium, stout. Coxal plates 2–4 relatively shallow, broader than deep; coxa 5 very deep; coxa 6, hind lobe deep, anterior margin vertical. Head, eye relatively small, subrectangular, black. Antenna 1 short, flagellum 4–5 segmented, tip not exceeding peduncle 4 of antenna 2. Antenna 2 (♂), peduncular segments 4 and 5 very stout, incrassate, flagellum short, 12-segmented, proximal segments broader than long. Maxilliped, palp segment 3 stout; 4th minute, masked by 3rd.

Gnathopod 1 (♂), propod shorter than carpus; palm relatively deep, merus not tumescent posteriorly; in ♀, propod much shorter than carpus, slender, posterior margin with 6–7 spines; palm very short, vertical, greatly exceeded by dactyl. Gnathopod 2 (♂), propod large, deep, margins subparallel; palm convex, nearly vertical, with sharp notch above posterior angle (mature specimens); palm oblique, notch weakly or not developed in subadult ♂; in ♀, basis moderately expanded, margin convex proximally, straight distally; propod shorter than carpus; merus not tumescent posteriorly.

Peraeopods 3 and 4, segments 4–6 relatively short, 5 shorter than 6 (esp. in peraeopod 4), slightly expanded. Peraeopods 5–7, segments 4 and 5 relatively broad, 6 slender. Peraeopod 7 (♂), segments 4 and 5 distinctly broadened, thick.

Abdominal side plates 2 and 3, hind corner acuminate, slightly produced. Pleopods normal. Uropod 1 short, not exceeding uropod 2; outer ramus lacking marginal spines; inner margin of inner ramus bare; interramal spine lacking. Uropod 2, rami with 1–2 marginal spines. Telson, dorsal and apical spines short.

Distribution: Nearly cosmopolitan in Atlantic and Pacific oceans. American Atlantic coast: from southern Newfoundland and the Gaspé Peninsula to Florida, the West Indies; and in South America to Patagonia.

Ecology: Almost ubiquitous; on rocky, gravelly, sandy, or muddy shores, and in salt marshes, under drift at high water level. A primary beach colonizer, prior to arrival of more specialized talitrids; often abundant on sand beaches where *Talorchestia* is not present.

Life Cycle: Annual; ovigerous females from May–September; several broods.

Orchestia uhleri Shoemaker 1936 Plate XLVI.1   (ūl' ĕr ī)

L. 10 mm ♂, 8 mm ♀. Body small, slender, strikingly mottled and banded on back and sides. Coxal plates 2–4 large, broader than deep; coxa 5 very large, deeper than others; coxa 6, posterior lobe deep. Head, eye large, rounded, black. Antenna 1 slender, exceeding peduncle 4 of antenna 2; peduncle segment 1 very short; flagellum 4–5 segmented; antenna 2 slender, peduncular segments 4 and 5 subequal; flagellum slender, 12-segmented. Maxilliped, palp segment 3 relatively small, narrow, 4th segment not discernible; inner plate small.

Gnathopod 1 (♂), propod shorter than carpus, posterior margin straight; merus with low tumid process on posterior margin. In ♀, similar, but palm shorter, carpus and merus lacking posterior tumid lobe; gnathopod 2 (♂), propod very large, long-oval, posterior margin very short; palm very oblique, with double toothlike processes and separating depression near hinge, into which fits the corresponding knob on the long sinuous dactyl; merus short. In ♀, basis not much expanded anteriorly, margin convex proximally, straight distally; propod about equal to carpus; merus with low tumid process posteriorly.

Peraeopods 3 and 4, segments 4–6 relatively long and slender; dactyl small. Peraeopods 5–7, basis very expanded posteriorly; segments 4–6 slender, dactyls short.

Abdominal side plates, hind corners acuminate, not produced. Pleopods slender, weak. Uropod 1 slender, slightly exceeding uropod 2, both rami with 3–4 marginal spines; interramal spine present, simple tip. Uropod 2, rami marginally spinose. Uropod 3, ramus slender, apex acute. Telson longer than broad, narrowing toward notched apex, spines slender.

Distribution: American Atlantic coast, from central Maine (Casco Bay?) south to Florida and the Gulf of Mexico.

Ecology: Salt marshes and estuaries, at and just below mean high water level; among *Spartina* roots and under debris. During high flood tides, animals crawl up on grass stems.

Life Cycle: Annual; ovigerous females June–August; more than one brood.

Genus *Talorchestia* Dana 1853 (tăl ör kĕst' ĭ ă)

Body heavy, broadened, appendages spinose, fossorial. Coxal plates large, deep, posterior marginal processes weakly developed. Eyes, very large. Antenna 2 (♂) elongate, peduncle powerful. Maxilliped palp 3-segmented, segments very broad and spinose.

Gnathopod 1 (♂) weakly subchelate, dactyl exceeding palm; in ♀, simple, dactyl powerful. Gnathopod 2 (♂) powerfully subchelate; in ♀, basis broadly expanded in front.

Peraeopods 3 and 4 fossorial, dactyls elongate (larger than one-half propod), dactyl of 2 with posterior notch. Peraeopods 5–7 fossorial; peraeopod 6 slightly longest; peraeopod 5 much the shortest. Pleopods slender, peduncular outer margin spinose. Uropods 1 and 2 powerfully developed, elongate, margins of rami and peduncle heavily spinose. Uropod 3, ramus laterally compressed, longer than peduncle, apex rounded. Telson short, broad, spines apical. Branchial lobes relatively short and saclike on peraeon segments 2–6.

KEY TO AMERICAN ATLANTIC SPECIES OF *TALORCHESTIA*

1. Coxal plate 1 acutely produced, reaching antennal sinus; peraeopod 5, segment 6 not longer than 5; anterior margin with 2–4 spine groups; uropod 3, ramus broad, distally with subapical group of long spines . *T. megalophthalma* (p. 162).
   Coxal plate 1 blunt, not reaching sinus; peraeopod 5, segment 6 distinctly longer than 5, both with 4–6 anterior groups of spines; uropod 3, ramus narrowing, apex with small spines only . . . . *T. longicornis* (p. 163).

Talorchestia megalophthalma (Bate) 1862
Plate XLVII.1 (mĕg ăl öf thălm' ă)

L. 24 mm ♂, 22 mm ♀. Body stout, broadest anteriorly, some peraeon and pleon segments dorsally transversely pigmented. Eyes extremely large, black, nearly contiguous dorsally, bulging out from sides of head (in dorsal view). Coxal plates 2–4 very large, deeper than wide; coxa 2 lacking posterior process; coxa 5 much less deep than 4. Antenna 1 short, peduncular segments 1, 2, and 3, subequal, flagellum with 4–6 segments. Antenna 2 (♂) up to one-half body length; peduncle powerful, 5th segment much the longest; flagellum of 25+ segments, distal segments not distinctly toothed. Maxilliped, palp segments 1 and 2 about equally wide.

Gnathopod 1 (♂), propod shorter than carpus, margins parallel; palm very short, posterior angle not tumescent; carpus with narrow subapical posterior lobe. In ♀, propod slightly shorter than carpus; carpus lacking posterior lobe. Gnathopod 2 (♂), propod very large and deep, broadening distally; palm gently convex, nearly vertical, about equal to posterior margin. In ♀, posterior lobe of carpus deepest proximally; merus with slender posterior lobe, basis very broad, anterior margin evenly convex.

Peraeopods 3 and 4 relatively short, dactyls long, nearly equal to propods. Peraeopod 4, segment 5 much shorter than segment 6. Peraeopod 5 very short, segment 6 not longer than 5, anterior margins with 2–4 groups of simple stout spines. Peraeopod 6, basis nearly circular in outline. Peraeopod 7, basis broader than deep (long). Abdominal side plates, hind corners weakly acuminate; pleopods moderately strong.

Uropod 2, outer ramus distinctly the shorter. Uropod 3, ramus distally broad, apex rounded, upper margin with subapical group of slender spines. Telson very short and broad.

Distribution: American Atlantic coast; north shore of Gulf of St. Lawrence to Georgia.
Ecology: Burrows in flat, surf-exposed, high-salinity (above 17 o/oo) clean sand beaches, at and somewhat below mean high water level; less frequently hides under logs and beach debris. Burrow openings are circular in outline.
Life Cycle: Annual; ovigerous females May–August; several broods per female. Animals burrow deep into high berm and overwinter.

Talorchestia longicornis (Say) 1818 Plate XLVII.2   (lŏnj ĭ kŏrn′ ĭs)

L. 27 mm ♂, 24 mm ♀. Body subfusiform, broadest medially; segments uniformly pigmented. Eyes very large, not bulging out from head margin (in dorsal view). Coxal plates 2–4 large, squarish or wider than deep; coxa 5 nearly as deep as 4. Coxa 1, anterodistal angle blunt-acute. Antenna 1, peduncular segments 2 and 3 subequal, longer than 1; flagellum 7–8 segmented. Antenna 2 (♂) very long, nearly equal to body length; peduncular segment 5 elongate. Maxilliped, palp segment 2 widest.

Gnathopod 1 (♂), propod widening distally, palmar angle tumescent; carpus with posterior distal lobe. In ♀, propod shorter than carpus, narrowing distally. Gnathopod 2 (♂), propod very large, long-ovate; palm oblique, much shorter than posterior margin, with large convex median prominence. In ♀, basis expanded, margin convex proximally, slightly concave distally; carpus posterior lobe deepest in midmargin; merus lacking posterior lobe.

Peraeopods 3 and 4, segments 4–6 relatively long; dactyls about one-half propods. Peraeopod 4, segment 5 nearly equal to 6. Peraeopod 5, segment 6 longer than 5, anterior margins with 4–6 groups of simple spines. Peraeopod 6, basis oval. Peraeopod 7, basis deeper than wide, posterior margin evenly fringed with medium spines.

Abdominal side plate 2 deeper than 3, hind corners weakly acuminate. Pleopods weak, rami shorter than peduncles.

Uropod 2, rami subequal, all margins spinose. Uropod 3, ramus long, tapering, apex sharply rounded, armed with short spines. Telson short, broad.

Distribution: American Atlantic coast, from Chaleur Bay, western Newfoundland, and Burin peninsula south to northern Florida.

Ecology: On fine sandy beaches, steep or flat, in more protected locations and in estuaries to 3 o/oo, at and above mean high water level, sometimes ranging more than 100 yards inland and to elevations of several feet in the dunes; frequently under logs and debris.

Life Cycle: Annual or biennial; ovigerous females from May–August; several broods per summer; juvenile reaching maturity the following summer.

Mostly epifaunal amphipods that construct tubes or nests of detritus or other inert materials stuck together by means of the animal's peraeopodal cement glands, or that cling to the substratum. Characterized by shallow, depressed, or cylindrical bodies with small coxal plates tending to separation at the base, 4th coxa not excavate (incised) behind; by very short rostrum, small eyes, but well developed, often densely setose antennae; by basic mouthparts that are adapted for feeding on algae or detritus; by reduced urosome and entire telson. Sexual dimorphism pronounced, especially in gnathopods and/or antenna 2. The group is further divisible into the Aoridae-Ampithoidae family complex, in which antennae and mouthparts are relatively weakly setose and unmodified. The remaining families have strongly setose antennae (esp. peduncles) and mouthparts, and a tendency to reduction in number of gills, and fusion or loss of abdominal segments and abdominal appendages.

Family AORIDAE (emended)   (ā ôr'ī dē)

Body usually smooth, slender, tending to depressed. Coxal plates medium to small, tending to separation at base; coxa 1 largest, produced anteriorly (esp. in ♂); 4th not excavate behind; coxa 5, anterior lobe deeper. Rostrum very short; anterior head lobe blunt, eye at its base, lateral. Antenna 1 slender, longer than 2, lightly setose; accessory flagellum almost always present. Antenna 2, peduncle tending to enlargement. Sexual dimorphism shown strongly in gnathopod 1 and antenna 2.

Mandible, molar strong; palp slender, segment 3 tapering distally. Maxilla 1, inner plate with 1–2 setae; outer plate with 7 spine-teeth; palp large. Maxilliped, plates strong, especially inner; palp medium. Upper lip rounded, broad. Lower lip with distinct inner lobes, mandibular processes attenuated.

Body often with midventral sternal processes, especially in ♂. Gnathopods 1 and 2 subchelate. Gnathopod 1 larger; in ♂, very much larger and often complexly subchelate. Gnathopod 2, segments 5 and 6 slender.

Peraeopods 3 and 4, basis usually not expanded, segment 5 normal; dactyls with gland ducts. Peraeopods 5–7 slender, 7th conspicuously longest; dactyls normal. Pleopods, peduncles normal; occasionally basal segment of outer ramus expanded.

Uropods 1 and 2 slender, peduncles produced distally as spine; rami subequal, usually biramous. Uropod 3 slender, peduncles short, inner ramus tending to reduction or fusion with peduncle; terminal spines of rami simple. Telson short, deep, entire, often with widely paired apical cusps. Coxal gills saclike on peraeon segments 2–6. Brood plates large, laminar, marginal setae long, smallest on 6.

Component Genera: All genera listed by Barnard (1969) in Aoridae plus: *Unciola, Pseudunciola, Unciolella, Parunciola, Comacho, Grandidierella, Chevreuxius,* and *Neohela.*

KEY TO AMERICAN ATLANTIC GENERA OF AORIDAE

1. Uropod 3 uniramous; abdominal side plates separated distally . . . . 2.
   Uropod 3 biramous; abdominal side plates overlapping distally . . . . . . 5.
2. Uropod 3, ramus (and peduncle) slender; gnathopod 1 (♂) may be complexly subchelate . . . . 3.
   Uropod 3, ramus short, broad; gnathopod 1 stoutly normally subchelate (both
   sexes) . . . . 4.
3. Accessory flagellum prominent, multisegmented; peraeopods 6 and 7, dactyls
   elongate, falcate; gnathopod 1 (♂) normally subchelate . . . . *Neohela.*
   Accessory flagellum minute, 1-segmented; peraeopods 6 and 7, dactyls normal;
   gnathopod 1 (♂) complexly subchelate. . . . . *Grandidierella.*
4. Peraeopods 5–7 similar, dactyls not reversed; uropod 2 biramous . . . . . .
   . . . . *Unciola* (p. 173).
   Peraeopods 5 and 6 similar, dactyls reversed (Fig. 11D); uropod 2 uniramous . . .
   . . . . *Pseudunciola* (p. 177).
5. Head, anterior lobe acute; antenna 2, peduncle weak, little exceeding peduncle
   of antenna 1; gnathopods 1 and 2 very slender, similar (♀, imm.) . . . .
   . . . . *Rudilemboides* (p. 170).
   Head, anterior lobe rounded or truncate; antenna 2, peduncle stout, usually exceeding that of antenna 1; gnathopods dissimilar in ♀ and im. . . . . 6.
6. Coxal plates 1–4, moderately deep, setose below; urosome with clusters of dorsal setae and/or spines; peraeopods 5–7, bases broadly expanded . . . . . .
   . . . . *Leptocheirus* (p. 166).
   Coxal plates 1–4, medium to shallow, nearly bare; urosome dorsally smooth or
   with simple seta only; peraeopods 5–7, bases little expanded . . . . 7.
7. Gnathopod 1 complexly subchelate in ♂; gnathopod 2, segment 6 slender, linear
   (both sexes) . . . . *Microdeutopus* (p. 171).
   Gnathopod 1 normally subchelate in ♂; gnathopod 2, segment 6 moderately
   strongly subchelate, segment 6 subrectangular (both sexes) . . *Lembos* (p. 168).

Genus *Leptocheirus* Zaddach 1844   (lĕpt ō kīr′ ŭs)

Body little compressed. Pleon segment 3 longest. Coxal plates 1–4, deep, lower margin richly setose; coxa 5, anterior lobe deeper than broad. Head, rostrum lacking, anterior head lobe rounded. Antennae 1 and 2 subequal; antenna 1, peduncular segment 3 shorter than 1; peduncle and flagellum weakly setose behind, accessory flagellum well developed.

Mandible, many blades in spine row; length of palp segment 3 equals 2, apex acute. Maxilla 1, inner plate elongate. Maxilla 2, outer plate broadened. Maxilliped, palp segment 2 strongly setose on inner margin.

Gnathopod 1 subchelate (stronger in ♂), segment 5 longer than 6. Gnathopod 2 simple, segment 5 elongate, slender; segments 2, 5, and 6, anterior margin with numerous long plumose setae. Peraeopods 5–7, basis strongly expanded, setose posteriorly.

Uropods 1 and 2 subequally biramous, peduncle produced distally as stout spine between spinose rami. Uropod 3, rami subequal, stout, spinose marginally. Telson short, broad.

KEY TO NORTH AMERICAN SPECIES OF *LEPTOCHEIRUS*

1. Coxa 1 narrow, directed anteriorly; coxa 5, anterior lobe narrowing distally; uropod 3, posterior margin of rami with 3–4 spine groups . . . . *L. pinguis* (p. 167).
   Coxa 1 broad, vertical; coxa 5, anterior lobe margins subparallel; uropod 3, rami with a few posterior spines and long apical spines . . . *L. plumulosus* (p. 168).

Leptocheirus pinguis (Stimpson) 1853 Plate XLVIII.2   (pǐng′ wǐs)

L. to 17 mm. Coxal plates 2–4, lower corners rounded; coxa 1 narrow and directed anteriorly; coxa 5, anterior lobe smaller than coxa 4, narrowing below. Urosome segments 1 and 2 with paired dorsolateral cusps and with paired groups of setae and short spines, respectively. Head, anterior lobe truncate; eye subreniform, black. Antenna 1 longer than antenna 2; accessory flagellum 5–7 segmented. Antenna 2, peduncle relatively short, segment 3 with only 3 posterior groups of setae.

Mandibular palp, segment 3 subfalcate, distal inner margin with 15–20 tall comb-setae. Maxilliped, palp segments 3 and 4 short.

Gnathopod 1 (♂), segment 5, length about twice segment 6; palm straight, slightly oblique, posterior margin with proximal tubercle; dactyl finely serrate posteriorly; in ♀, segment 6, palm is vertical and posterior margin lacks tubercle. Gnathopod 2, segments 5 and 6 slender; segment 5 about twice length of 6; dactyl short.

Peraeopods 5–7, bases moderately expanded, nearly bare anteriorly, finely long-setose posteriorly; segment 6 (in all) slightly longer than 5; dactyls slender, normal; peraeopod 7 distinctly the longest.

Abdominal side plate 3 very broad, shallow, lower margin shallow-concave. Uropod 1, rami longer than peduncle, with numerous short marginal spines. Uropod 2, rami longer than peduncle, strongly spinose. Uropod 3, rami with 3–4 transverse rows of posterior facial spines and short slender apical spines. Telson short, very broad, apex broadly rounded, posterodorsal corners very acute, cuspate.

Distribution: American Atlantic coast, from Labrador south to Virginia. Common along Gulf of Maine shores and in Cape Cod Bay.
Ecology: Occurs from low intertidal, subtidally to more than 250 m, on sand and sandy mud or even muddy bottoms, especially in channels of estuaries.
Life Cycle: Probably two years; ovigerous females in April–June.

Leptocheirus plumulosus Shoemaker 1932 Plate XLVIII.1   (plōōm ūl ōs'ŭs)

L. 10–13 mm. Coxal plates 2–4 subrectangular; coxa 1 rounded below, not directed anteriorly; coxa 5, anterior lobe large, as deep as coxa 4. Urosome segments 1 and 2 with paired dorsolateral groups of longish setae, but lacking cusps. Head, anterior lobe rounded; eye oval, black. Antenna 1 slightly shorter than 2; accessory flagellum 3–4 segmented. Antenna 2, peduncle stout; segment 3 with 4–5 posterior setal groups.

Mandibular palp, segment 3 with 10–12 inner distal comb setae. Maxilliped, palp and outer plate relatively short; palp segments 3 and 4 combined about equal to segment 2.

Gnathopod 1 (♂), segments 5 and 6 subequal, stout; palm vertical, excavate near hinge; dactyl with fine posterior serrations, closing at posterior angle, posterior margin straight; in ♀, segment 5 longer than 6, palm not excavate, slightly convex. Gnathopod 2, segment 5 relatively stout, plumosesetose anteriorly; dactyl nearly as long as segment 6.

Peraeopods 5–7, bases progressively larger and broadly laminar, anterior margins with plumose setae; segment 6 slender, distinctly longer than segment 5.

Abdominal side plate 3, lower margin nearly straight. Uropod 1 extending beyond 2, rami and peduncle subequal, spines slender. Uropod 2, distal process of peduncle less than one-half length of outer ramus. Uropod 3 short, rami with a few stout lateral spines and longish slender apical setae. Telson short, rounded, with dorsolateral setae.

Distribution: From Cape Cod peninsula south through Chesapeake Bay and the Carolinas to Georgia and northern Florida. Recorded in New England from Bass River and Pocasset estuaries of Cape Cod; also taken on Nantucket Island (Quaise stream) and on Martha's Vineyard (Lake Tashmoo).
Ecology: Intertidal and shallow water to about 5 m in brackish estuaries, on mud, detritus, and sandy mud bottoms in areas with a current. Animal constructs tubes of sand grains and debris.
Life Cycle: Annual; ovigerous females May–September; two broods per female.

Genus *Lembos* Bate 1856   (lĕm' bŏs)

Coxal plates 1–5 shallow, tending to separation. Antenna 1, peduncle strong; accessory flagellum prominent. Antenna 2, peduncle strong. Mouthparts similar to *Microdeutopus.*

Gnathopod 1 (♀), segment 6 strongly subchelate, longer than 5. In ♂, gnathopod 1 normally subchelate, powerful, larger than in ♀; segment 6 longer than 5. Gnathopod 2 smaller, segment 5 not prolonged distally (both sexes); segment 6 subrectangular, longer than segment 5 in ♀, shorter in ♂. Peraeopods 5–7, basis not much broadened.

Pleon side plates tending to separation distally. Uropod 1, peduncle prolonged as spine between subequal rami. Uropod 2, outer ramus shorter than inner. Uropod 3, peduncle short, rami subequal. Telson with pair of terminal dorsal hooks.

KEY TO NEW ENGLAND SPECIES OF *LEMBOS*

1. Coxal plates 1–4 separated at base; antenna 2, peduncle extending much beyond peduncle of antenna 1; gnathopod 1, segment 5 much shorter than 6 . . . .
   . . . . *L. smithi* (p. 170).
   Coxal plates 1–4 overlapping basally; antenna 2, peduncle not longer than peduncle of antenna 1; gnathopod 1, segment 5 slightly shorter than 6 . . . .
   . . . . *L. websteri* (p. 169).

Lembos websteri Bate 1856 Plate L.2   (wĕb' stër ĭ)

L. 5 mm ♀. Peraeon segments 2–5 with midventral (sternal) spines (♂). Coxal plates 1–4 overlapping, 1st anteriorly acute, 5th and 6th with deeper anterior lobe. Head, eye large, roundish, occupying most of anterior lobe. Antenna 1 much longer than 2, peduncle 1 and 2, subequal; accessory flagellum 3–4 segmented. Antenna 2, peduncle 5 slightly longest, flagellum of 5–6 distinct segments, shorter than peduncle 5.

Mandible, palp segment 3 strongly narrowing, subfalcate. Maxilliped, basal segments and palp segment 1 with outer marginal winglike expansion.

Gnathopod 1 (♂), segment 6 slightly broader than 5, anterior margin with dense setae, not swollen above posteriorly serrated dactyl; palm with strong tooth above posterior angle spine; in ♀, less strongly subchelate; segment 6 larger than 5, palm evenly convex, posterior angle exceeded by serrated dactyl. Gnathopod 2 (♂), segment 2 with prominent tooth at anterodistal angle; segment 5 much longer and slightly deeper than 6, anterior margins of both thickly setose; in ♀, segments 5 and 6 subequal in length, not narrow, sparsely setose anteriorly.

Peraeopods 5 and 6, proximal posterior margin of basis rounded. Peraeopod 5, segment 5 only slightly shorter than 4. Peraeopod 7, basis posteriorly setose. Peraeopod dactyls strong, length about one-third respective segment 6.

Pleon side plates overlapping, hind corners rounded or obtuse. Uropod 3, both rami distinctly longer than peduncle, terminal and marginal spines relatively short. Telson short, broad.

Distribution: Amphi-Atlantic; south side of Cape Cod to Georgia and northern Florida. Common in Vineyard Sound (Hadley Harbor).
Ecology: Subtidal to over 100 ft.; among algae, in moderate current.
Life Cycle: Annual; ovigerous females from May–September; several broods.

Lembos smithi Holmes 1905 Plate LI.1   (smĭth' ī)

L. 5–6 mm. Coxae 1–4 very shallow, not contiguous, posterior margins of 2–4 setose. Eye relatively small, black, oval. Antenna 1, peduncle slender; accessory flagellum 3–4 segmented; antenna 2, peduncle elongate, strong (♂); segments 4 and 5 with posterior marginal clusters of long setae; flagellar proximal segments fused.

Mandible, palp segment 3 not falcate. Maxilliped, dactyl very strong; basal segments lacking winglike expansion. Outer plate with long setae on outer margin.

Gnathopod 1 (♂), basis short, stout; segment 5 short, deep; segment 6, large, deeper, anterodistal margin swollen above base of dactyl, margin long setose; in ♀, similar, slightly smaller. Gnathopod 2, segment 5 slightly longer than 6, anterior margins armed with simple setae, 4–8 clusters of facial setae; similar in ♀.

Peraeopods 5 and 6, proximal posterior angle of basis sharply rounded. Peraeopod 5, segment 5 short. Peraeopod dactyls relatively short, slender, less than about one-fourth length of segment 6. Pleon segments 1–3 separated distally, hind corners obtuse. Uropod 3, terminal spines of rami long, slender.

Distribution: South side of Cape Cod to northern Florida. Recorded in Vineyard Sound (Hadley Harbor); Buzzards Bay.
Ecology: Low intertidal to moderate depths (20 m); on pilings, wharves, eelgrass, in moderately strong tidal currents. At base of sponges and *Amourecium*.
Life Cycle: Annual; ovigerous females from May–September; several broods per female.

Genus *Rudilemboides* Barnard 1959   (rōō dĭ lĕm boid' ēz)

Body slender, small. Coxal plates shallow, tending to separation; coxae 5 and 6, anterior lobe deep. Short midventral spines on peraeon segments 3, 5, and 6. Head short, anterior lobe prominent, acute, bearing eye basally. Antennae slender, not elongate. Antenna 1, peduncle 1 with median posterior marginal spine group; accessory flagellum short.

Mandible, palp short, segment 3 setose apically. Maxilla 2, inner plate, inner margin setose throughout. Maxilliped, outer plate narrowing distally.

Gnathopods 1 and 2 (♀) slender, subequal; segment 5 longer than 6; palm short; segment 4 short; in ♂, gnathopod 1 complexly subchelate, segment 5, posterodistal angle not strongly produced; gnathopod 2, segment 5 stout, segment 6 palm short.

Peraeopods 3 and 4, basis slightly swollen, segment 4 broadest. Peraeopod 5 much shorter than 6 and 7, dactyl reversed.

Pleon side plates overlapping. Uropod 2, outer ramus shorter. Uropod 3, rami subequal, peduncle shorter. Telson short, paired terminal dorsal hooks.

KEY TO NORTH AMERICAN SPECIES OF *RUDILEMBOIDES*

1. Gnathopod 1 (♂), segment 5 with palm and teeth . . . . . . *R. naglei* (p. 171).
   Gnathopod 1 (♂), segment 5 lacking palm and teeth . *R. stenopropodus* Barnard.

Rudilemboides naglei n. sp. Plate L.1   (nā' gĕl ī)

L. 3.0 mm ♀. Antenna 1, accessory flagellum 2-segmented, equal to about one-half peduncular segment 3. Antenna 2, flagellum 3–4 segmented, terminal segment with 2 clawlike spines.

Gnathopod 1 (♂) complexly subchelate; segment 2 short, expanded, anterior margin straight; segment 5 with short palm, and 3 distal marginal teeth; segment 6 nearly straight. Gnathopod 2 (♂) weakly subchelate, palm short, oblique, exceeded by dactyl; segment 5 inflated; segment 2, anterodistal angle produced into short lobe.

Peraeopods 3 and 4, segment 4 expanded anterodistally, slightly overhanging short segment 5. Peraeopod 5, segment 2 slighted expanded, posterior margin slightly concave, segment 6 and dactyl directed posteriorly.

Pleon side plate 3, hind corner subquadrate, not produced or toothed. Uropod 3, outer ramus longer, with paired marginal spines; inner ramus with 1–2 marginal spines.

Distribution: South side of Cape Cod and southern New England to Georgia, and eastern Gulf of Mexico. Recorded in Hadley Harbor, Vineyard Sound, Buzzards Bay.
Ecology: Low water level to about 10 m; in eelgrass.
Life Cycle: Annual; ovigerous females May–August; several broods per female.

Genus *Microdeutopus* Costa 1853 (including *Coremapus*)
                                 (mī krö dūt' ö pŭs)

Head with rounded lateral lobes; eye at base, rounded. Coxal plates 1–4 relatively shallow, 1st produced anteriorly (in ♂); coxa 5, anterior lobe deeper than posterior. Antennae slender; antenna 1 longer than antenna 2; accessory flagellum distinct, short. Mandibular palp, segment 3 longest, distally narrowing. Maxilla 1, inner plate small, with single apical seta. Maxilliped, palp slender, outer plate large.

Gnathopod 1 (♂) powerfully and complexly subchelate; carpus very large, posterior distal angle produced; segment 4 not excessively elongate; propod much smaller than carpus. Gnathopod 1 (♀) normally subchelate, larger than gnathopod 2. Gnathopod 2 (both sexes), segments 5 and 6 slender, subequal, small, posterior lobe of carpus shallow.

Peraeopods slender, 5 much shorter than 7. Uropod 1, peduncle prolonged as stout spine between rami. Uropod 3, rami short, subequal, linear, about equal to peduncle. Telson with paired dorsal apical cusps. Coxal gills large, broad.

KEY TO NEW ENGLAND SPECIES OF *MICRODEUTOPUS*

1. Antenna 1, accessory flagellum 2-segmented; uropod 3, rami with 3 spines across apex, and paired and solitary spines on margins; gnathopod 2 (♂), basis broadly expanded, front margin convex, crenulated; gnathopod 1 (), dactyl tip closing on palmar angle . . . . *M. gryllotalpa* (p. 172).
   Antenna 1, accessory flagellum 4–5 segmented; uropod 3, rami with 2 apical spines, and singly inserted marginal spines; gnathopod 2 (♂), basis slightly expanded, anterior margin straight, smooth; gnathopod 1 (♀), tip of dactyl exceeding palm . . . . *M. anomalus* (p. 173).

Microdeutopus gryllotalpa Costa 1853 Plate XLIX.1   (grĭl ö tălp′ ă)

L. 8 mm ♂, 10 mm ♀. Peraeon segments 2–7 with short median sternal process. Coxal plate 1 moderately produced (in ♂), not reaching antennal sinus. Coxa 5, anterior lobe slightly deeper than posterior. Antenna 1, flagellum up to 22 segments, accessory flagellum of 2 segments, 2nd segment minute. Antenna 2, peduncle 4 and 5 subequal, flagellum of 6–9 segments, shorter than peduncle 5.

Mandibular palp, segment 3 with comb setae along distal half of inner margin. Maxilliped, basal segments lacking outer marginal processes.

Gnathopod 1 (in ♂), segment 2 (basis) short, expanded distally; carpus immensely expanded, about as deep as long, posterior distal margin with 2–4 teeth, most distal longest; propod short, posterior margin irregularly lobed. Gnathopod 2 (in ♀), carpus and propod subequal in length, propod with parallel margins, dactyl about equal to angular palmar margin. Gnathopod 2 (♂), basis greatly expanded, anterior margin convex and crenulated; slightly expanded in ♀, anterior margins of segments 4, 5, and 6 (both sexes) with numerous long plumose setae; segment 4 not greatly prolonged distally behind segment 5; segments 5 and 6 subequal.

Peraeopods 3 and 4, segments 2, 4, and 5 moderately dilated. Uropod 3, rami about equal in length to peduncle, outer ramus slightly the longer. Rami apices with transverse group of three spines; margins with paired and solitary spines; terminal setae long.

Distribution: Coasts of northwestern Europe; Norway south to the Mediterranean and Black Sea; western Atlantic from Cape Cod and southern Mass., Conn., Long Island Sound to Chesapeake Bay.
Ecology: Intertidal, among *Chaetomorpha* and other algae; among *Zostera,* in salt marshes associated with *Ruppia,* in somewhat brackish (to 15 o/ᴏᴏ) water; also sublittoral down to 150 m. On oyster beds, tunicates, *Mytilus,* in

high detrital accumulations. Frequently around docks and man-made installations; associated with *Corophium acherusicum.*
Life Cycle: Annual; ovigerous females from May–September; several broods per female.

Microdeutopus anomalus (Rathke) 1843 Plate XLIX.2   (ă nŏm′ ăl ŭs)

L. 8 mm ♂, 10 mm ♀. Peraeon segments 2–4 with strong spine on midventral line (in ♂). Coxal plate 1 (♂) strongly produced anteriorly, tip acute, reaching antennal sinus; coxa 5, anterior lobe deep. Antenna 1, flagellum of about 23 segments; accessory flagellum with 4–5 segments, terminal segment minute. Antenna 2, flagellum of 6–8 segments, nearly equal to peduncle 5. Mandibular palp, segment 3 with comb setae along distal two-thirds of posterior margin. Maxilliped, basal segments with outer marginal winglike processes.
      Gnathopod 1 (in ♂), basis elongate, often with proximal anterior marginal process; carpus very large, long, with posterior distal angle produced into strong acute tooth, having one (or two) accessory teeth arising at its base; propod slender, posterior margin concave except near hinge of dactyl. Gnathopod 1 (in ♀), propod longer than carpus, distally broadening, dactyl tip closely well beyond posterior angle of oblique palm. Gnathopod 2, basis slightly expanded (in ♂), anterior margin about straight; propod and carpus (in both sexes) elongate, 5th slightly broader; propod more than twice as long as broad; anterior margins of segments 4, 5, (6), with simple setae only.
      Peraeopods 3 and 4 slender, segments 2, 4, and 5 not inflated. Uropod 3, peduncle slender; dorsal and inner flanges not pronounced; rami slender, slightly longer than peduncle; outer ramus shorter, with 2 outer marginal spines, inner ramus with outer and inner marginal spines; no spines paired; terminal setae short.

Distribution: Coasts of northwestern Europe, from northern Norway to the Canary Islands, the Mediterranean and Black seas. In the western Atlantic coastal region, from Cape Ann, Mass., and southern New England to the Middle Atlantic states; also Bermuda.
Ecology: An essentially sublittoral species, in depths to 200 m, but occasionally intertidal. Occurring in areas affected by detritus, often among arborescent weeds, also in *Zostera* beds, among shells, sponges, tunicates, and *Mytilus;* in relatively high salinities (over 25 o/oo).
Life Cycle: Annual; ovigerous females from May–September; several broods per female.

Genus *Unciola* Say 1818   (ŭns ĭ ōl′ ă)

Coxal plates very shallow, subequal, separated; coxa 1 acute anteriorly, coxa 5 with deeper anterior lobe. Peraeon segment 1 longer in ♂ than in ♀.

Head, rostrum distinct, acute, anterior lobe prominent, rounded, bearing eye. Antenna 1 longer than 2; accessory flagellum well developed. Antenna 2, peduncle strongly developed (esp. in δ); flagellum short.

Upper lip slightly notched apically. Mandible, palp strong, segments 2 and 3 long-setose. Maxilla 1, inner plate small, with 1 apical seta. Maxilliped, inner plate broadened, palp dactyl strong.

Gnathopod 1 normally and powerfully subchelate (both sexes), stronger in δ; segments 3, 4, and 5, short, deep; dactyl serrate or setose behind. Gnathopod 2 subchelate or parachelate, small; segments 5 and 6 slender, 6 narrowing distally.

Peraeopods 3 and 4 slender, segments 4 and 5 long, dactyls glandular. Peraeopods 5–7 slender, bases little expanded, dactyls directed forward.

Pleosome side plates small, separated. Pleopods powerful; peduncles short, not expanded. Uropods 1 and 2 biramous, rami short, spinose. Uropod 3 uniramous, ramus very short; peduncle with lobelike inner distal angle. Telson entire, rounded.

KEY TO NEW ENGLAND SPECIES OF *UNCIOLA*

1. Abdominal side plate 3, hind corner strongly produced beyond posterior margin; peraeopod 7 distinctly longer than 6 (segment 4 differing in length) . . . . . 2.
   Abdominal side plate 3, hind corner short, acute, not extending beyond posterior margin; peraeopods 6 and 7 subequal (segment 4 about equal in length) . . . 3.
2. Abdominal side plate 3, hind corner process upcurved or recurved, hooklike; gnathopod 2 basically subchelate, palm nearly vertical . . . *U. irrorata* (p. 174).
   Abdominal side plate 3, hind corner process nearly straight; gnathopod 2 basically parachelate . . . . *U. inermis* (p. 175).
3. Uropod 1, rami with 6–8 spines along outer margin; peraeopod dactyls with 4–6 inner marginal setae; antenna 2 (δ), posterior inner margin of segment 4 not strongly serrate. . . . . *U. dissimilis* (p. 176).
   Uropod 1, rami with 1–3 lateral marginal spines; peraeopod dactyls with 2–4 inner marginal setae; antenna 2 (δ), segment 4, inner posterior margin strongly serrate . . . . *U. serrata* (p. 176).

Unciola irrorata Say 1818 Plate LI.2   (ĭr ŏr āt' ă)

L. 13 mm δ, 10 mm ♀. Coxal plates 2 and 5 deepest, anterior lobe of coxa 5 produced downwards, sharply rounded. Head, rostrum long, acute; anterior lobe, corners quadrate. Eye small, oval, black. Antenna 1, peduncle 2 much longer than 1; accessory flagellum 5-segmented. Antenna 2 (δ), peduncular segments 3 and 4 strongly broadened, powerful, not serrate posteriorly; segment 4 narrowing distally to less than one-half proximal margin, peduncle 5 slender, broader than segmented flagellum; in ♀, peduncular segments 3 and 4, slightly expanded. Mandibular palp long, relatively slender, segments 2 and 3, subequal. Maxilliped, inner plate short.

Gnathopod 1 (♂), segment 6 very powerful, palm straight, with small prominence near hinge; posterior angle, prominence with stout apical spine; dactyl setose (not serrate) behind; in ♀, smaller, less powerful; palm slightly convex; dactyl serrate behind. Gnathopod 2, segments 5 and 6 subequal in length; segment 6, narrowing and bent downwards distally, palm nearly vertical or slightly parachelate, gently convex; dactyl finely serrate and short setose behind.

Peraeopods 3–7, dactyls 2–6 setose on inner margin; peraeopod 7 distinctly longer than 6; margins of segment 4 with plumose setae. Abdominal side plates 1, 2, and 3, hind corners acutely produced, prominent, that of 3 upturned, hooklike, or even recurved. Uropod 1, peduncle outer margin with 8–10 short spines and a few long setae. Uropod 3, ramus elongate, setose, exceeding acute distal lobe of peduncle. Telson subovate.

Distribution: Eastern Canada (Strait of Belle Isle), Gulf of St. Lawrence and Gulf of Maine, sporadically southward to off Chesapeake Bay. Common in Cape Cod Bay and estuaries northward.

Ecology: In coarse to medium sand; from lower intertidal to over 30 fathoms, living in tubes of other organisms.

Life Cycle: Annual. Ovigerous females from March–July; one brood per female.

Unciola inermis Shoemaker 1942 (no illustration)   (ĭn ĕrm′ ĭs)

L. 11–12 mm. Coxal plates shallow, anterior angles acute; coxa 5, anterior lobe produced acutely downward, less so in ♀. Head, rostrum strong, continuing as keel down front of head; anterior lobe angles subquadrate; eye oval, black. Antenna 1, segment 2 longest; segment 2, spinose and setose behind; accessory flagellum with 4–5 segments. Antenna 2 (♂), segments 3 and 4 very expanded; 4 narrowing distally to one-third width of proximal end; posterior margin not serrate, but with double row of small spines; in ♀, segments 3 and 4 little expanded; segment 3 with 2–3 small posterior spines. Maxilla 1, inner plate with 2–3 setae.

Gnathopod 1 (♂), palm nearly straight, without protuberance; in ♀, smaller, palm slightly convex; dactyl weakly or finely serrate behind. Gnathopod 2, segment 6 shorter than 5, narrowing to distinctly chelate (parachelate) palm; dactyl nearly straight, finely setose behind. Peraeopods 3–7, dactyls with 4–5 inner marginal setae.

Pleon side plates 1–3, posterior corner strongly and acutely produced, 3rd not upturned. Uropod 1, peduncular outer margin strongly spinose and setose; outer ramus, outer margin with 2–3 spines and several setae. Uropod 2, outer ramus, outer margin with several setae. Uropod 3, ramus short, scarcely exceeding distal lobe of peduncle bearing 3 apical setae. Telson circular.

Distribution: American Atlantic coast, from southern Newfoundland, Gulf of St. Lawrence, and Cape Cod region southward to North Carolina. Common in Bay of Fundy and southern Gulf of Maine region.
Ecology: On sandy mud and sand, from low water level to 30 fathoms and deeper.
Life Cycle: Annual; ovigerous females from April–June.

Unciola dissimilis Shoemaker 1942 Plate LII.1   (dĭ sĭm′ĭl ĭs)

L. 16 mm ♂, 12 mm ♀. Coxa 5, anterior lobe shallow, rounded below. Head, rostrum relatively long, acute; anterior lobe angular above. Eye brownish, oval. Antenna 1, peduncular segment 2 much longer than 1, posterior margin with 10–14 groups of setae; accessory flagellum 5-segmented. Antenna 2 (♂), peduncular segments 3 and 4 broad and deep, inner margin not serrate; in ♀, peduncle segments not expanded. Mandibular palp, segment 3 slightly shorter than segment 2, margins heavily setose. Maxilliped, palp segments 2 and 3 broad; outer plate very broad.

Gnathopod 1, palm nearly straight; posterior angle produced as prominent single-toothed knob; dactyl very finely serrate and long-setose on hind margin; in ♀, segment 6 smaller, palm with median concavity and rounded prominence toward hinge. Gnathopod 2, parachelate, segment 5 longer than 6; segment 6 distally narrowing, palm short, convex, anteriorly oblique; dactyl finely serrate behind.

Peraeopods 3–7, dactyls with 4–6 anterior marginal setae. Peraeopod 5 distinctly shorter (esp. segment 5) than subequal 6 and 7.

Abdominal side plates 2 and 3, hind corners acute, short, not upturned, not extending beyond posterior margin. Uropod 1, peduncle large, with strong distal process, margins with about 8 spines; outer ramus with 6–8 outer marginal spines. Uropod 3, ramus longer than wide, strongly setose apically and laterally, extending well beyond strongly setose distal lobe of peduncle. Telson broader than long.

Distribution: American Atlantic coast, from Cape Ann, Mass., to North Carolina. Recorded at Quick's Hole, Mass.
Ecology: On sandy mud and coarse sand, from low water level to over 500 fathoms.
Life Cycle: Unknown.

Unciola serrata Shoemaker 1942 Plate LII.2   (sĕr āt′ ă)

L. 8 mm ♂, 6 mm ♀. Coxal plate 5, anterior lobe rounded below. Head, rostrum prominent, acute; anterior lobe rounded; eye dark brown, oval. Antennae subequal. Antenna 1, peduncular segments 1 and 2 subequal; accessory flagellum 3–4 segmented. Antenna 2 (♂), segments 3 and 4 very broad and deep, 4th not greatly narrowing distally, lower (inner) margin with 5–7 heavy even serrations; in ♀, segments 3 and 4 not strongly expanded, hind margins regularly setose, not serrate. Mandibular palp stout; segment 3 much shorter than 2, modestly setose. Maxilliped, palp segments 2 and 3 and outer plate not unusually expanded.

Gnathopod 1 (♂), palm convex, with low prominence near hinge, posterior angle prominent, with two spines; dactyl serrate posteriorly; in ♀, similar, smaller. Gnathopod 2 (♂), segments 5 and 6 subequal, palm very short, somewhat parachelate; in ♀ and imm., palm is nearly vertical, convex, dactyl finely serrate behind.

Peraeopods 3–4, dactyls with 2–3 posterior marginal setae; peraeopods 5–7, dactyls with 3–4 posterior marginal setae. Abdominal side plates 2 and 3, hind corner forming a small tooth, not level with posterior margin. Uropod 1, peduncular margins with 3–6 spines; outer ramus with 2 outer marginal spines. Uropod 3, ramus short, with 2 lateral marginal setae, apex about level with that of peduncle inner distal lobe. Telson subovate.

Distribution: Vineyard Sound southward to New Jersey, Chesapeake Bay, and Georgia.
Ecology: Inhabits tubes in sandy mud from low water level to about 20 m.
Life Cycle: Annual; ovigerous females from May–August; several broods per female.

Pseudunciola new genus   (sūd ŭns ĭ ōl′ ă)

Coxal plates shallow, separated; coxa 1 produced anteriorly, 5 with deep anterior lobe. Head shallow; rostrum short; eyes lacking. Antennae short, peduncular segments stout; accessory flagellum small.

Mandibular palp slender, 3-segmented. Maxilliped, palp short. Lower lip, mandibular lobes short.

Gnathopod 2 powerfully and normally subchelate (both sexes); segment 2 stout. Gnathopod 2 small, weakly subchelate.

Peraeopods 3 and 4, basis dilated; segments 4 and 5 not reduced. Peraeopods 5 and 6, segment 4 expanded, segment 5 short; segment 6 and dactyl directed posteriorly. Peraeopod 7, dactyl directed anteriorly.

Pleon side plates separated distally. Pleopods short, strong, segments of rami expanded. Uropod 1 unequally biramous. Uropod 2 uniramous. Uropod 3 very short, peduncle mediodistally expanded; ramus shorter than peduncle. Telson broad, rounded.

Relationships: A remarkable monotypic genus, derived from a *Lembos smithi*-like ancestor, but evolving in parallel to true corophiids (*Siphonoecetes*) the reversed dactyls of peraeopods 5 and 6, and reduced uropods 1 and 2; pleopods gaining width by enlargement of basal ramal segments rather than inward expansion of peduncle.
Sole Species: *Unciola obliquua* Shoemaker 1949.

Pseudunciola obliquua (Shoemaker) 1949 Plate LIII.1   (ŏb lēk′ wă)

L. 6.0 mm ♀. Blind. Rostrum very short. Antenna 1, accessory flagellum 1-seg-
mented; peduncular segment 2 shorter than 1. Antenna 2, peduncle 3 larger
than peduncle 4. Mandibular palp, segment 3 shorter than 2.

Gnathopod 1, palm of segment 6 convex; dactyl posteriorly serrate.
Gnathopod 2, segment 6 short, smaller than segment 5. Peraeopods 3–7,
dactyls with posterior setae or stiff pectinations. Peraeopods 5 and 6, segment 4, margins with plumose setae.

Pleon side plates 1 and 2, hind corners acute, produced; pleon 3, hind
corner obtuse, margin oblique. Uropod 1, inner ramus bare, narrowing,
length about one-half outer ramus. Uropod 2, peduncle with outer marginal
setae; ramus with stout terminal spines. Uropod 3, peduncular mediodistal
lobe larger than ramus, having 4 long apical setae, slender spine. Telson
short, broader than long.

Distribution: Bay of Fundy (St. Mary Bay) and Gulf of Maine (Cape Cod
Bay), Vineyard Sound to Middle Atlantic states (off Sandy Hook, N.J.)
Ecology: Living in tubes in medium fine to coarse sand, from just below
low water level to more than 50 m in depth.
Life Cycle: Presumably annual; ovigerous females from April–August; 4–6
relatively large eggs per brood.

Family AMPITHOIDAE   (ăm pĭ thō′ĭ dē)

Body smooth, little compressed. Strongly sexually dimorphic. Coxal plates
deep, smooth or lightly setose below, 4th and 5th largest. Head, rostrum
lacking, anterior lobe short, inferior antennal sinus shallow, eyes lateral,
rounded. Antenna 1, longer than 2, peduncular segment 3 short; accessory
flagellum lacking or very short. Antenna 2, peduncle strong, sparsely setose
behind.

Upper lip rounded or very slightly notched; epistome not produced.
Lower lip, inner lobes distinct; outer lobes with distinct notch or emargination on "shoulders." Mandible, molar strong; palp slender, reduced, or lacking. Maxilla 1, inner plate small; outer plate with 9 spine-teeth. Maxilla 2,
outer plate somewhat broadened, inner plate strongly setose. Maxilliped,
outer plate large; palp strong, dactylate.

Gnathopods 1 and 2 strongly subchelate, 2nd usually the larger. Peraeopods 3 and 4, basis somewhat inflated, dactyls with glandular ducts. Peraeopod 5, segment 6 and dactyl reversed; basis more broadly expanded than
in 6 and 7.

Uropods 1 and 2 subequally biramous, peduncle produced distally as a
spine between rami. Uropod 3, rami shorter than peduncle, inner tapering,
outer with 2 hooked spines near apex.

Coxal gills saclike on peraeon segments 2–6. Brood plates laminar, margins lined with long hooked setae.

KEY TO NEW ENGLAND GENERA OF AMPITHOIDAE

1. Uropod 3, outer ramus distinctly shorter than inner; accessory flagellum present ·
   . . . . *Cymadusa* (p. 182).
   Uropod 3, outer ramus about equal to or longer than inner ramus; accessory flagellum lacking . . . . 2.
2. Mandible without palp; peraeopods 3 and 4, segment 2 very broadly expanded in front; peraeopod 5, segments 4, 5, and 6 short, stout . . . *Sunamphitoe* (p. 183).
   Mandible with palp; peraeopods 3 and 4, basis moderately broadened; peraeopod 5, distal segments slender, normal. . . . . *Ampithoe* (p. 179).

Genus *Ampithoe* Leach 1813–14   (ăm pĭ thō′ ē)

Coxal plates deep, smooth below or nearly so; coxa 5 deepest; coxa 1, expanded anterodistally. Head, inferior antennal sinus shallow. Antenna 1, accessory flagellum lacking. Upper lip rounded. Lower lip, notch on inner margin of shoulder of outer lobes. Mandible, palp slender, 3-segmented; molar strong.

Gnathopods 1 and 2 normally subchelate, 2nd larger than 1st, esp. in ♂; basis with anterodistal lobe. Gnathopod 1, segments 5 and 6 subequal; gnathopod 2, segment 5, posterior lobe short. Peraeopods 3 and 4, basis slightly expanded anteriorly. Peraeopods 5–7 slender, segment 6 not distally expanded.

Uropods 1 and 2, rami subequal; peduncle of 1 prolonged strongly as spine between posteriorly spinose rami. Uropod 3, rami subequal. Telson with pair of apical dorsal hooks.

KEY TO NEW ENGLAND SPECIES OF *AMPITHOE*

1. Coxal plate 2 larger and broader than 1 or 3; peraeopod 7, posterodistal angle of basis rounded; body segments often with middorsal light spots. . . . .
   . . . . *A. rubricata* (p. 180).
   Coxal plate 2 not broader than 3; peraeopod 7, posterodistal angle of basis sharply or obtusely cornered; dorsal body coloration uniform . . . . 2.
2. Gnathopod 1, segment 6 deep, palm slightly exceeded by dactyl; coxal plates 1–4, posterodistal margin setose; peraeopod 5, basis wider than long . . . . .
   . . . . *A. valida* (p. 180).
   Gnathopod 1, palm short, distinctly exceeded by dactyl (esp. in ♂); coxal plates 1–4, lower margins bare; peraeopod 5, basis not as wide as deep . . . .
   . . . . *A. longimana* (p. 181).

Ampithoe rubricata Montagu Plate LIV.1   (rōō brĭ kāt′ ă)

L. 14–20 mm ♂. Coxal plates moderately large, smooth below; plate 2, broader than 1 or 3 (♂), plate 1 narrow; anterior lobe of coxa 5 as large as coxa 4. Head, inferior antennal sinus shallow-convex; eye smallish, sub-ovate, dark red. Antennae 1 and 2 subequal, about one-half length. Antenna 2, peduncle stout, especially in ♂, flagellum 8–10 segmented. Mandibular palp strong, well developed; segment 3 about equal to 2, distally expanded. Maxilliped palp stout; outer plate broad.

Gnathopod 1 (♂), segments 5 and 6 subequal in length and moderately deep; palm oblique, slightly convex; in ♀, similar but smaller. Gnathopod 2 (♂) larger than 1; segment 6 larger than segment 5, palm irregularly concave, oblique, with stout tooth at posterior angle; segment 5, posterior lobe narrow; in ♀, segment 6 is smaller and palm less oblique. Peraeopods 3 and 4, segment 5 shorter than segment 4; bases slightly inflated. Peraeopod 5, basis about as long as wide. Peraeopods 6 and 7, bases lacking posterodistal lobe, angles rounded, margin bare.

Abdominal side plate 3, hind corner subquadrate, sharply rounded. Uropod 1, outer ramus, outer margin with 8–12 short spines. Uropod 2, outer ramus with 4–5 short spines. Uropod 3, peduncle short, with distal row of 5 stout spines; outer ramus, outer margin setose. Telson basally broadest, apex broadly convex, between stout dorsolateral cusps.

Distribution: Amphi-Atlantic boreal; in North America, from Labrador south to Cape Cod Bay and Long Island Sound.
Ecology: Intertidal, mainly along rocky coasts and outer parts of estuaries; constructs nest among fucoids and in mussel beds.
Life Cycle: Annual; ovigerous females from April–July; one brood per female.

Ampithoe valida Smith 1873 Plate LV.1   (văl′ ĭ dă)

L. 10–12 mm. Coxal plates 1–4 deep, with short row of slender setae at posterodistal margins; coxa 5 longest. Head, inferior sinus small; eye small, round, black. Antenna 1 longer than antenna 2; peduncle short, about equal to head and first peraeon. Antenna 2 (♀), peduncle short, segments 4 and 5 subequal, little inflated. Mandible, palp short, scarcely exceeding incisor, segment 3 with apical setae only. Maxilla 1, inner plate bare. Maxilliped, palp segment 2 very short, scarcely longer than 1. Inner plate short, not extending beyond palp segment 1.

Gnathopod 1 (♂), segment 5 longer and deeper than segment 6, posterior lobe of 5 broad, setose below; in ♀, segments 5 and 6 subequal in length and depth. Gnathopod 2 (♀) slightly larger than 1; segment 5 shorter than 6, hind lobe narrow; in ♂, segment 6 very large, stout, subrectangular, palm vertical, with rounded median tooth and prominent posterior angle, hind margin long.

nearly bare; dactyl very heavy. Peraeopods 3 and 4, basis slightly inflated; segment 5 not shorter than 4. Peraeopod 5, basis broader than deep. Peraeopods 6 and 7, posterodistal lobes of basis sharply quadrate.

Abdominal side plate 3, hind corner obtuse, hind margin convex. Uropod 1, peduncle strong, with 8–10 longish outer marginal spines; outer ramus with about 10 marginal spines. Uropod 2, outer ramus shorter than that of uropod 1, outer margin with 3–4 spines. Uropod 3, peduncle short, not twice length of rami; with two groups of distal spines; outer ramus with one group of outer marginal setae. Telson short, narrowing; apical cusps low.

Distribution: American Pacific coast; American Atlantic coast, from Middle Atlantic states (Chesapeake Bay) to Cape Cod, Cape Ann, and New Hampshire (Piscataqua estuary).
Ecology: Found mainly in estuaries and brackish-water habitats, usually nestling among Ulvacea, from lower intertidal to depths of a few meters.
Life Cycle: Annual; ovigerous females from May–September; several broods per female.

Ampithoe longimana Smith 1873 Plate LIV.2   (lŏnj ĭ măn' ă)

L. 10 mm ♂ and ♀. Coxal plates relatively shallow, little deeper than wide, smooth below; coxa 1, broadest. Head, inferior antennal sinus broad, shallow. Eye medium large, rounded, black. Antennae 1 and 2 elongate (esp. in ♂), nearly equal to body; peduncles long and slender (less in ♀); flagellum of antenna 2 with bristly whorls of setae. Mandibular palp slender; segment 3, setose on inner distal margin and at apex. Maxilla 1, inner plate with 2 short setae. Maxilliped, palp slender; segment 2 about twice length of slender segment 3.

Gnathopod 1 (♂), segments 5 and 6 elongate, shallow, subequal, strongly setose posteriorly, nearly equal in length to basis; segment 6, palm short, greatly exceeded by dactyl; in ♀, segments 5 and 6 shorter but stout, palm longer, distinctly exceeded by dactyl. Gnathopod 2 (♂), segments 5 and 6 shorter but heavier than gnathopod 1; palm slightly oblique, irregular; in ♀, segment 6 is less robust, shorter, palm convex. Peraeopods 3 and 4, segments 4 and 5 subequal, basis slightly inflated. Peraeopod 5, basis deeper than broad. Peraeopods 6 and 7, posterodistal lobe shallow, hind corner sharply quadrate.

Abdominal side plate 3 broad, hind corner broadly rounded. Uropod 1, peduncle slender, outer margin with 6 spines, outer ramus with 5–6 spines. Uropod 2, outer ramus with 3–4 spines. Uropod 3, peduncle relatively long, about twice rami; outer ramus (hooked) not setose marginally. Telson short, apex rounded, cusps low.

Distribution: American Atlantic coast, from Florida, southeastern states and Bermuda north to Chesapeake Bay, Cape Cod, and southern Maine; an isolated relict population occurs in southwestern Gulf of St. Lawrence around Prince Edward Island and the Magdalen Islands.

Ecology : Constructs nests from secretion and bits of seaweed on algae and *Zostera,* in shallow water from lower intertidal to about 10 m; in higher salinities (down to 22 o/oo). Feeds essentially on diatoms. Usually in medium-current areas, often in very dense populations.

Life Cycle: Annual; ovigerous females from May—September; several broods per female.

Genus *Cymadusa* Savigny 1816   (sī mă dūs'ă)

Coxal plates 1–5, deep, distal margin posteriorly fringed with setae; 1st anterodistally expanded. Head, eye relatively large, extending onto anterior head lobe. Antenna 1, peduncle 3 about one-third segment 2; accessory flagellum distinct.

Mandible, palp short, 3-segmented. Maxilla 1, inner plate with several marginal setae.

Gnathopod 1 subchelate. Gnathopod 2 slightly stronger than 1 (both sexes). Peraeopods 3 and 4, basis slightly expanded anteriorly. Peraeopod 5, basis about as wide as long. Peraeopods 5–7 distal segments slender, 6 slightly broadened distally.

Uropods 1 and 2, rami subequal; peduncle of 1 distally produced as spine between rami. Uropod 3, inner ramus distinctly longer than outer.

KEY TO NORTH AMERICAN SPECIES OF *CYMADUSA*

1. Gnathopod 2 (both sexes), segment 6 longer and broader than segment 5; peraeopod 7, proximal posterior angle of basis sharply acute . . *C. filosa* (Savigny).
   Gnathopod 2 (both sexes), segment 5 longer and as broad as segment 6; peraeopod 7, posteroproximal angle of basis sharply rounded . . . *C. compta* (p. 182).

Cymadusa compta (Smith) 1873 Plate LV.2   (kŏmpt' ă)

L. 12–15 mm ♀, 7–11 mm ♂. Coxal plates 1–5 deep, lower margins sparingly setose, coxa 5 deepest. Eye large, encapsulated, light reddish, occupying most of head lobe. Antenna 1, accessory flagellum 1½-segmented, longer than flagellar segment 1. Antenna 2, peduncle strong, sparsely setose.

Gnathopod 1, segments 5 and 6 somewhat elongate (♂); segments 2, 5 and 6, anterior margins with feathery setae. Gnathopod 2 little larger, similar, segments 2, 5, and 6, margins similarly setose. In ♀, segments 5 and 6 subequal in length. Peraeopod 7, proximal posterior angle of basis sharply rounded.

Uropod 3, outer margin of peduncle with 2–3 groups of slender setae. Pleon side plate 3, hind corner acute, not produced.

Distribution: Central Maine and sporadically to Cape Cod, south to northern Florida; Gulf coast states. Common in the Cape Cod region: Pleasant Bay, Nantucket, Martha's Vineyard, Buzzards Bay.

Ecology: Common in eelgrass, in shallow water, low intertidal to over 30 ft, often in brackish water (to 5 o/oo).

Life History: Annual; ovigerous females from May–September. Several broods per female.

Genus *Sunamphitoe* Bate 1856 (sŭn ăm fĭ tō′ ĕ)

Coxal plates increasing posteriorly to 5; lower margins with posterodistal setae; coxa 1, not expanded, anterior distal corner acute. Accessory flagellum lacking.

Lower lip, outer lobe notched on center of distal margin. Mandible, palp lacking; molar reduced, incisor and lacinia strong. Maxilla 1, inner plate bare.

Gnathopod 1, segment 6 longer than 5, palm short (both sexes). Gnathopod 2 (♀) similar, larger; in ♂, very large and strongly subchelate; segment 5, posterior lobe narrow. Peraeopods 3 and 4, basis strongly dilated anteriorly. Peraeopod 5, basis expanded anteriorly and posteriorly; segments 4–6, short, stout; dactyl reversed. Peraeopods 6–7, segment 6, distally broadest, but not forming a subchela with dactyl.

Uropods 1 and 2, outer ramus slightly shorter, peduncle of 1 acutely produced between rami. Uropod 3, rami subequal, outer ramus with two large hook spines. Telson with pair of terminal dorsal hooks.

Sunamphitoe pelagica (M.E.) 1830 Plate LIII.2 (pĕl ăj′ ĭk ă)

L. 9 mm ♂. Coxal plates 1–5, progressively large, deep. Head, inferior antennal sinus relatively deep and concave; eye small, almond-shaped, black. Antenna 1 distinctly longer than 2; peduncle of antenna 1 short, scarcely reaching peduncle 5 of antenna 2. Mandible, molar grinding surface relatively small. Maxilla 1, inner plate unarmed. Maxilliped, palp short, segment 2 about 1½ segment 3, inner plate narrow.

Gnathopod 1 (both sexes), basis with rounded anterodistal lobe; segment 6 subrectangular, dactyl about twice length of short vertical palm. Gnathopod 2 (♀) similar, larger than 1; segment 5 shorter, posterior lobe narrow; in ♂, segment 5 very short, lobe deep, narrow; segment 5 very large, palm very long, oblique, shallow concave, with tubercle and low prominence near hinge, posterior margin very short; dactyl as long as segment 5, inner margin sinuous, spinulose. Peraeopods 3 and 4, segment 5 slightly shorter, less inflated than 4. Peraeopod 5, basis much expanded, subcircular. Peraeopods 6 and 7, basis without posterodistal lobe, rounded angle marked by a spine.

Abdominal side plate 3, hind angle sharply obtuse, margins smooth.

Uropods 1 and 2 not reaching end of uropod 3. Uropod 1, outer ramus shorter, outer margin with 2–3 spines. Uropod 2, rami as long as uropod 1. Uropod 3, peduncle stout, about twice length of rami; outer ramus, outer margin finely denticulate; inner ramus with short terminal spine and seta. Telson short and broad, apex truncate between cusps.

Distribution: Epibiotic-pelagic in warm water, North Atlantic region, northward to Cape Cod and Sable Island; European coast north to western Britain.
Ecology: Found mainly on *Sargassum* drifting in from the Gulf Stream and North Atlantic Drift, at or near the surface.
Life Cycle: Probably reproducing year-round.

Family PHOTIDAE (Isaeidae of Barnard, 1969)   (fōt' ĭ dē)

Body smooth. Coxal plates deep, setose below; 4th not excavate behind; 5th with deep anterior lobe. Head, interantennal lobe prominent, bearing eye on proximal portion; inferior antennal sinus deep. Antennae slender, subequal, peduncles elongate; posteriorly setose. Accessory flagellum present, vestigial, occasionally lacking.

Mandible, molar strong; palp slender, 2nd segment longest, 3rd terminally setose. Upper lip slightly notched, epistome variously produced. Lower lip, inner lobes present, mandibular processes not attenuated. Maxilla 1, inner plate small, with 1 apical seta. Maxilliped, outer plate large; palp segment 4 short.

Gnathopods 1 and 2 subchelate; gnathopod 2 larger. Peraeopods 3 and 4, basis narrow, dactyls with gland ducts emptying terminally. Peraeopods 5–7 increasingly longer, bases expanded behind; 7th not very elongate.

Uropods 1 and 2 subequally biramous. Uropod 3 usually biramous, inner ramus shorter or lacking; terminal spines simple, not hooked. Telson short, entire.

Coxal gills simple sacs on peraeon segments 2–6. Brood plates lamellate, with numerous long marginal setae.

KEY TO AMERICAN ATLANTIC GENERA OF PHOTIDAE

1. Head, anterior lobe subtruncate; peraeopod 3 (occ. 4), segments 2–5 long-setose anteriorly; uropod 3, rami spinose, not acutely tapering . . . . . *Protomedia.*
Anterior head lobe sharply acute or strongly produced; peraeopod 3 weakly setose; uropod 3, rami slender, tapering to apex . . . . 2.
2. Antenna 1, peduncular segment 3 distinctly shorter than segment 1; antennal peduncles weakly setose behind; uropod 3, uniramous. . *Microprotopus* (p. 188).
Antenna 1, peduncular segment 3 equal to or longer than segment 1; antennal peduncles strongly setose behind; uropod 3, biramous . . . . 3.

3. Uropod 3, inner ramus very short; eye located on strongly produced anterior head lobe; coxal plates large, very deep . . . . *Photis* (p. 186).
Uropod 3, rami subequal; eye lateral; coxal plates not exceptionally large or deep. . . . . 4.

4. Accessory flagellum present, 3+-segmented; gnathopod 2, carpus (segment 5) a little shorter than 6, carpal lobe shallow . . . . . *Gammaropsis (Eurystheus)*.
Accessory flagellum lacking; gnathopod 2, segment 5 very short, posterior lobe deep. *Podoceropsis* (p. 185).

Genus *Podoceropsis* Boeck 1871   (pŏd ö sēr ŏp' sĭs)

Head, anterior lobe acute. Coxal plates increasingly deep posteriorly, 4th deepest. Antenna 1, peduncular segment 3 longer than 1; posterior margins of antennae strongly setose; accessory flagellum lacking or very rudimentary.

Upper lip, epistome conical. Mandible, palp large; terminal segment rounded, strongly setose. Maxilliped, palp segments not expanded, dactyl medium long.

Gnathopod 1 subchelate, palm oblique; dactyl large; segment 6 long. Gnathopod 2 subchelate, larger than 1; carpus (segment 5) very short, lobes deep behind. Peraeopods 6 and 7 subequal, longer than 5; bases moderately expanded.

Uropods 1 and 2, peduncle prolonged distally between rami as sharp process. Uropod 3, rami slender, inner ramal spines simple, shorter. Telson small, short.

KEY TO AMERICAN ATLANTIC SPECIES OF *PODOCEROPSIS*

1. Gnathopod 1, dactyl large, greatly exceeding palm; coxa 5 almost as deep as 4; uropod 3, rami subequal; telson apex blunt . . . . *P. nitida* (p. 185).
Gnathopod 1, dactyl little exceeding palm; coxa 4 much deeper than 5; uropod 3, inner ramus about one-half outer; telson apex acute . . . *P. inaequistylis* Shoem.

Podoceropsis nitida (Stimpson) 1853 Plate LVII.2   (nĭt' ĭd ă)

L. 7.5 mm. Coxal plates 1–5, lightly setose below; coxa 5, anterior lobe much larger and deeper than posterior lobe. Antenna 1, slightly shorter than 2. Antenna 2, peduncular segment 3 about twice as long as deep.

Epistome conical and strongly produced. Mandibular palp segment 3 nearly as long as segment 2. Maxilla 1, inner plate with single apical seta. Maxilliped, palp segment 3 slender, longer than dactyl.

Gnathopod 1 shallow-subchelate (both sexes); segment 5 longer than 6, posterior lobe shallow; palm very oblique, convex, shallow, posterior angle greatly exceeded by stout dactyl, serrate behind. Gnathopod 2 (♂), segment 6 stout, with two strong palmar teeth; in ♀, palmar margin deeply excavate near posterior angle, with shallow tooth near hinge.

Peraeopods 3 and 4, segment 5 short, less than one-half segment 4; dactyls short. Peraeopods 5–7, bases moderately expanded, evenly convex, and lightly setose behind.

Abdominal side plates 2 and 3, hind corners acute, not produced. Uropod 1, rami slender, subequal; peduncle with acute distal interramal process. Uropod 2, outer ramus shorter than inner. Uropod 3, rami slender, acute, inner ramus slightly the longer, with two long apical setae. Telson short, apex blunt or broadly acute, with pair of dorsolateral setal groups.

Distribution: Amphi-Atlantic boreal; in America, southward from the Gulf of St. Lawrence and Gulf of Maine to Rhode Island and Connecticut. Common in Buzzards Bay.
Ecology: Mainly on rocky bottom, subtidal, to depths of more than 50 m.
Life Cycle: Annual; ovigerous females April–July.

Genus *Photis* Krøyer 1842   (fōt′ ĭs)

Coxal plates 1–4 very large, deep; coxa 5, anterior lobe equal to coxa 4, posterior lobe small. Head, anterior lobe projecting, acute, eyes subapical. Antennae slender, subequal. Antenna 1, peduncle 3 about equal to 1; peduncle and flagellum weakly setose behind; accessory flagellum very rudimentary.

Mandible, palp segment 3 shorter than 2 apex acute. Upper lip, epistome conical.

Gnathopods 1 and 2 subchelate, 2nd larger (esp. in ♂); segment 5 short, posterior lobe narrow, short. Peraeopod 5, dactyl reversed; peraeopods 6 and 7 subequal, basis less broad than 5.

Uropods 1 and 2 subequally biramous, weakly spinose. Uropod 3 biramous, inner ramus very short, outer ramus 2-segmented; peduncle long. Telson small, triangular.

KEY TO NEW ENGLAND SPECIES OF *PHOTIS*

1. Coxa 3, lower margin, and gnathopod 2, basis with stridulating ridges (in ♂)
. . . . *P. reinhardi* (p. 186).
   Coxa 3 and gnathopod 2 basis lacking stridulating ridges . . . . 2.
2. Coxa 3 largest, deepest; eye fully pigmented . . . . . . *P. macrocoxa* (p. 187).
   Coxae 2–4 subequal; eye with a few pigmented facets only . . *P. dentata* Shoem.

Photis reinhardi Krøyer 1842 Plate LVI.1   (rīn′ hârd ī)

L. to 5 mm. Coxal plates 1 and 2 (both sexes) deep and regular, lower margin with long setae; coxae 3 and 4 deeper, 3rd (♂) with row of oblique stridulating ridges along lower margin; coxa 5 not larger than 4. Head, anterior lobe narrowly rounding; eye large, oval, amber colored. Antennae 1 and 2 subequal, falgellum of 6–8 segments; accessory flagellum rudimentary.

Gnathopod 1 (δ), basis short, broad, with anterior distal lobe; segment 6, palm deeply excavate, lobate at posterior angle; dactyl with 4–5 inner marginal teeth; in $\female$, palm oblique and nearly straight. Gnathopod 2 (δ), basis stout, curved, with strong anterodistal lobe and oblique row of stridulating ridges; segment 5, posterior lobe narrow, deep; segment 6, palm deeply and sharply excavate above long tooth at palmar angle; dactyl with rounded spinose protruberance on inner margin; in $\female$, palm is less deeply excavate. Peraeopods 3 and 4 subequal and alike; segment 4 (of peraeopod 3 only) with long setae at anterodistal angle.

Uropod 3, outer ramus with lateral setae, shorter than peduncle. Telson about as long as wide.

Distribution: An arctic and boreal amphi-Atlantic species, ranging south in America to the Gulf of Maine and Long Island Sound. Recorded in Cape Cod Bay.
Ecology: Subtidally from about 5 to over 50 m; mainly on soft bottoms.
Life Cycle: Annual; ovigerous females from April–July; one brood.

Photis macrocoxa Shoemaker 1945 Plate LVI.2   (măk rō kŏks′ ă)

L. 2.5–3.5 mm. Coxal plates 1 and 2 (δ) narrowing and rounded below, marginal setae long. Coxa 3 largest; broadest distally; coxae 4 and 5 subequal in width, hind lobe of 5 small. Head, anterior lobe sharply angular. Eye, subovate, dark brown, almost apical on head lobe. Antenna 1 slightly shorter than antenna 2; flagellae 5-segmented; accessory flagellum minute.

Mandibular palp, segment 3 with 4–5 inner marginal pectinate setae. Maxilliped, inner plate short, not extending beyond palp segment 1.

Gnathopod 1 (δ), palm excavate; palmar angle a rounded prominence, with stout spine; dactyl with 1–2 inner distal teeth; in $\female$, palm very oblique, not excavate, angle weakly defined. Gnathopod 2 (δ), segment 3 with prominent outer distal lobe; segment 6, palm very oblique and deeply excavate, with long sharp tooth at posterior angle and stout double-toothed prominence near hinge; dactyl with small inner distal tooth.

Abdominal side plates 2 and 3, lower hind corner about quadrate. Uropod 3, outer ramus shorter than peduncle; inner ramus longer than terminal segment of outer ramus. Telson triangular, apex subacute.

Distribution: North American coastal waters, from the Gulf of St. Lawrence and Bay of Fundy to southern New England, and the Middle Atlantic states (Virginia).
Ecology: Subtidally to depths of 58 fathoms, on muddy and sandy mud bottoms.
Life Cycle: Annual; ovigerous females from April–July.

Genus *Microprotopus* Norman 1867   (mī krō prŏt′ ō pŭs)

Head, anterior lobe moderately produced. Coxal plates deep, 2 broadest. Antennae 1 and 2 short, subequal; posterior margins weakly setose. Antenna 1, segment 3 shorter than 2, accessory flagellum present.

Mandible, palp slender, sparsely setose. Maxilliped, palp segment 3 inflated beyond base of short dactyl; inner plate tall. Upper lip, epistome not produced.

Gnathopod 1 subchelate, segment 6 shorter than 5. Gnathopod 2 usually subchelate, very large in ♂, much stronger than gnathopod 1; segment 5 short, posterior lobe narrow. Peraeopods 3 and 4 short, segments 2, 4, and 5, variably dilated. Peraeopods 5–7 successively longer, bases broadly expanded, with distal free lobes.

Uropods 1 and 2, peduncle not prolonged distally, rami subequal. Uropod 3 uniramous; ramus 1-segmented. Telson small, truncate.

KEY TO NORTH ATLANTIC SPECIES OF *MICROPROTOPUS*

1. Antenna 1, acessory flagellum 1½ -segmented; European. . . . . 2.
   Accessory flagellum 2½ –3½ segmented; American . . . . 3.
2. Antennal flagella 3–4 segmented; gnathopod 2 (♀) simple . . . .
   . . . . *M. longimanus* (du Croisic).
   Antennal flagella 5–8 segmented; gnathopod 2 (♀) strongly subchelate . . . . .
   . . . . *M. maculatus* Norman.
3. Antenna 2, length 1½ times antenna 1; uropod 1, inner ramus lacking marginal spines; peraeopods 3 and 4 (♂), segments 2 and 4 strongly plumose-setose anteriorly
   . . . . *M. shoemakeri* Lowry.
   Antennae 1 and 2 subequal in length; uropod 1, inner ramus marginally spinose; peraeopods 3 and 4 (♂), segments 2 and 4 weakly setose (not plumose) anteriorly
   M. raneyi (p. 188).

Microprotopus raneyi Wigley 1966 Plate LVII.1   (rān′ ē ī)

L. 2–4 mm. Coxal plate 1 expanded distally, corners broadly rounded; coxae 1–4 deep, lower margins with numerous setae; coxa 5 as deep as 4. Head, anterior lobe acute. Eye oval, dark brown. Antenna 1 equal to antenna 2; accessory flagellum 2½ -segmented.

Mandibular palp segment 3 with apical and subapical setae only. Maxilliped, inner and outer plates narrow.

Gnathopod 1 (both sexes), segment 5 slightly longer than 6; segment 6 expanding distally, palm convex. Gnathopod 2 (♂) very large, segment 6 subovate, palm elongate, v. oblique, nearly straight, with 2 unequal teeth near hinge, dactyl long, falcate, tip closing on strong, inward-projecting tooth at base of palm (mature ♂); in ♀, segment 6 much smaller, palm with shallow median excavation; segment 5 relatively larger, lobe deeper, margins (inner distal) with long plumose setae.

Peraeopod 3 slightly heavier than 4; segments 2 and 4 short-setose, slightly inflated. Peraeopods 5–7 slightly increasing in length; basis of peraeopod 5 suborbicular, of peraeopod 6 subovate, and of peraeopod 7 broad ovate, posterodistal lobe extending beyond segment 3. In all, segment 5 is shorter than segments 4 and 6.

Abdominal side plate 2, hind corner quadrate. Side plate 3, posterior margin convex, hind corner obtuse.

Uropods 1 and 2 extending slightly beyond uropod 3. Uropod 3, ramus slender, longer than peduncle, lacking lateral spines, with 2 longish apical spines. Telson broadest basally, apex subtruncate.

Distribution: Southwestern Cape Cod Bay, Vineyard Sound and Buzzards Bay south to Georgia, northern Florida and Gulf of Mexico.
Ecology: Builds small tubes in sandy substrata, from lower intertidal to several meters in depth.
Life Cycle: Annual; ovigerous females from May–September; several broods per female.

Family ISCHYROCERIDAE   (ĭs kĭr ŏ sĕr' ĭ dē)

Body smooth, slightly depressed. Coxal plates shallow, contiguous, smooth below; 2 larger than 1, 4th not excavate; 5th with deep anterior lobe. Strongly sexually dimorphic. Head, rostrum short; inferior head lobe strong, bearing eye basally; inferior antennal sinus deep. Antenna 2, longer than 1, peduncles of both long, strongly setose posteriorly; flagella short, proximal segments fused. Accessory flagellum small.

Upper lip rounded, epistome produced anteriorly. Lower lip with inner lobes. Mandible, molar strong; palp large, terminal segment subclavate, heavily setose. Maxilla 1, inner plate small, with 1 apical seta; outer plate with 9 terminal spine-teeth. Maxilliped basic.

Gnathopod 1 subchelate, segment 6 longer than 5. Gnathopod 2 subchelate, larger than gnathopod 1, usually very large in ♂; segment 5 short, posterior lobe narrow. Peraeopods 3 and 4, basis expanded; segment 5 short; dactyl with gland duct to tip. Peraeopods 5–7 regularly increasing in size posteriorly, basis expanded.

Uropods 1 and 2, inner ramus longer; peduncle prolonged as spine between rami. Uropod 3, peduncle strong; rami very short, outer curved, with hooked teeth near apex. Telson short, apex acute.

Coxal gills saclike on 2–6, small on 2. Brood plates large, laminar, margins with long hook-tip setae.

KEY TO AMERICAN ATLANTIC GENERA OF ISCHYROCERIDAE

1. Coxal plate 5, distinctly deeper than 6. . . . . 2.
   Coxal plate 5, only slightly deeper than 6 . . . . 3.

2. Head, anterior lobe acute; peraeopods 3 and 4, segment 5 not grossly overhung by segment 4; gnathopod 2 (♂) without palmar thumb. . . *Ischyrocerus* (p. 191).
Head, anterior lobe rounded or blunt; peraeopods 3 and 4, segment 5 small, largely overhung by segment 4; gnathopod 2 (♂) with strong palmar thumb . . .
. . . . *Jassa* (p. 190).
3. Coxal plates 3 and 4 large; 4th wider than deep, excavate behind; accessory flagellum distinct; antenna 2, flagellum weakly setose . . . . *Microjassa*.
Coxal plates 3 and 4 small; 4th coxal margins subparallel; accessory flagellum minute; antenna 2, flagellum very richly long-setose . . . . *Parajassa*.

Genus *Jassa* Leach 1813–14   (yă's ă)

Head shallow, weakly rostrate; anterior head lobe large, rounded or blunt, bearing eye. Coxal plates medium, tending to separation; 4th deepest, not excavate behind; 5th much deeper than 6. Antenna 1, peduncle 3 much longer than 1; accessory flagellum small but distinct. Antenna 2 very powerful in ♂, peduncle longer than flagellum, proximal segments of flagellum fused, apical segment with hooked spines. Maxilla 1, inner plate lacking setae. Maxilla 2, inner plate shorter, setose.

Gnathopod 1 subchelate. Gnathopod 2 subchelate, much larger than 1 (both sexes); in ♂, palmar angle with thumblike process; posterior lobe of segment 5 short. Peraeopods 3 and 4, segment 5 reduced; anterior margin partly or nearly completely overhung by distal process of 4. Peraeopods 1–4, anterior distal margin with longest setae. Peraeopods 6–7, distal margin of basis excavate at limb junction.

Uropod 3, outer ramus equal to inner. Telson triangular, apex acute.

Jassa falcata (Montagu) 1818 Plate LVIII.2   (făl kāt' ă)

L. 9.0 mm ♂. Coxal plates 2, 3, and 4 deeper than 1 and 5; coxa 2 (mature ♂) often narrow and deep below. Head, anterior lobe smoothly rounded; eye near its extremity, oval, black. Antenna 1, segment 2 longer than 3, longer than 1; flagellum with about 5 segments; accessory flagellum with 1 very short segment. Antenna 2, peduncle stout, much stronger than antenna 1, posterior margin strongly setose, distally plumose; flagellum of 2–3 segments with pair of terminal hook spines. Mandibular palp, terminal segment short, expanded, apex rounded. Maxilliped, inner plate not extending beyond segment 1 of palp.

Gnathopod 1 (both sexes), basis with a few short anterodistal setae; segment 5, posterior lobe distinct, not narrow; segment 6, palm convex, oblique. Gnathopod 2, basis with several anterodistal setae; segment 5 very short, posterior lobe very small, narrow; segment 6 (♂) very large, powerfully subchelate; palm deeply excavate, with strong, simple-tipped thumblike process at posterior angle and irregular tooth near hinge; dactyl very long with inner proximal protuberance; in ♀ and imm. ♂ strongly subchelate, weakly concave, palmar angle with low tooth and short spines; dactyl serrate behind.

Peraeopods 3–4, basis with anterodistal lobe and marginal setae; segment 4 strongly produced anterodistally, and short-setose, nearly masking short segment 5. Peraeopods 5–7, segment 4 moderately expanded, with several posterior marginal setae.

Abdominal side plate 3, hind corner subquadrate or slightly acuminate. Uropod 2 extending beyond uropod 1. Uropod 3 longest; peduncle stout, outer ramus equal to or slightly longer than inner, with terminal hook and 2 subterminal denticles. Telson subtriangular.

Distribution: Cosmopolitan in temperate—warm temperate areas; on the American Atlantic coast northward to Chaleur Bay and southern Newfoundland.

Ecology: Builds open-ended mud and silt tubes, on solid substrata in strong tidal and wave currents such as on pilings, wharves, aids-to-navigation, and ships' hulls. May reach high densities on eelgrass and on soft sediments along channels. A dominant fouling organism. A suspension feeder and predator on small crustaceans and ostracods.

Life Cycle: Annual; ovigerous females from May–September; several broods per female.

Genus *Ischyrocerus* Krøyer 1838   (ĭs kĭr ŏ sēr′ ŭs)

Coxal plates medium deep; 4th largest; 5th much deeper than 6. Head, anterior lobe acute; eye lateral. Antenna 1, peduncular segment 3 longer than 1; accessory flagellum small, distinct. Antenna 2, peduncle 5 longest, flagellum shorter, segments distinct, not heavily setose.

Gnathopod 2 little larger than 1 in ♀, powerfully subchelate in ♂, palmar angle without thumblike process; segment 5, posterior lobe small. Peraeopods 3 and 4, segment 5 medium small, anterior margin free. Peraeopods 5–7, distal margin regularly rounding to junction anteriorly.

Uropod 3, inner ramus longer than outer. Telson broadly acute.

KEY TO NEW ENGLAND SPECIES OF *ISCHYROCERUS*

1. Uropod 3, outer ramus lacking posterodistal denticles; peraeopods 5–7, basis lacking posterodistal lobes; antennae elongate, flagellum of 6–8 segments; gnathopod 2 (♂), palm lacking toothed prominence near hinge . . *I. megacheir.*
   Uropod 3, outer ramus with posterodistal denticles; peraeopods 5–7, basis expanded, with distinct posterodistal lobes; antennae short, flagella with 4–5 segments; gnathopod 2 (♂), palm with toothed prominence near hinge . . . . . 2.
2. Uropod 3, peduncle lacking marginal spines, outer ramus with 7 blunt denticles; gnathopod 2 (both sexes), palmar process with 2 teeth; telson, apex broad . . . .
   . . . . *I. commensalis.*
   Uropod 3, peduncle with marginal spines, outer ramus with 4–5 blunt denticles; gnathopod 2, palmar process with 1 tooth ♂, lacking in ♀; telson acute . . . .
   . . . . *I. anguipes* (p. 192).

Ischyrocerus anguipes Krøyer 1838 Plate LVIII.1   (ăng′ wĭ pās)

L. 8.0–12.0 mm ♂. Coxal plates relatively large and deep; coxa 5, anterior lobe deep, much larger than posterior lobe. Head, anterior angle acute, produced beyond roundish black eye. Antennae 1 and 2 relatively short, subequal; flagella 5-segmented; accessory flagellum short, not reaching end of first flagellar segment. Mandibular palp, segment 3 much shorter than 2, distally broadest. Maxilliped, palp segment 4 short; inner plate extending beyond palp segment 1.

Gnathopod 1 (both sexes), segment 5, posterior lobe broad; segment 6 little expanded; palm oblique, convex, merging smoothly with posterior margin; dactyl toothed behind. Gnathopod 2 (♂) much larger than 1; segment 5, posterior lobe small; segment 6 large, elongate palm nearly horizontal, concave, richly setose, with single-toothed prominence near hinge, rounding proximally to very short posterior margin; dactyl strong, with median posterior marginal prominence. Gnathopod 2 (♀) similar to but slightly larger than gnathopod 1; segment 5, posterior lobe narrower. Peraeopods 3 and 4, basis somewhat inflated, margins bare; segment 4 little produced anterodistally and nearly bare of setae. Peraeopods 5–7 increasing successively, segment margins nearly bare; basis, posterodistal lobes sharply rounding.

Abdominal side plate 3, hind corner rounded. Uropod 2, not extending beyond uropod 1 but shorter than 3. Uropod 3, posterior margin of peduncle spinose; outer ramus slightly shorter than inner, bearing 4–5 sharp denticles. Telson triangular, apex acute.

Distribution: Subarctic and boreal; American Atlantic coast from Hudson Strait south to New England and in deeper waters to Cape Hatteras.
Ecology: A common shallow-water tube-building amphipod, occurring from the low tide level, especially along rocky shores, to depths of over 50 m. A common fouling organism on pilings, aids-to-navigation, etc. in cold-water areas.
Life Cycle: Annual; ovigerous females from March–July; several broods per female.

Family COROPHIIDAE (emended)   (kör öf ē′ ĭ dē)

Body smooth; urosome visibly depressed; segments tending to fusion. Coxal plates usually short, shallow, separated at the base. Head, rostrum short; interantennal (anterior) lobe prominent, bearing eye; inferior antennal sinus deep. Antenna 1, slender, equal to or shorter than 2; peduncular segment 3 medium to long; accessory flagellum lacking. Antenna 2, peduncle segments 3–5 often stoutly developed (esp. in ♂); flagellum short, segments tending to fusion.

Upper lip notched; epistome slightly produced anteriorly. Lower lip with inner lobes; mandibular processes of outer lobes short. Mandible, molar strong; palp weak, 1-, 2-, or 3-segmented. Maxilla 1, inner plate small, bare; outer plate with 7 terminal spine-teeth. Maxilla 2, outer plate larger. Maxilliped, inner plate small; outer large; palp dactylate.

Gnathopod 1 subchelate (primarily). Gnathopod 2 subchelate or simple, larger than 1. Peraeopods 3 and 4, segment 2, expanded anteriorly; segment 5 short, usually overhung by segment 4. Peraeopod 5 short; segment 6, dactyl directed backwards. Peraeopod 6 variable. Peraeopod 7 elongate, slender, dactyl directed anteriorly.

Pleopods short, powerful, peduncle tending to inward expansion. Uropod 3 short, typically uniramous, ramus usually shorter than peduncle. Telson short, broad. Coxal gills saclike, on peraeon segments 3–6, occasionally on 7, lacking on 2. Brood plates narrow, smallest on peraeon 2, large on peraeon 5. Sexual dimorphism prominent in head shape, antenna 2, and/or gnathopod 2.

Component Genera: *Cerapus, Cerapopsis, Concholestes, Corophium, Dryopoides, Erichthonius, Gaviota, Kamaka, Paracorophium, Pseuderichthonius, Ranunga, Siphonoecetes.*

KEY TO NEW ENGLAND GENERA OF COROPHIIDAE

1. Telson and uropod 3 ramus marginally with dorsally directed hooked spines; antennae 1 and 2 subequal, peduncles posteriorly strongly setose; pleopod peduncles normal; gnathopod 2 (♂) complexly subchelate . . . . 2.
   Telson and uropod 3, ramus with simple setae marginally; antenna 1 shorter than 2 not strongly setose behind; pleopod peduncles inwardly expanded; gnathopod 2 (♂) normally subchelate or simple . . . . 3.
2. Coxa 5, anterior lobe about subequal to coxae 1–4; uropod 2 biramous, large; pleopods subequal in size . . . . *Erichthonius* (p. 193).
   Coxa 5, anterior lobe very much larger than coxae 1–4; uropod 2 uniramous, small; pleopod 1 much stronger than 2 and 3 . . . . *Cerapus* (p. 196).
3. Gnathopod 2 subchelate (both sexes); peraeopods 3 and 4, anterior and posterior margins of segment 5 overlapped by distal expansion of segment 4
   . . . . *Siphonoecetes* (p. 198).
   Gnathopod 2 simple; peraeopods 3 and 4, segment 5 margins free or with only anterior margin overlapped by segment 4 . . . . *Corophium* (p. 198).

Genus *Erichthonius* Milne-Edwards 1830   (ĕr ĭk thŏn' ĭ ŭs)

Coxal plates relatively deep, slightly separated; coxa 5 with deep anterior lobe. Head, anterior lobe moderately produced. Antennae 1 and 2 slender, subequal; peduncular posterior margins strongly setose. Antenna 2 inserted well posterior to antenna 1.

Upper lip, epistome acutely produced anteriorly. Mandible, palp strong, 3-segmented, distally setose. Maxilla 1, inner plate small, margin with setae. Maxilliped normal.

Gnathopod 1 normally subchelate; segment 5 longer than 6; hind lobe long. Gnathopod 2 (♂) complexly subchelate, very powerful; segment 5, posterodistal angle produced as strong tooth; in ♀, gnathopod 2 normally subchelate, stronger than 1, posterior lobe of segment 5 large. Peraeopods 3 and 4, segment 5, margins free, basis dilated. Peraeopod 5 much shorter than 6 and 7; segment 6 and dactyl reversed.

Pleopod peduncles stout, ramal segments broad. Uropods 1 and 2 normally biramous. Uropod 3 uniramous; peduncle strong, ramus very short, curved, with hooked spines near apex. Telson short, broad, bilobed; upper surface with rows of hooked spines.

KEY TO NEW ENGLAND SPECIES OF *ERICHTHONIUS*

1. Peraeopods 3 and 4, basis moderately expanded, widest distally, dactyls short, about one-half length of segment 6; eye small, almost entirely located on anterior head lobe; peraeopod 7, basis posterodistal margin distinctly excavate
. . . . *E. difformis* (p. 194).
Peraeopods 3 and 4, basis strongly expanded, widest medially, dactyl more than two-thirds length of segment 6; eye large, about one-half on anterior head lobe; peraeopod 7, posterodistal angle quadrate, distal margin about straight . . . 2.
2. Uropod 3, peduncle about equal to urosome 3; coxa 2, lower margin convex, lined with close-set vertical stridulating ridges (♂); gnathopod 2 (♂), segment 5, distal process strongly double-toothed . . . . *E. brasiliensis* (p. 195).
Uropod 3, peduncle longer than urosome 3; coxa 2, lower margin straight or concave, lacking vertical ridges (♂); gnathopod 2 (♂), segment 5, distal process essentially single-toothed . . . . *E. rubricornis* (p. 196).

Erichthonius difformis (Milne-Edwards) 1830 (no illustration)   (dĭ fŏrm′ ĭs)

L. 5 mm. Coxal plate 5, anterior margin rounded, setose below; coxa 2 (♂) concave below, margin smooth. Head elongate, shallow, length about twice depth; eye small, more than one-half situated on elongate rounded anterior head lobe. Antenna 1, peduncle 1 with 4–6 groups of posterior marginal setae.

Gnathopod 1 (♂), basis stout, hind margin rounded, with strong anterodistal lobe; more slender in ♀. Gnathopod 2 (♂), segment 5 slightly excavate above long single-tipped posterodistal tooth, 4–5 clusters of setae below; segment 6, lower margin concave between low proximal and distal processes; dactyl strongly setose apically and posteriorly; in ♀, segment 5, posterior lobe deeper than segment proper, curving distally toward palmar angle. Peraeopods 3 and 4 small, well separated; basis little expanded, widest distally; dactyls about one-half length of segment 6. Peraeopod 5, basis with quadrate posterodistal lobe, distal margin not excavate. Peraeopod 7, posterodistal margin of basis excavate, hind corner sharply quadrate.

Abdominal side plate 3, hind margin very slightly crenulate. Uropod 2, margins of subequal rami smooth. Uropod 3, peduncle short and stout, about twice length of ramus. Telson very short and broad; dorsal hooked spines in two widely separated clusters, 3 rows per cluster.

Distribution: Circum-Atlantic; in North America southward to Newfoundland, outer coast of Nova Scotia, and the Bay of Fundy.
Ecology: Cold-water areas, from extreme low water level to more than 100 m in depth.
Life Cycle: Presumably annual; ovigerous females from March–July.

Erichthonius brasiliensis (Dana) 1853 Plate LIX.2   (brǎ sǐl ǐ ěn′ sǐs)

L. 4–6 mm. Coxal plate 5, anterior lobe rounded, with lower fringe of long setae; coxa 2 deep, rounded below; in ♂, margin lined with about 25–30 fine stridulating ridges. Head, length about 1½ times depth; eye large, about one-half situated on large, anteriorly mucronate head lobe. Antenna 1, peduncle 1 with 4–5 groups of posterior marginal setae.

Gnathopod 1 (♂), basis with knoblike posteroproximal process; smoothly convex in ♀. Gnathopod 2 (♂), segment 5 with long double-toothed posterodistal projection armed behind with 6–10 groups of setae; segment 6 with low posterodistal tubercle; dactyl short, setose behind and long-setose at apex; in ♀, segment 5, posterior lobe narrow, deeper than segment proper.

Peraeopods 3 and 4, basis strongly expanded, widest medially, margins nearly touching; dactyls about three-fourths length of segment 6. Peraeopod 5, basis posterodistal margin truncate, hind lobe not produced. Peraeopod 7, basis posterodistal corner sharply rounded, distal margin not excavate.

Abdominal side plate 3, hind margin rounded, weakly crenulated. Uropod 1 extending beyond uropods 2 and 3; peduncle inner margin with fine serrations. Uropod 2, margins of rami very finely serrate. Uropod 3, peduncle short, little more than twice length of ramus, posterior margin with median and distal group of setae. Telson twice as wide as long; dorsal hooked spines stout, in two approximated clusters, 3 rows per cluster.

Distribution: Western Atlantic; in North America from Cape Cod and Vineyard Sound to Chesapeake Bay, Florida, Gulf states and West Indies.
Ecology: A very common tube-building species, mainly in shallow water, from low water level to over 200 m. Constructs tubes on stems and branches of hydroids and ectoprocts. Tubes are composed of mud and secretory material. In winter, found exclusively on *Sertularia argentea*. In bays and mouths of estuaries; polyhaline down to salinities of about 15 o/oo.
Life Cycle: Annual; ovigerous females from May–September; several broods.

Erichthonius rubricornis Smith 1873 (= *E. hunteri* Bate)
Plate LIX.1   (rōō brĭ kŏrn' ĭs)

L. 7–9 mm. Coxa 5, anterior lobe relatively large, margin straight or concave, lower margin setose (♀); coxa 2, lower margin smooth, concave (♂) or straight (♀). Head, inferior antennal sinus about one-half length of head; anterior lobe strongly produced, acute; eye moderately large, black, about one-half on anterior lobe. Antenna 1, peduncle 1 with 4–5 posterodistal groups of setae. Epistome very strongly produced in front. Urosome segments 1 and 2 with paired dorsolateral setae.

Gnathopod 1 (♂), basis stout, rounded behind, more slender in ♀; segment 4 produced below as sharp triangular process. Gnathopod 2 (♂), segment 5 with long acute, single-toothed posterodistal process, sharply excavate above in most mature stages; segment 6 heavy, convex below, with low distal prominence; dactyl nearly straight, inner margin lined with short setae, closing on distal process of segment 5; in ♀, segment 5, posterior lobe deep, curving distally to posterior angle of segment 6. Peraeopods 3 and 4, bases broadly expanded, widest medially, margins approximated; dactyls slender, more than three-fourths length of segment 6. Peraeopod 5, basis broad, posterodistal lobe pronounced, broadly rounding. Peraeopod 7, basis with sharply quadrate posterior lobe, distal margin not excavate.

Abdominal side plate 3, hind margin rounded, usually very weakly crenulate. Uropods 1 and 2 extending beyond uropod 3, uropod 1 slightly beyond 2; peduncle (inner margin) and rami (both margins) lined with median stout spines. Uropod 3, peduncle elongate, about 2½ times ramus, ramus with 1–2 posterior marginal setae and terminal group of 3 small hooked spines. Telson, width about twice length, dorsal hooked spines small, numerous, in separated clusters, several rows per cluster.

Distribution: Amphi-Atlantic; in North America: Labrador, Gulf of St. Lawrence, outer coast of Nova Scotia, Gulf of Maine, south to Long Island Sound.
Ecology: Lives in flexible sandy mud tubes, usually in colonies, attached to rocks and other bottom materials, from low water level to depths of over 200 m.
Life Cycle: Annual; ovigerous females from April–July; one brood per year.

Genus *Cerapus* Say 1818   (sĕr' ă pŭs)

Body elongate, subcylindrical. Coxal plates 1–4 shallow, separated; coxa 5, anterior lobe very large and deep. Head elongate, rostrum thornlike; anterior lobe prominent. Antennae 1 and 2 stout, subequal, strongly setose posteriorly; flagella 3–6 segmented. Antenna 1, segment 1 expanded laterally. Epistome not produced in front. Mandible, palp slender, 3-segmented. Maxilla 1, inner plate with 1 apical seta. Maxilliped normal.

Gnathopod 1 normally subchelate (both sexes); segments 5 and 6 sub-equal. Gnathopod 2 (♀) normally subchelate, slightly larger than 1, similar; in ♂, complexly subchelate; segment 5, posterodistal angle produced as tooth, distal margin also with tooth. Peraeopods 3 and 4 short; segments 4 and 5 short and broad; margins free. Peraeopod 5 very short, dactyl bicuspate, reversed. Peraeopods 6 and 7 much longer than 5, linear, dactyls hooked.

Pleon side plates shallow, separated distally. Pleopods unequal, 1 much the largest, basal segments of outer ramus broadened. Uropod 1 unequally biramous. Uropod 2 uniramous. Uropod 3 uniramous, ramus with hooked spines. Telson bilobed, short, with dorsal rows of hooked spines.

KEY TO NORTH AMERICAN SPECIES OF *CERAPUS*

1. Antennae, flagella of 3 segments; gnathopod 2 (♂), segment 5, distal marginal spine acute; uropod 1, terminal spine of outer ramus set in fan of short blade spines . . . . *C. tubularis* (p. 197).
   Antennae, flagella of 6 segments; gnathopod 2 (♂), segment 5, distal marginal spine blunt-tipped; uropod 1, outer ramus rosette spines long, about one-half length of terminal spine . . . . *Cerapus* sp. (Florida).

Cerapus tubularis Say 1818 Plate LX.1   (tū būl âr' ĭs)

L. 3.5–4.5 mm. Antennae 1 and 2 relatively short, flagella 3-segmented. Antenna 1, peduncular segment 1 strongly broadened medially, forming plug for tube. Peduncular segments, posterior margins with 6–9 groups of long setae. Mandibular palp, inner margin of segment 2 nearly devoid of setae.

Gnathopod 1 (♂), propod slightly longer than carpus, posterior margin with long pectinate setae. Gnathopod 2 (♂), anterior margin of basis nearly bare; inner palmar tooth of segment 5 acute, not blunt-tipped; dactyl, inner margin lined with low cusps; distally and submarginally with 3–4 long setae. Gnathopod 2 (♀), propod more slender than in gnathopod 1. Peraeopod 5, anterior coxal lobe longer than deep; dactyl bicuspate. Peraeopod 7, posterior margin of basis proximally with many fine setae.

Uropod 1, peduncle produced in short soft apically ciliated lobes at base of each ramus; outer ramus with 16–18 marginal cusps; large pectinate terminal spine of inner and outer ramus set in a rosette of flat spines all less than one-half its length. Uropod 2, ramus with two outer cusps and subapical spine. Uropod 3, peduncle with 1–2 inner marginal setae.

Distribution: American Atlantic coastal waters, Cape Cod to eastern Florida.
Ecology: Shallow, muddy sand, and in deeper channels, from immediately subtidal to more than 100 feet. Portable case rectangular in X-section.
Life Cycle: Annual; ovigerous females from June–September.

Genus *Siphonoecetes* Krøyer 1845   (sī fŏn ē kĕt′ ēz)

Head large, long; rostrum strong; interantennal lobe produced, bearing eye terminally. Antenna 1 short, peduncular segments 1 and 3 subequal. Antenna 2 longer; flagellum shorter than peduncular segment 5.

Mandible, palp short, 1–3 segmented. Maxilla 1, inner plate very short, not setose.

Gnathopods 1 and 2 well developed, weakly subchelate, palm exceeded by dactyl. Peraeopods 3 and 4, segment 4 very expanded distally, overlapping or masking very small segment 5. Peraeopods short, subequal, dactyls reversed, bidentate. Peraeopod 7 elongate, dactyl normal, bidentate.

Pleopods powerful, peduncles expanded medially. Uropod 1 biramous, inner ramus shorter. Uropod 2 with one or two rami. Uropod 3, peduncle medially expanded, ramus shorter than peduncle. Telson broadly rounded.

Siphonoecetes smithianus Rathbun 1908 Plate LX.2   (smĭth ĭ ān′ŭs)

L. 6.0 mm. Coxal plate 1 deepest, anteroventral angle produced, rounded. Head, rostrum strongly produced, acute; anterior lobe narrow, sharply rounded; eye oval, black, terminal. Antenna 1 short, less than one-half body length, not reaching antenna 2, peduncle segment 5. Antenna 2 (♂), peduncle very stout, segment 3 broadest, flagellum of one stout segment and minute terminal segment. Mandibular palp apparently of 3 segments, 3rd minute; segment 2, outer margin strongly setose. Maxilla 1, inner plate bare. Maxilliped, palp segment 1 with outer marginal group of plumose setae.

Gnathopod 1 (both sexes), segments 5 and 6 subequal in length, 6th narrower; palm very oblique, with 2–3 strong stout spines near posterior angle; dactyl long, serrate behind. Gnathopod 2 (♂), segment 6 stout, palm oblique, slightly concave toward hinge, with 4–5 stout spines near posterior angle, posterior margin short. Peraeopods 3 and 4, basis very broadly expanded, subovate; segment 6 short, nearly as broad as long. Peraeopods 5, 6, and 7, bases subequal in width.

Abdominal side plates 2 and 3, hind corners quadrate. Uropod 1, outer margin of peduncle lined with stiff setae. Uropod 2, inner margin of inner ramus bare. Uropod 3, inner distal lobe of peduncle small, margin nearly bare, as broad as long, ramus not extending beyond peduncle lobe. Telson as long as broad, apex rounded.

Distribution: Known only from a few localities south of Cape Cod, Long Island Sound, and off New Jersey.
Ecology: Subtidal, 20–40 m, on sandy mud bottoms.
Life Cycle: Probably annual; ovigerous females in August.

Genus *Corophium* Milne-Edwards 1830   (kŏr ōf′ ĭ ŭm)

Body very depressed. Urosome segments tending to fusion. Coxal plates small and separated. Head, anterior lobes prominent, bearing eyes. Antenna

1 shorter than antenna 2; flagellum multiarticulate. Antenna 2, peduncular segments 3–5 stout, often pediform (esp. in ♂), flagellum short, segments fused.

Upper lip rounded or flat. Mandible, palp weak, 2-segmented. Maxilla 1, inner plate bare. Maxilliped, row of plumose setae crossing basal segments and palp segment 1.

Gnathopod 1 subchelate, weak; segment 5 elongate. Gnathopod 2 simple; segment 4 prolonged behind 5, long-setose; dactyl strong, variously toothed on posterior margin. Gnathopods not sexually dimorphic. Peraeopods 3 and 4, segments 2 and 4 dilated; segment 5 short; dactyl elongate, nearly straight. Peraeopods 5 and 6 short; segment 5 with 2 rows of comb spines; segment 6 and dactyl directed posteriorly. Peraeopod 7 elongate; segment 2 broad, margins fringed with plumose setae.

Pleon side plates small, separated. Pleopods short, powerful. Peduncles broadened inwardly. Uropods 1 and 2 biramous. Uropod 3 uniramous, ramus lamellate. Telson small, broad, not conspicuously uncinate.

KEY TO NEW ENGLAND SPECIES OF *COROPHIUM*

1. Segments of the urosome separate; antenna 1, peduncular segment 2 short, about one-half length (or less) of segment 1 (both sexes); peraeopods 3 and 4, segment 5 normal, anterior margin free; gnathopod 2, dactyl simple
   . . . . *C. volutator* (p. 200).
   Segments of urosome fused; antenna 1, peduncular segment 2 nearly equal to segment 1 (in ♂); peraeopods 3 and 4, segment 5 very short, anterior margin partly or completely overhung by segment 4; gnathopod 2, dactyl with one or more posterodistal accessory teeth. 2.
2. Urosome with distinct lateral marginal ridge; uropods 1 and 2 arising ventrally . 3.
   Urosome lacking lateral ridge; uropods 1 and 2 arising laterally or ventrolaterally 5.
3. Peraeopods 3 and 4, anterior margin of segments 2 and 4 lined with long plumose setae; gnathopod 2, segment 5 about equal to 2; antenna 2 (♂), peduncular segment 5 lacking posterior marginal tubercle. . . . . . . *C. lacustre* (p. 204).
   Peraeopods 3 and 4, anterior margin of segments 2 and 4 lacking long setae; gnathopod 2, segment 5 distinctly shorter than 2; antenna 2 (♂), peduncular segment 5 with posterior marginal tubercle . . . . 4.
4. Antenna 2, peduncular segment 4 alike in ♂ and ♀, both with strong posterodistal curved process; peduncular segment 5, posterior marginal tubercle small, set proximally; gnathopod 1, palm nearly vertical, strongly convex . *C. simile* (p. 205).
   Antenna 2, peduncular segment 4 differing markedly in ♂ and ♀, without distal process in ♀; peduncular segment 5, posterior marginal tubercle strong, set towards mid-margin; gnathopod 1, palm oblique, gently convex . . . *C. acutum* (p. 205).
5. Peraeopods 3 and 4, dactyls long, nearly straight, usually exceeding length of segments 4 and 5 combined; uropods 1 and 2, rami with one apical spine much longer than the others; uropod 2, inner ramus with stout lateral marginal spine(s) 6.
   Peraeopods 3 and 4, dactyls normal, curved, not longer than segments 5 and 6 combined; uropods 1 and 2, rami with apical spines not greatly differing in size; uropod 2, inner ramus lacking stout lateral marginal spine(s) . . . . 8.

6. Gnathopod 2, segment 4 longer than segments 2 and 6; dactyl with one postero-distal (subapical) tooth; antenna 2 (♀), segments 3 and 4 very broadly expanded, lower margin heavily spinose; uropod 3, ramus narrow, oval *C. crassicorne* (p. 201). Gnathopod 2, segment 4 shorter than segments 2 and 6; dactyl usually with two posterodistal teeth; antenna 2 (♀), segments 3 and 4 not unusually expanded or spinose; uropod 3, ramus broad, subcircular . . . . 7.

7. Head (♀), rostrum on prominence well in front of eyes; dorsal pigment in broad band between eyes, extending onto posterior part of head; uropod 1, inner margin of peduncle with 2–3 spines; males present, rostrum minute
. . . . *C. acherusicum* (p. 201). Head (♀), rostrum little produced in front of eyes; dorsal head pigment in narrow band between the eyes; uropod 1, inner margin with 3–4 spines; males apparently lacking . . . . *C. bonelli* (p. 202).

8. Peraeopods 3 and 4, anterior margin of segment 2 bearing long plumose setae in mature animals; uropod 1, inner margin with one spine (at apex); gnathopod 2, dactyl extending greatly beyond palm of segment 6; rostrum short (both sexes); antenna 2 (♂), peduncle 5 with strong posterior marginal tubercle
. . . . *C. tuberculatum* (p. 203). Peraeopods 3 and 4, anterior margin of segment 2, short-setose; uropod 1, inner margin with 2 spines (distally); gnathopod 2, dactyl slightly exceeding palm; rostrum (♂) very long, acute, exceeding anterior-lateral lobes; antenna 2 (♂), peduncular segment 5, posterior marginal tubercle vestigial . . . *C. insidiosum* (p. 203).

Corophium volutator (Pallas) 1766 Plate LXI.1   (vō lūt āt'ör)

L. 5–6 mm. (excl. antennae). Urosomal segments separate; uropods 1 and 2 arise laterally. Head, rostrum short (both sexes). Antenna 2, segment 4 alike in ♂ and ♀, powerful, elongate; segment 5 about equal to segment 4; posterior margin of 5 without tubercle. Antenna 1, peduncular segment 2 much shorter than 1; segment 1, inner margin (dorsal view) scalloped or crenulated (weakly in ♀); flagellum long, 10–12 segments. Mandibular palp lacking distal setal lobe; segment 2 longer than 1.

Gnathopod 1, segment 5 longer than 6; dactyl tip not exceeding palmar angle. Gnathopod 2, segments 5 and 6 subequal, both slightly shorter than 2; segment 5 with distinct posterior distal free margin; dactyl simple, lacking posterior distal tooth. Peraeopods 3 and 4, basis little inflated, margins not setose; segment 5 not very small, anterior margin free, not overhung by segment 4. Peraeopod 5, margins of basis lined with plumose setae. Peraeopod 7 elongate, segments 4 and 6, about equal to basis.

Uropod 1, peduncle outer margin with 10 or more singly inserted spines; rami heavily spinose, apices rounded. Uropod 2, rami longer than peduncle, with 3–5 stout lateral spines. Uropod 3, ramus not set off eccentrically; peduncle subequal, ramus ovate, armed with setae only, much smaller than telson.

Distribution: Amphi-Atlantic; in Europe, from southern Norway throughout the North Sea region to the Bay of Biscay and the Adriatic; American Atlantic coast; Gulf of Maine only, widely throughout the Bay of Fundy south to Casco Bay.

Ecology: Forms tubes intertidally in the mud of estuarine mud flats, salt-marsh pools and brackish ditches, in almost fresh waters, and vertically almost to mean high water level. Occasionally taken planktonically.

Life Cycle: Annual; ovigerous females from May–June; reproducing the following spring.

Corophium acherusicum Costa 1857 Plate LXII.2   (ă chĕr ōōs′ ĭk ŭm)

L. 4.0 mm. Urosome segments fused, without distinct lateral ridge; uropods 1 and 2 arising ventrolaterally. Head, rostrum minute, deeply recessed in ♂, moderately strong, on prominence ahead of eyes in ♀. Head, (ocular) lobe very large, extended anteriorly in ♂. Antenna 2, peduncle segment 4 unlike in ♂ and ♀; segment 5 with small tubercle proximally on posterior margin. Antenna 1 (♂), peduncular segment 1 elevated proximally; flagellum 6–8 segmented. Antenna 2 (♀), segment 4 with 3–4 stout posterior marginal spines. Mandibular palp, distal lobe of segment 1 moderate, distal margin straight; segment 2 larger than 1.

Gnathopod 1, segment 5 greater than 6; tip of dactyl exceeding palmar angle. Gnathopod 2, segment 5 longer than 2; segment 6 equal to or longer than segment 2; dactyl with 2 posterodistal teeth. Peraeopods 3 and 4, anterior margins of segments 2 and 4 lacking plumose setae; segment 5 small, partly overhung anteriorly by segment 4; dactyls very long, nearly straight. Peraeopod 5, margins of basis bare. Peraeopod 7, segment 6 shorter than basis.

Uropod 1, inner margin of peduncle distally with 2–3 stout spines. Uropod 2, outer ramus shorter than inner; margins with stout spines. Uropods 1 and 2, apex of all rami with one very long and 1–3 much shorter spines. Uropod 3, ramus longer than peduncle, relatively broad, but narrower than telson.

Distribution: Virtually cosmopolitan in warm temperate coastal waters. American Atlantic coast north to central Maine.

Ecology: Commonly in shallows, in protected and estuarine situations, in somewhat reduced salinities, in lotic water areas.

Life Cycle: Annual; ovigerous females from May–September.

Corophium crassicorne Bruzelius 1859 Plate LXI.2   (krăs ĭ körn′ ē)

L. 3.5 mm. Body short and very broad. Urosome segments fused, without distinct lateral ridge; uropods arising laterally. Head, rostrum (both sexes) weak. Antenna 2, segment 4 unlike in ♂ and ♀. Antennae 1 and 2 (♂) more or less "normal"; antenna 2, segment 5 with very small tubercle proximally on posterior margin. Antenna 1 (♀), segment 1 very broadly expanded, proximal dorsal margin with 4–5 spines; flagellum with 5–6 segments. Antenna 2 (♀), peduncular segments 3 and 4 very broad, powerful; inner lower margin lined with stout spines; segment 5 very short. Mandibular palp, distal setal lobe of segment 1 prominent, distal margin concave; segment 2 much longer than 1.

Gnathopod 1, segment 5 much longer than 6; dactyl short, tip reaching palmar angle. Gnathopod 2, segments 2, 5, and 6 subequal; dactyl with 1 posterior marginal (subapical) tooth. Peraeopods 3 and 4, margins of segments 2 and 4 virtually bare; segment 5 very short, completely overhung anteriorly by segment 4; dactyls very long and nearly straight, longer than segments 5 and 6 combined. Peraeopod 5, posterior margin of basis with a few plumose setae. Peraeopod 7, segment 6 slightly shorter than basis.

Uropod 1, inner margin of peduncle distally with 2–3 spines. Uropod 2, rami and peduncle subequal, rami with stout marginal spine(s). Uropods 1 and 2, apex of rami with one very long and 2–3 much smaller spines. Uropod 3, ramus narrow, oval, apical setae very long.

Distribution: Arctic and North Atlantic, south along North American coast to Long Island Sound. Common along Massachusetts and in Cape Cod Bay.
Ecology: In consolidated sandy bottoms, in protected cold-water areas; subtidally to depths of more than 200 m.
Life Cycle: Annual.

Corophium bonelli (M.-E.) 1830* Plate LXII.1   (bŏn ĕl′ ī)

L. 4.5 mm. Males apparently never discovered (unless misidentified with *C. acherusicum*). Body relatively long and narrow. Head, rostrum short. Dorsal pigment in narrow band between eyes, not on posterior part of head. Antenna 2, peduncle 4, with 3 posterior marginal spines. Antenna 1, peduncular segment 2 extends beyond peduncle 4 of antenna 2, flagellum 6–7 segmented. Mandibular palp, distal setal lobe prominent, margin slightly concave.

Gnathopod 1, segment 5 longer than 6, about equal to segment 2; tip of dactyl closes on palmar angle. Gnathopod 2, segment 4 longer than segment 2; segment 6 about equal to 2; dactyl with 2 (occ. only 1) posterodistal teeth. Peraeopods 3 and 4, anterior margins of segments 2 and 4 nearly bare; segment 5 small, anteriorly overhung by segment 4; dactyls very long, nearly straight, greater than segments 5 and 6 combined. Peraeopod 5, margins of basis lacking plumose setae. Peraeopod 7 elongate; segment 6 longer than basis.

Uropod 1, inner margin of peduncle distally with 4–5 spines. Uropod 2, outer ramus shorter, margins of rami with stout spine(s). Uropods 1 and 2, apex of rami with one very long spine and 2–3 short spines. Uropod 3, ramus large, broad, longer than telson.

---

* This species has been further distinguished from the arctic *Corophium clarencense* Shoemaker, in which males are known (Just, 1970).

Distribution: Bipolar, subarctic and boreal in North Atlantic; in North America, south throughout eastern Canada to Massachusetts (Cape Cod), and Connecticut.
Ecology: Shallows and low intertidal, along rocky cold-water coasts, building tubes under stones, in *Mytilus* beds, etc.
Life Cycle: Annual; ovigerous females from March–July.

Corophium insidiosum Crawford 1937 Plate LXIII.1   (ĭn sĭd ĭ ōs' ŭm)

L. 4 mm. Urosome segments fused, lacking distinct lateral ridge; uropods arising laterally. Head, rostrum (♂) very long, exceeding produced anterior lobes; in ♀, rostrum short; dorsal pigment occurring broadly between eyes, onto rostrum and extending back onto posterior part of head. Antenna 2, segment 4 unlike in ♂ and ♀; segment 5 (♂) lacking distinct posterior marginal tubercle. Antenna 1 (♂), peduncular segments 1 and 2 long, subequal; segment 1 with inner marginal tubercle, at about level of rostrum tip. Mandible palp segment 1, setal lobe moderate, distal margin straight; segment 2 slightly longer than 1.

Gnathopod 1, segment 5 longer than 6, tip slightly exceeding posterior angle. Gnathopod 2, segment 5 shorter than 2; segment 6 longer than 2; dactyl with 2–4(3) posterodistal teeth. Peraeopods 3 and 4, anterior margin of segment 2 with short setae only; anterior margin of segment 4 with 8–12 long simple setae; segment 5 short, completely overhung anteriorly by segment 4. Peraeopod 5, margins of basis lacking plumose setae. Peraeopod 7, segment 6 shorter than basis.

Uropods 1 and 2, apex of rami with spines of two sizes; long spine is not more than twice shorter spines. Uropod 1, inner margin distally with 2 spines. Uropod 2, outer ramus with 1 stout marginal spine, inner ramus lacking marginal spines. Uropod 3, ramus broad, about equal to peduncle.

Distribution: Endemic to the North Atlantic region; western Europe, south to Mediterranean; American Atlantic, from Chaleur Bay to Cape Breton Island and outer coast of Nova Scotia, New Hampshire to Long Island Sound; absent from upper Bay of Fundy, introduced into San Francisco Bay.
Ecology: Common in shallows and low intertidal, on sandy mud flats and among eelgrass; estuarine to about 15 o/oo salinity. Builds mud and detritus tubes, on oysters, aids-to-navigation. Sometimes common in plankton.
Life Cycle: Annual; ovigerous females from April–August; several broods per summer. Migrates into and overwinters in deep channels, moving inshore in late spring.

Corophium tuberculatum Shoemaker 1934 Plate LXIII.2   (tū bër kū lāt'ŭm)

L. 3 mm. Body short, relatively broad. Urosome segments fused, lateral ridge not distinct; uropods 1 and 2 arising ventrolaterally. Head, rostrum very short (both sexes). Antenna 2, peduncular segment 4 unlike in ♂ and ♀; antenna 2 (♂) enormously developed, about as long as body; peduncular segment 5 with strong tubercle proximally on posterior margin; in ♀, peduncle segment 4 with 2 mid posterior marginal spines. Antenna 1 (♂), segment 2 long, equal to segment 1; flagellum 6–7 segmented. Mandibular palp, setal lobe moderate, distal margin straight; segment 2 longer than 1.

Gnathopod 1, segment 5 longer than 6; tip of dactyl distinctly exceeding palm. Gnathopod 2, segment 5 shorter than 2; segment 6 longer than 2; dactyl with 2 posterodistal teeth. Peraeopods 3 and 4, anterior margins of segment 2 with long plumose setae, of segment 4 with 6–8 shorter, simple setae; segment 5 small, partly overhung anteriorly by segment 4. Peraeopod 5, margins of basis lacking long setae. Peraeopod 7 elongate, segment 6 about equal to basis.

Uropod 1, inner margin of peduncle with distal angle spine only. Uropod 2, rami subequal, marginal spines slender. Uropods 1 and 2, apex of rami, long spine not more than twice length of short spines. Uropod 3, ramus broad, orbicular, narrower than telson.

Distribution: American Atlantic coast, from upper Bay of Fundy (Minas Basin); south side of Cape Cod to Florida, and the eastern Gulf of Mexico.
Ecology: In shallow bays and estuaries, sandy mud bottoms.
Life Cycle: Annual.

Corophium lacustre Vanhoffen 1911 Plate LXIV.1   (lă kŭst' ĕr)

L. 3.5 mm. Urosome segments fused, lateral ridge distinct; uropods 1 and 2 arising ventrally. Head, rostrum (♂) prominent, acute; in ♀, short, broad. Antenna 2, segment 4 unlike in ♂ and ♀; segment 4 (♀) has only single short distal process; segment 5 (♂), posterior margin lacking tubercle, but armed with clusters of long simple setae. Antenna 1 (♂), segments 1 and 2 subequal in length, inner dorsal margin densely lined with long simple setae. Mandible, setal process of palp segment 1 small, distal margin convex; segment 2 equals segment 1.

Gnathopod 1, segment 5 longer than 6; tip of dactyl not exceeding palm. Gnathopod 2, segment 5 equals segment 2; segment 6 longer than 2; dactyl with 2 posterodistal teeth. Peraeopods 3 and 4, anterior margin of segments 2 and 4 with long plumose setae; segment 5, very small, completely overhung anteriorly by segment 4; dactyls short, curved, about equal to segment 6. Peraeopod 5, margins of basis with long plumose setae. Peraeopod 7, segment 6 shorter than basis.

Uropod 1, inner margin of peduncle lacking spines, distal angle with minute spine. Uropod 2, marginal spines of rami slender. Uropod 3, ramus moderately broad, set at angle to body axis, smaller than telson.

Distribution: Brackish estuaries of the American Atlantic coast, from the Bay of Funday (St. John estuary), south to Florida (St. John's system); dominant

in the Chesapeake Bay estuaries. Also in western Europe (southern North Sea region), possibly introduced.
Ecology: Estuarine, brackish water (25 o/oo to almost fresh), shallow and lower intertidal, in marshy banks, on pilings, aids-to-navigation.
Life Cycle: Annual. Ovigerous females May–September.

Corophium acutum Chevreux 1908 Plate LXIV.2   (ă kūt' ŭm)

L. 2–3 mm. Body short and broad. Urosome segments fused, lateral ridge prominent; uropods 1 and 2 arising ventrally. Head, rostrum short (both sexes). Antenna 2, peduncle segment 4 unlike in ♂ and ♀; in ♀, posterior margin with 3 stout spines; segment 5 (♂) with strong tubercle towards midpoint of posterior margin. Antenna 1, peduncular segments 1 and 2 subequal in ♂; flagellum of 6 segments. Mandibular palp, setal lobe of segment 1 moderately large, distal margin concave; segment 2 longer than 1.

   Gnathopod 1, segments 5 and 6 subequal; dactyl slightly exceeding palmar angle. Gnathopod 2, segment 5 much shorter than segment 2; segment 6 about equal to 2; dactyl with 2 posterodistal teeth. Peraeopods 3 and 4, anterior margins of segments 2 and 4 lacking long plumose setae; segment 5 very short, partly overhung anteriorly by segment 4; dactyl short, curved. Peraeopod 5, margins of basis without long plumose setae. Peraeopod 7, segment 6 shorter than basis.

   Uropod 1, inner margin of peduncle with spine at distal angle only. Uropod 2, rami shorter than peduncle, inner ramus lacking marginal spines. Uropod 3, ramus large, about as broad as telson, set at angle to body axis.

Distribution: Nearly cosmopolitan in warm temperate-tropical regions; American Atlantic coast from Florida north to Cape Cod.
Ecology: Common inshore in very lotic (current swept) areas; on pilings and aids-to-navigation, in channels and on open surf coast.
Life Cycle: Annual; ovigerous females June–September; several broods per female.

Corophium simile Shoemaker 1934 Plate LXV.1   (sĭm' ĭl ē)

L. 3.5 mm. Body medium broad. Urosome segments fused, lateral ridge distinct; uropods 1 and 2 arising ventrally. Head, rostrum medium strong, broadly acute, tip level with anterolateral head lobes. Antenna 2, peduncular segment 4 essentially alike in ♂ and ♀, with strong posterodistal curved spine; segment 4 stronger and more setose in ♂; peduncular segment 5 lacking posterior marginal tubercle. Antenna 1 (both sexes), segment 2 long, equal to segment 1; flagellum of 5–6 segments.

   Mandibular palp, setal lobe short, distal margin straight; segment 2 slightly longer than 1.

   Gnathopod 1, segment 5 longer than 6; dactyl strongly exceeding palm.

Gnathopod 2, segment 5 distinctly shorter than 2, posterior margin completely occluded by segment 4; segment 6 about equal to segment 2; dactyl with 3 posterior marginal teeth.

Peraeopods 3 and 4, anterior margin of segment 2 with a few short setae; anterior margin of segment 4 with simple setae at distal apex; segment 5 very short, partly overhung by segment 4; dactyls nearly straight, longer than segments 5 and 6 combined. Peraeopod 5, margins of basis nearly bare. Peraeopod 7 medium long, segment 6 not longer than basis.

Uropod 1, inner margin of peduncle lacking spines; outer margin of inner ramus lined with 3–5 variable spines. Uropod 2, rami subequal; outer margin of outer ramus with 4 slender spines. Uropod 3, ramus moderately broad, set at an angle to body axis, smaller than telson.

Distribution: American Atlantic coast: from south side of Cape Cod and Long Island Sound, south through the Middle Atlantic states to Florida and the eastern Gulf states.
Ecology: Builds tubes at bases of sponges, tunicates, etc., on sandy and sandy mud bottoms, in shallows to depths of over 15 m; relatively uncommon.
Life Cycle: Presumably annual; ovigerous females from June–September in New England.

Family CHELURIDAE   (kēl ūr' ǐ dē)

Body depressed, setose dorsally. Coxal plates short, separating distally. Urosome segments fused. Head, frontal margin continuous; rostrum lacking, eyes lateral. Antenna 1 short; accessory flagellum present. Antenna 2 short, squamiform; flagellum clavate. Sexual dimorphism strong in urosome and appendages.

Upper lip truncate apically. Lower lip lacking inner lobes. Mandible, molar strong, palp 3-segmented. Maxilla 1, inner plate moderate, apically setose; outer plate with 6 apical spine-teeth. Maxilliped, outer plate not enlarged.

Gnathopods weakly subchelate or parachelate (both sexes); gnathopod 1 slightly the stronger. Peraeopods 3 and 4 short, basis not dilated. Peraeopods 5–7 short, successively longer; basis narrow.

Pleon side plates shallow, overlapping. Pleopods powerful, peduncle broadened inwardly. Uropods 1–3 dissimilar in shape and size. Telson entire. Coxal gills saclike, on peraeon segments 2–6. Brood plates sublinear, short-setose.

KEY TO WORLD GENERA OF CHELURIDAE

1. Uropod 3, inner ramus small but distinct; pleon segment 1, with strong dorsal spine (esp. in ♂) . . . . *Chelura* (p. 207).
   Uropod 3, inner ramus lacking; pleon dorsally smooth . . . . 2.

2. Uropod 2 biramous; gnathopod 1 powerful in ♂ . . . . *Tropichelura* Barnard.
   Uropod 2 lacking rami; gnathopod 1 weak . . . . *Nippochelura* Barnard.

Genus *Chelura* Philippi 1839   (kēl ūr′ ă)

Superior antennal line absent, except to define lateral eye lobes.
Gnathopods small; parachelate. Pleosome 1 dorsally spinose.
Uropod 2 biramous, short. Uropod 3 biramous, inner ramus small.

Chelura terebrans (Philippi) 1839 Plate LXIX.1   (tĕr′ĕ brănz)

L. 4–6 mm. Pleosome segment 1 bearing large middorsal posteriorly directed spine, large in ♂. Posterior margins of body segments and margins of
coxal plates lined with setae. Head with prominent frontal process. Eye oval,
black, lateral. Antenna 1, flagellum of 5–6 segments; accessory flagellum 2-
segmented, terminal segment very short. Antenna 2 longer than 1, heavily
setose; flagellum consisting of one large squamiform segment and two minute terminal segments.

Gnathopod 1, segment 5 shorter than 6. Gnathopod 2, segments 5 and 6
subequal, more slender than in gnathopod 1; anterior margins of segments
2, 5, and 6 lined with plumose setae. Peraeopods 3 and 4 subequal; bases
slightly expanded, anterior margins lined with plumose setae. Peraeopods
5–7, bases moderately expanded, posterior margins lined with long plumose
setae; segment 4, posterodistal angle prolonged halfway along segment 5;
dactyls short, recurved.

Abdominal side plates 1–3 shallow, margins setose, hind corner sharply
rounded. Pleopods, inner ramus shorter than outer, segments broad. Uropod
1, peduncle more than twice rami; inner ramus expanding distally, with 3
apical spines. Uropod 2 short, peduncular inner margin with winglike lobe,
larger and more setose in ♂; rami short, subequal, apex truncate, serrate, setose (♂). Uropod 3, peduncle short, broad; outer ramus very large, margin
dentate, paddlelike in ♀; elongate, rudderlike in ♂; inner ramus rudimentary.
Telson short, subtriangular, with midventral carina.

Distribution: Cool-temperate (nontropical) waters of northern and southern
hemispheres; in North Atlantic, from Cape Cod and Vineyard Sound south to
Chesapeake Bay and northern Florida.
Ecology: Constructs shallow surface burrows in wooden pilings, floats,
stakes, etc., already occupied by *Limnoria* (gribble), from just above low
water level to a few meters in depth.
Life Cycle: Annual; ovigerous females from May–September; several broods
per female.

Family PODOCERIDAE   (pŏd ö sēr′ĭ dē)

Body slender, depressed. Urosome 1 elongate; 3 very small or fused with 2.
Coxal plates very small, widely separated. Sexual dimorphism strongly ex-

pressed in gnathopods. Head, rostrum lacking; eyes lateral. Antennae elongate (esp. peduncles); strongly setose posteriorly. Accessory flagellum present, rudimentary or occasionally lacking.

Upper lip bilobed. Lower lip with inner lobes. Mandible, molar strong; palp strong, 3-segmented. Maxilla 1, inner plate vestigial; outer plate with about 6 apical spine-teeth.

Gnathopods 1 and 2 normally subchelate, 2nd largest (esp. in ♂). Peraeopods 3 and 4, bases inflated. Peraeopods 5–7 increasing successively in length; basis narrow.

Pleon side plates small, separated. Pleopods, 2 or 3 pairs present. Uropods 1 and 2 biramous, slender. Uropod 3 abnormal, rudimentary or lacking. Telson entire, small, oval. Coxal gills saclike, from 2–4 pairs, absent from peraeon 7. Brood plates large, lamellate, with numerous simple marginal setae.

KEY TO AMERICAN ATLANTIC GENERA OF PODOCERIDAE

1. Peraeon segments 6 and 7 fused; urosome 2-segmented; uropod 3 lacking . . 2.
   Peraeon segments 6 and 7 separated; urosome 3-segmented, 1st normal; uropod 3 very small . . . . *Podocerus.*
2. Uropod 2 distinct, biramous; accessory flagellum 1–2 segmented *Dulichia* (p. 208).
   Uropod 2 very small, uniramous; accessory flagellum 3-segmented . *Paradulichia.*

Genus *Dulichia* Kroyer 1845   (dū līk′ ĭ ă)

Peraeon segments 6 and 7 coalesced. Urosome segments 2 and 3 coalesced. Head and rostrum elevated anterodorsally; eyes lateral, round. Coxal plates small, separated. Antennae slender, 1st longer. Accessory flagellum short. Mandible, palp slender, terminal segment short.

Gnathopod 1 simple or subchelate; gnathopod 2, subchelate, more powerful in ♂. Peraeopods 3 and 4 slender. Peraeopods 5–7 slender; segment 4 elongate; dactyls curved.

Pleopods strong; peduncles elongate. Uropods 1 and 2 biramous, inner ramus shorter. Uropod 3 lacking. Telson entire, rounded. Coxal gills, 2–4 pairs, lacking on peraeon 6–7.

KEY TO NEW ENGLAND SPECIES OF *DULICHIA*

1. Some body segments dorsally and laterally processiferous; peraeopods 3 and 4 much weaker than 5–7, basis little expanded; coxal gills present on peraeon segments 2–5 inclusive . . . . *D. spinosissima* (p. 209).
   Body segments dorsally and laterally smooth; peraeopods 3 and 4 nearly as large as 5 and 6, bases stoutly expanded; coxal gills lacking on peraeon 2 . . . . . 2.
2. Coxal gills present (small) on peraeon 5; coxa 1 (♂) produced anterodistally as long slender spine; gnathopod 2 (♀), segment 6 slightly longer than segment 5
   . . . . *D. monacantha* Metzger.

Coxal gills present only on peraeon 3 and 4; coxa 1 (♂), anterodistal angle not produced; gnathopod 2 (♀), segment 6 nearly twice length of segment 5
. . . . *D. porrecta* (p. 209).

Dulichia porrecta (Bate) 1857 Plate LXIX.2   (pŏ rĕkt′ ă)

L. 5–7 mm. Head and body segments smooth dorsally and laterally. Coxal plates 1–5 shallow; coxa 2 deepest, anteroventral angle acutely produced (♂). Head, rostrum short. Antennae 1 and 2, peduncles slender, lacking distal processes; accessory flagellum 3-segmented.

Gnathopod 1 (both sexes), segment 5 slightly broader proximally; segment 6 shorter, lacking palm. Gnathopod 2 (♂) powerfully subchelate; basis slender, not produced anterodistally; segment 5 small, deep; segment 6 much larger, palm very oblique, long-setose, with acute tooth near hinge and long sharp thumblike process at posterior angle; dactyl very large, sinuous, setose behind, with small knob near hinge. Gnathopod 2 (♀) simple; segment 5, markedly shorter, deeper than segment 6; dactyl short-setose behind. Peraeopods 3 and 4 nearly as strong as 5 and 6; bases expanded, broadest medially. Peraeopods 5 and 6 subequal, much shorter than 7; segments 4–6 in all weakly armed.

Uropod 1, peduncle with 1 strong outer marginal spine and numerous fine closely set spinules. Telson ovate, margins smooth. Coxal gills on segments 3 and 4 only.

Distribution: Arctic-boreal, North Atlantic and Pacific; in North America south to the Gulf of St. Lawrence, Gulf of Maine, and Cape Cod Bay.
Ecology: In cold high-salinity waters, subtidally to more than 100 m. Clings to hydroids and bryozoans; occasionally taken in plankton.
Life Cycle: Probably annual; ovigerous females March–July.

Dulichia spinosissima Krøyer (no illustration)   (spĭn ōs ĭs′ ĭm ă)

L. 10–14 mm. Peraeon segments 1–5 with lateral spinose processes; peraeon segment 7 and pleosome segments 1 and 2 with paired middorsal knoblike processes; pleon 3 with single strong middorsal spinous process. Coxal plates 1–4 as deep as long, acutely produced anteroventrally. Head, rostrum large, laminar, deep, extending back onto head as median carina. Antenna 1, peduncle segments 2 and 3, and antenna 2, peduncle segments 4 and 5 each with blunt anterodistal extension.

Gnathopod 1 (both sexes), segment 5 longer than 6; posterior lobe deepest proximally; segment 6, palm short but distinct, greatly overlapped by dactyl, posteriorly short setose. Gnathopod 2 (♂) larger but powerfully subchelate; segment 2 with large spinose anterodistal lobe; segment 5 short and deep; segment 6, palm concave, with sharp tooth near hinge and at posterior angle; dactyl, both margins setose; gnathopod 2 (♀) less powerful, palm smoothly convex, evenly merging with posterior margin. Peraeopods 3 and 4 much weaker than 5 and 6, bases little expanded, margins bare. Peraeopods 5–7 progressively larger and heavier; segments 4–6 margins strongly spinose.

Pleopods elongate, peduncles longer than bases of peraeopods 5–7. Uropods 1 and 2, peduncular outer margin strongly spinose throughout. Telson linguiform, broadest basally, with paired fine lateral setae. Coxal gills present on peraeon segments 2–5, small on segment 5.

Distribution: Amphi-Atlantic boreal; in North America, south from Labrador and the Gulf of St. Lawrence to Long Island Sound.
Ecology: In cold waters, subtidally to more than 100 m.
Life Cycle: Probably annual.

PLATES

1. _____

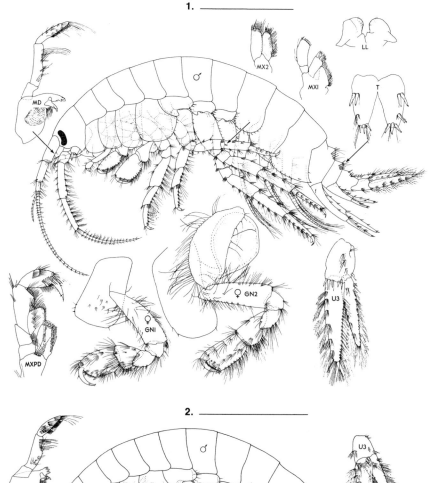

2. _____

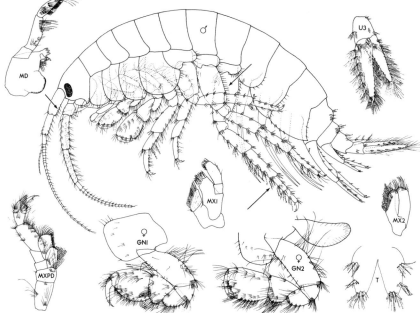

1. *Gammarus oceanicus* Segerstråle   **2.** *Gammarus setosus* Dementieva

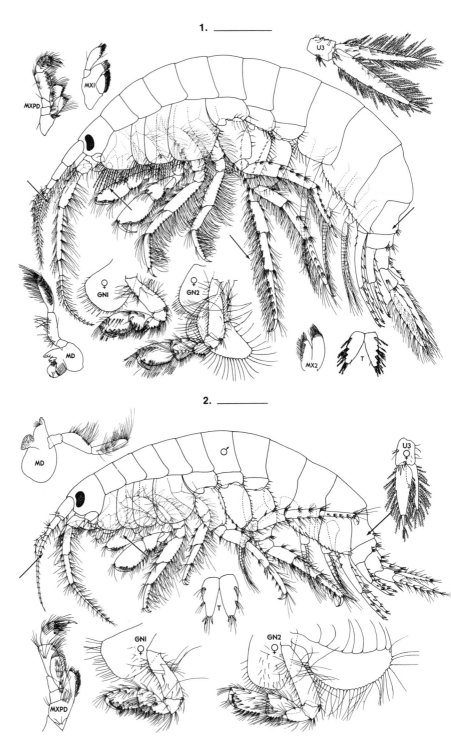

1. *Gammarus annulatus* Smith

2. *Gammarus lawrencianus* Bousf.

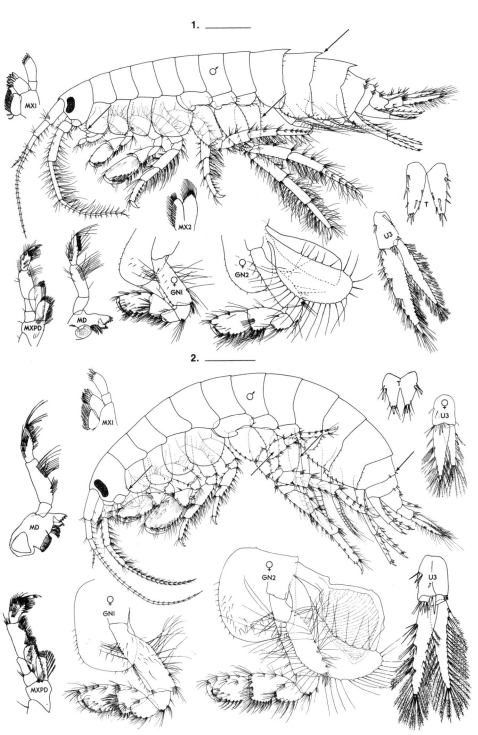

1. *Gammarus mucronatus* Say

2. *Gammarus palustris* Bousf.

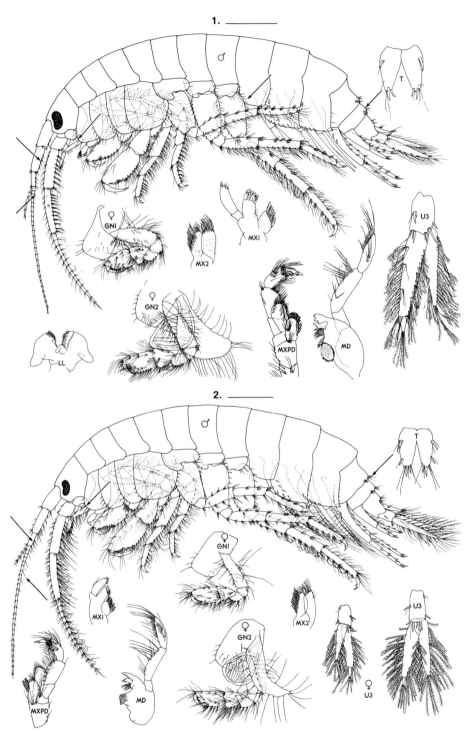

1. *Gammarus tigrinus* Sexton   **2. *Gammarus daiberi* Bousf.**

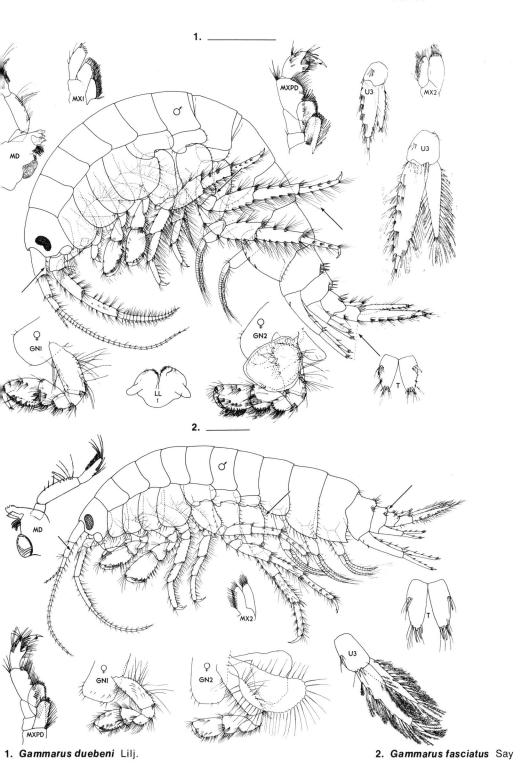

1. *Gammarus duebeni* Lilj.   **2.** *Gammarus fasciatus* Say

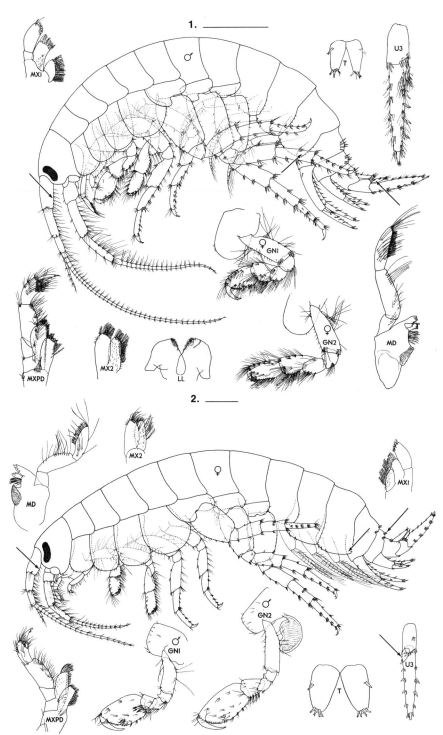

1. *Marinogammarus obtusatus* Dahl

2. *Marinogammarus stoerensis* (Reid)

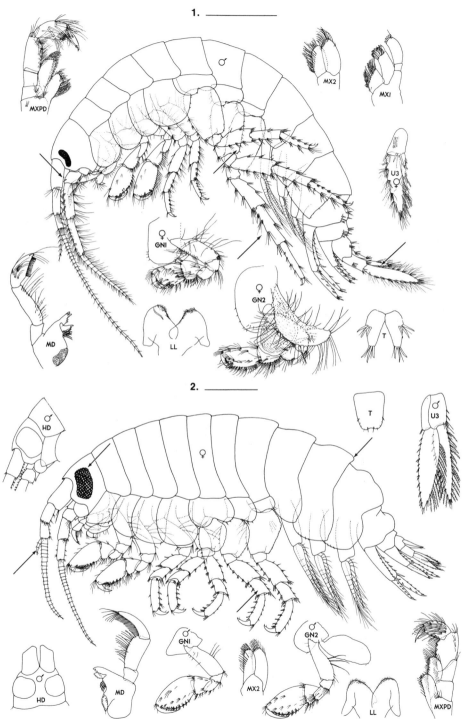

1. Marinogammarus finmarchicus Dahl   **2. Gammarellus angulosus** (Rathke)

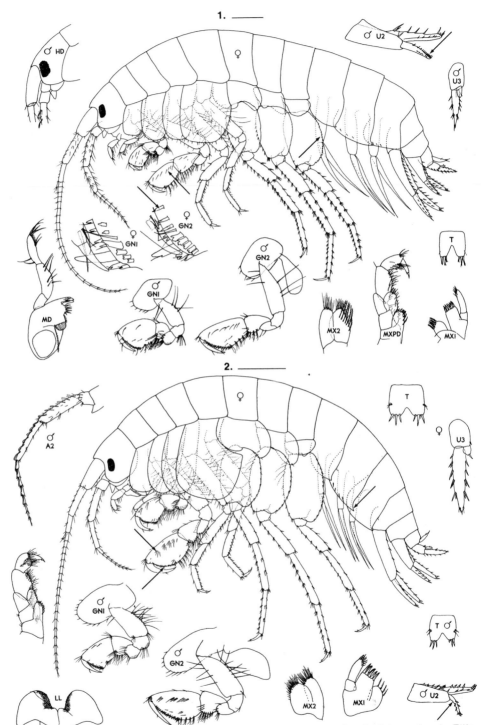

1. *Crangonyx pseudogracilis* Bousf.   2. *Crangonyx richmondensis richmondensis* Ellis

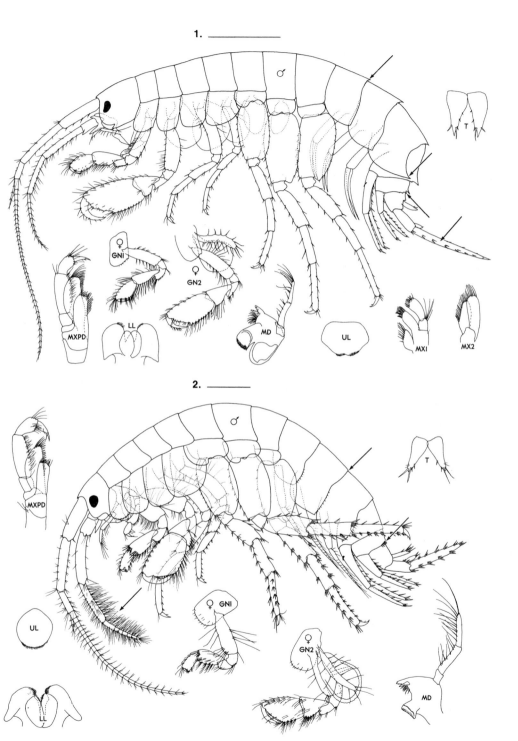

1. **Melita dentata** (Kr.)

2. **Melita nitida** Smith

PLATE X

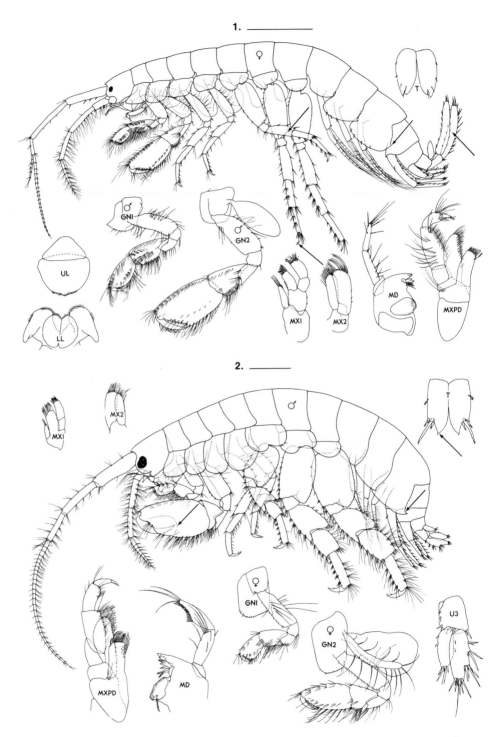

1. *Maera danae* Stimpson

2. *Elasmopus levis* Smith

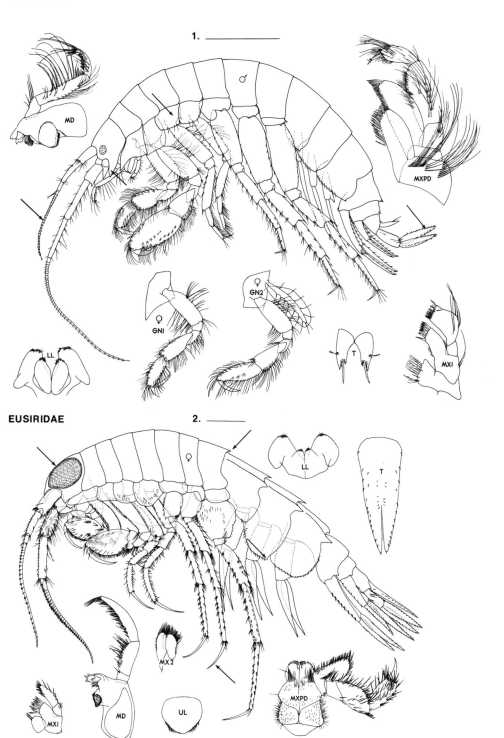

EUSIRIDAE

1. *Casco bigelowi* (Blake)   2. *Rhachotropis oculata* (Hansen)

1. ——

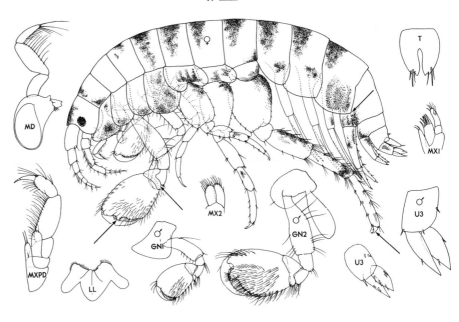

2. ——

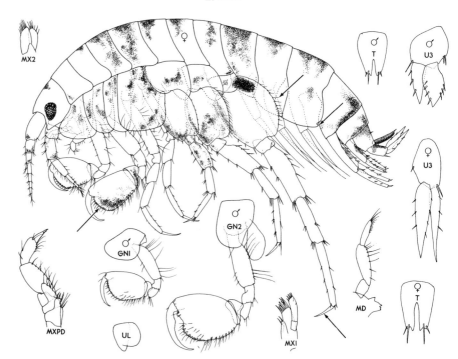

1. *Listriella clymenellae* Mills

2. *Listriella barnardi* Wigley

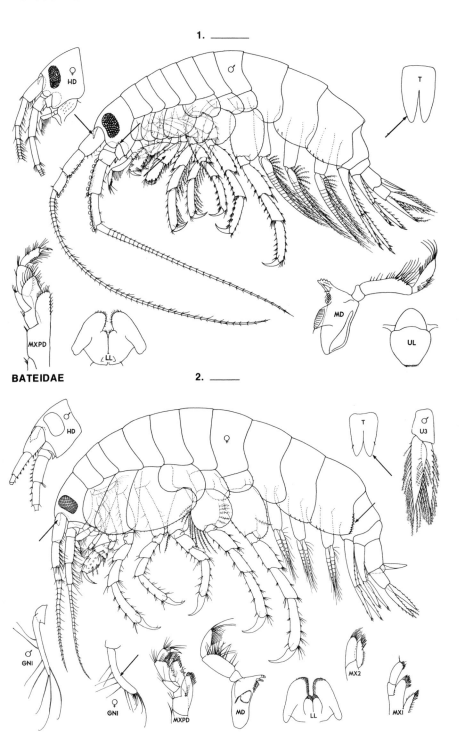

BATEIDAE

1. *Pontogeneia inermis* (Kr.)

2. *Batea catharinensis* Muller

1.

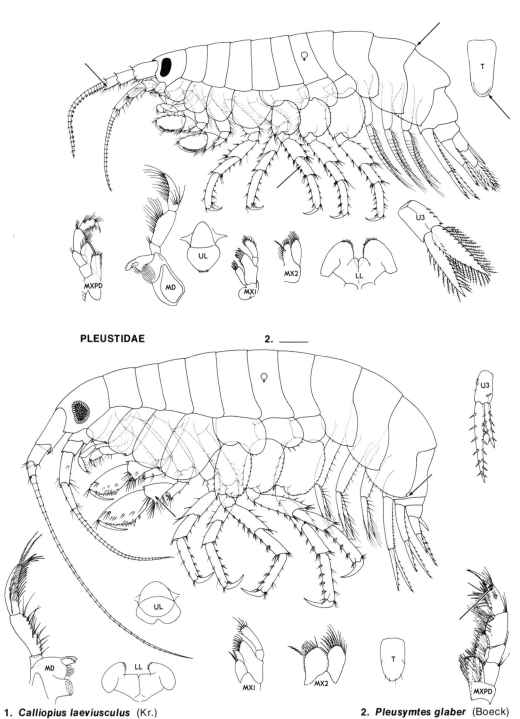

PLEUSTIDAE 2.

1. *Calliopius laeviusculus* (Kr.)

2. *Pleusymtes glaber* (Boeck)

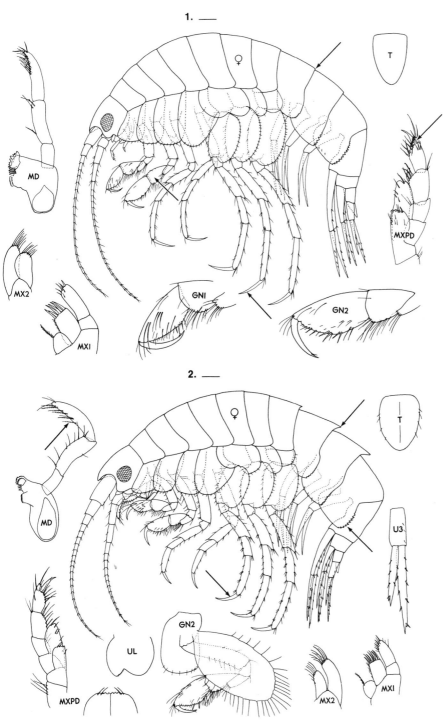

1. *Stenopleustes inermis* Shoem.

2. *Stenopleustes gracilis* Holmes

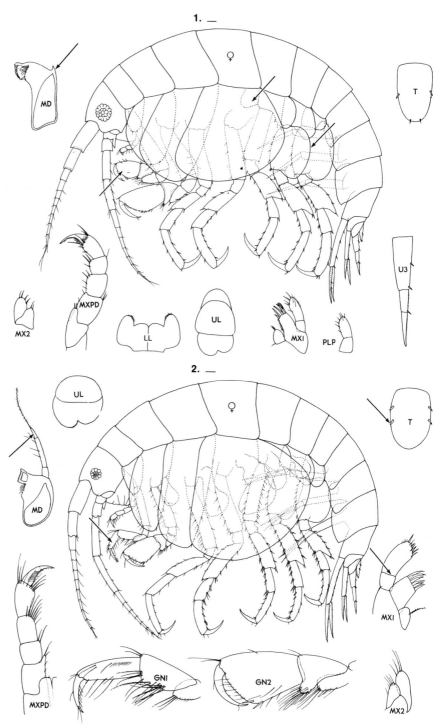

1. *Stenothoe minuta* Holmes

2. *Proboloides holmesi* n. sp.

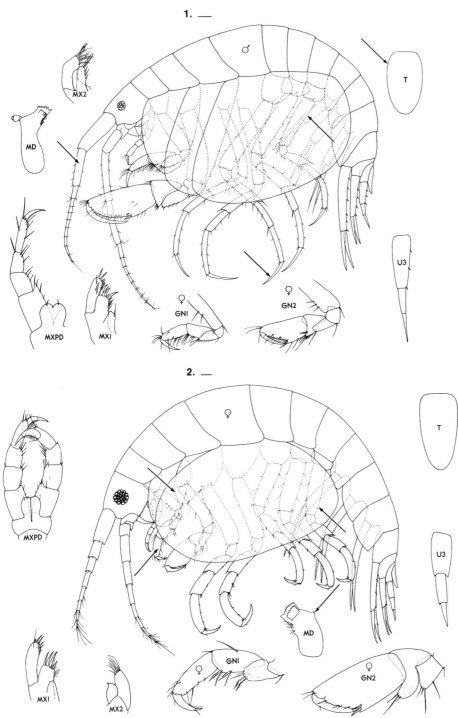

1. *Metopella angusta* Shoem.

2. *Parametopella cypris* (Holmes)

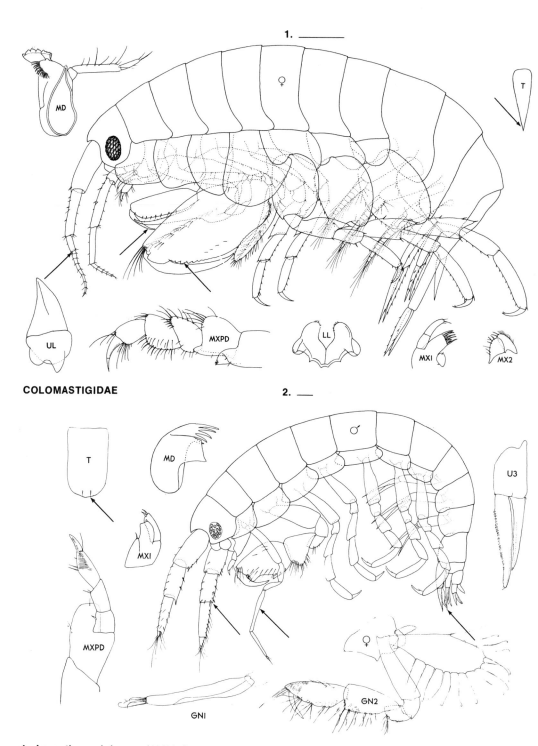

1. *Leucothoe spinicarpa* (Abildg.)

2. *Colomastix halichondriae* n. sp.

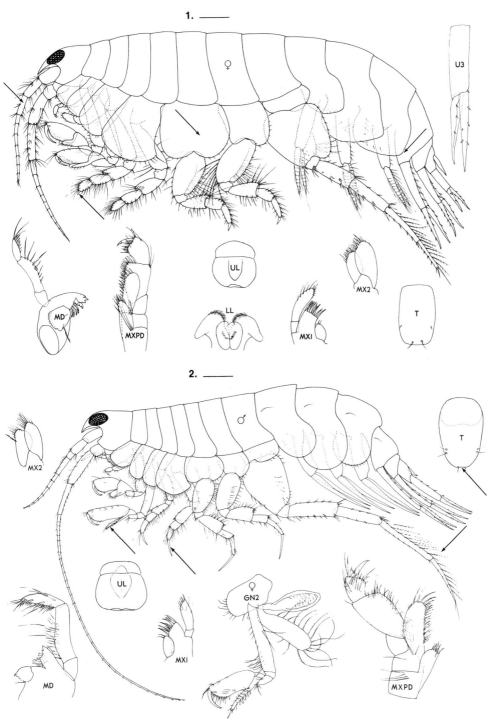

1. *Monoculodes edwardsi* Holmes

2. *Monoculodes intermedius* Shoem.

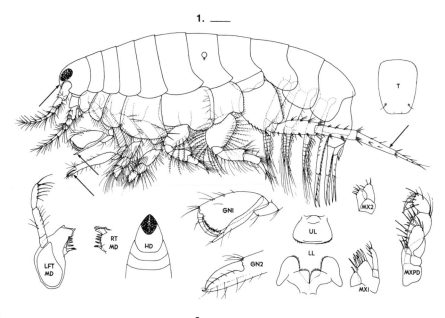

ARGISSIDAE

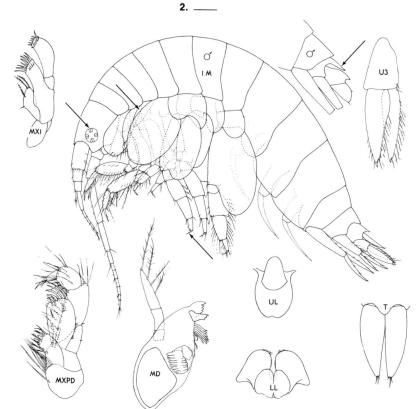

1. *Synchelidium americanum* n. sp.   2. *Argissa hamatipes* (Norman)

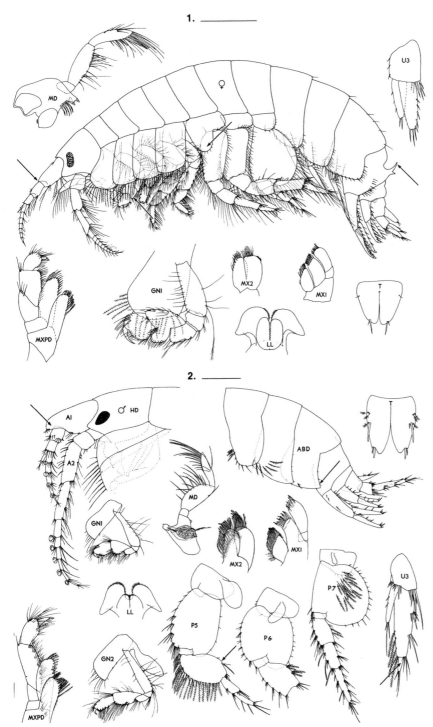

1. *Pontoporeia femorata* Kr.   2. *Amphiporeia lawrenciana* Shoem.

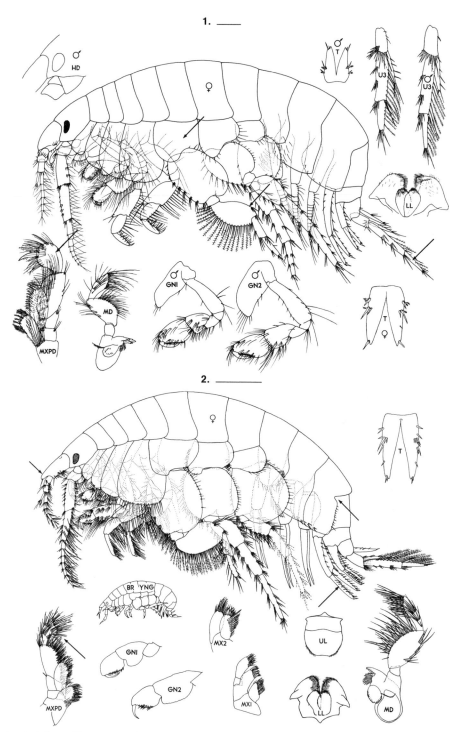

1. *Amphiporeia virginiana* Shoem.   2. *Amphiporeia gigantea* n. sp.

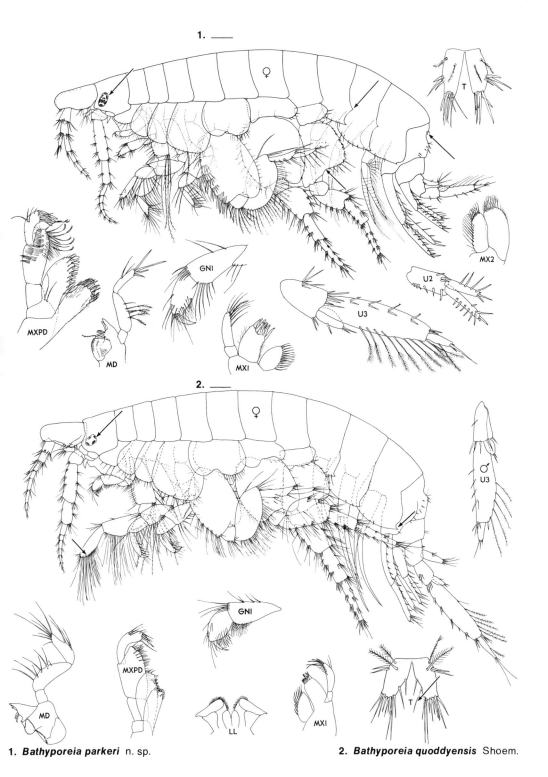

1. *Bathyporeia parkeri* n. sp.   **2. *Bathyporeia quoddyensis*** Shoem.

1. *Protohaustorius wigleyi* Bousf.   **2.** *Protohaustorius deichmannae* Bousf.

1. *Parahaustorius longimerus* Bousf.   **2.** *Parahaustorius holmesi* Bousf.

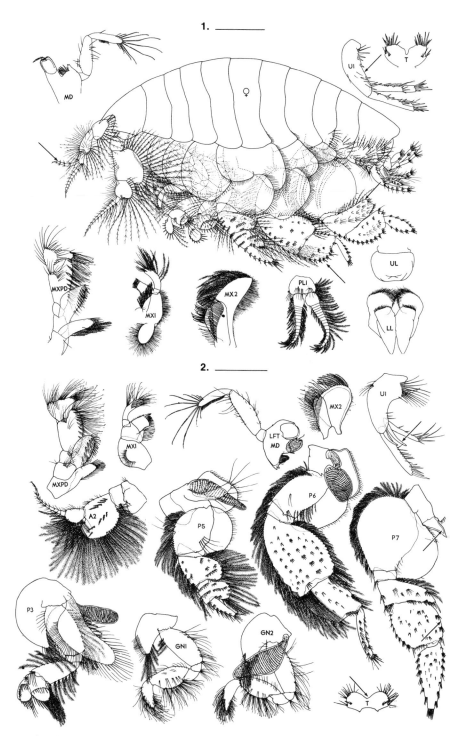

1. Haustorius canadensis Bousf.   **2. Parahaustorius attenuatus** Bousf.

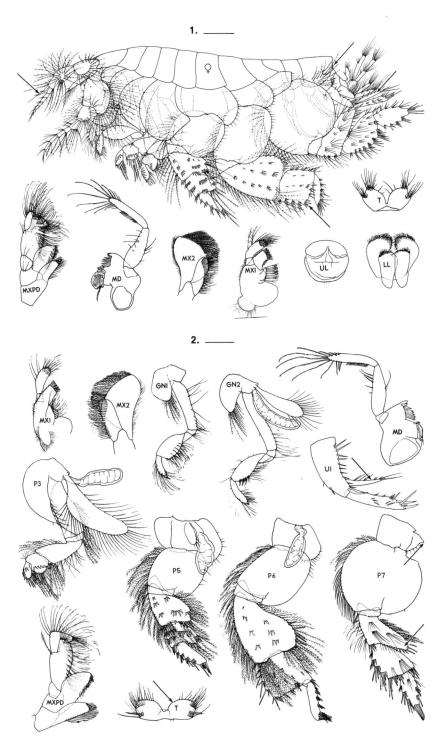

1. *Acanthohaustorius millsi* Bousf.   2. *Acanthohaustorius shoemakeri* Bousf.

1. *Acanthohaustorius intermedius* Bousf.   **2.** *Acanthohaustorius spinosus* Bousf.

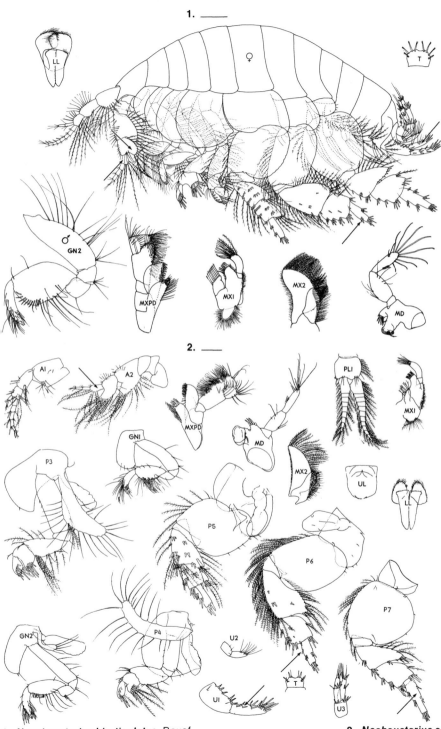

1. *Neochaustorius biarticulatus* Bousf.

2. *Neohaustorius schmitzi* Bousf.

1. *Pseudohaustorius caroliniensis* Bousf.   2. *Pseudohaustorius borealis* Bousf.

Head (dorsal view)

a. *Protohaustorius deichmannae*   b. *Protohaustorius wigleyi*   c. *Parahaustorius longimerus*   d. *Parahaustorius holmesi*   e. *Pseudohaustorius caroliniensis*   f. *Pseudohaustorius borealis*   g. *Parahaustorius attenuatus*   h. *Acanthohaustorius millsi*   i. *Acanthohaustorius shoemakeri*   j. *Acanthohaustorius intermedius*   k. *Acanthohaustorius spinosus*   l. *Haustorius canadensis*   m. *Neohaustorius biarticulatus*   n. *Neohaustorius schmitzi*

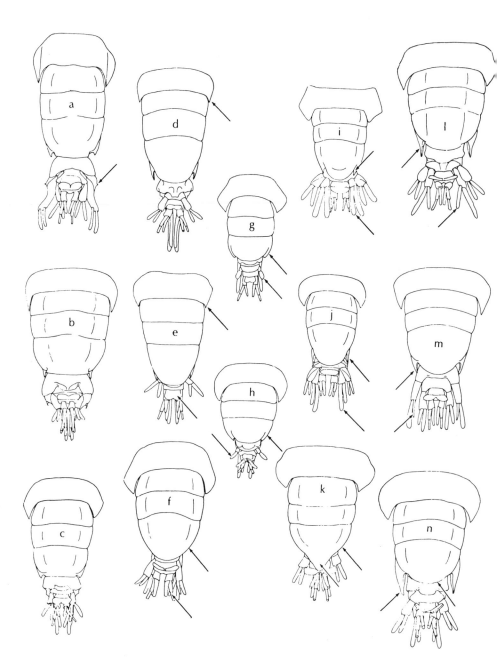

Abdomen (dorsal view)

a. *Parahaustorius holmesi*   **b.** *Parahaustorius attenuatus*   **c.** *Parahaustorius longimerus*   **d.** *Protohaustorius wigleyi*   **e.** *Protohaustorius deichmannae*   **f.** *Haustorius canadensis*   **g.** *Neohaustorius schmitzi*   **h.** *Neohaustorius biarticulatus*   **i.** *Pseudohaustorius caroliniensis*   **j.** *Pseudohaustorius borealis*   **k.** *Acanthohaustorius intermedius*   **l.** *Acanthohaustorius millsi*   **m.** *Acanthohaustorius shoemakeri*   **n.** *Acanthohaustorius spinosus*

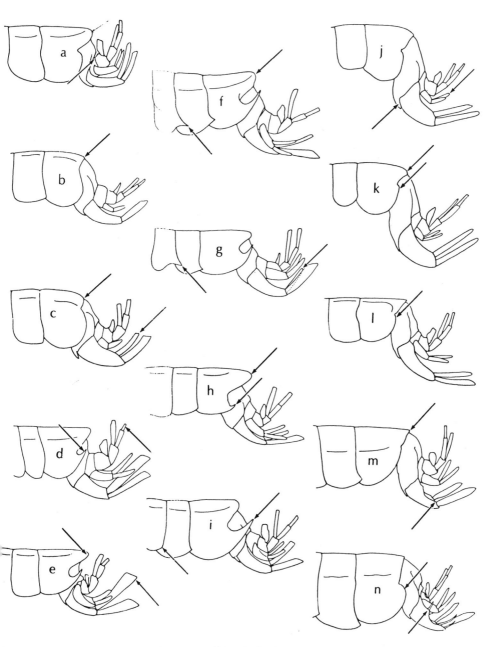

Abdomen (lateral view)

a. *Haustorius canadensis*   **b.** *Neohaustorius schmitzi*   **c.** *Neohaustorius biarticulatus*
d. *Pseudohaustorius borealis*   **e.** *Pseudohaustorius caroliniensis*   **f.** *Acanthohaustorius millsi*   **g.** *Acanthohaustorius shoemakeri*   **h.** *Acanthohaustorius intermedius*
i. *Acanthohaustorius spinosus*   **j.** *Protohaustorius wigleyi*   **k.** *Protohaustorius deichmannae*   **l.** *Parahaustorius longimerus*   **m.** *Parahaustorius holmesi*   **n.** *Parahaustorius attenuatus*

PLATE XXXIV

1. ____

2. ____

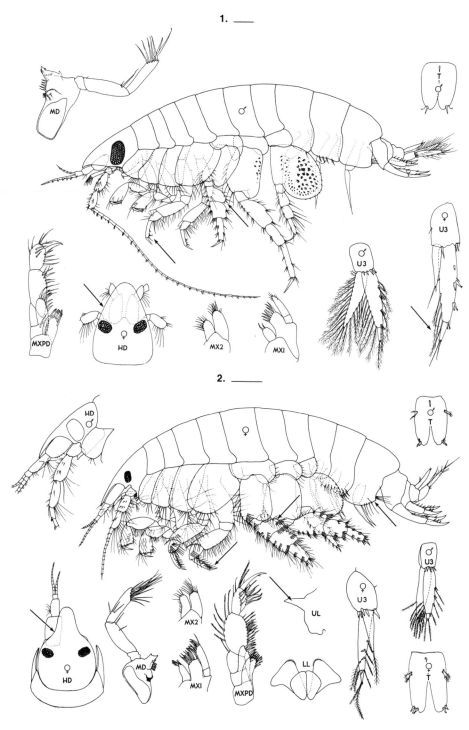

1. *Paraphoxus spinosus* Holmes

2. *Trichophoxus epistomus* (Shoem.)

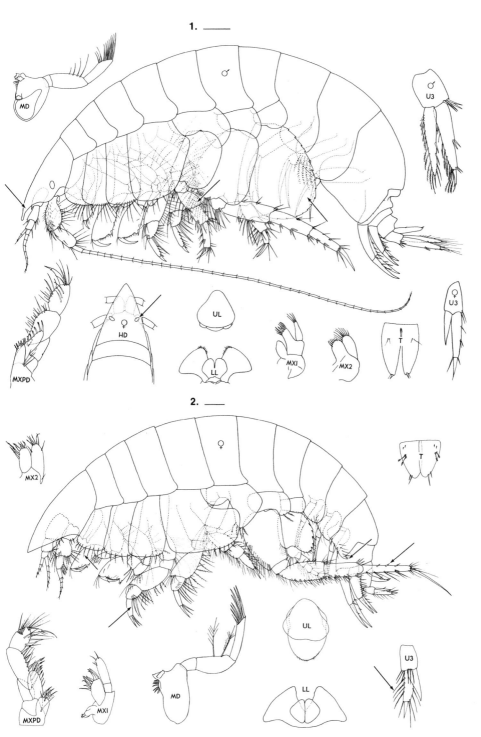

1. *Phoxocephalus holbolli* Kr.

2. *Harpinia propinqua* Sars

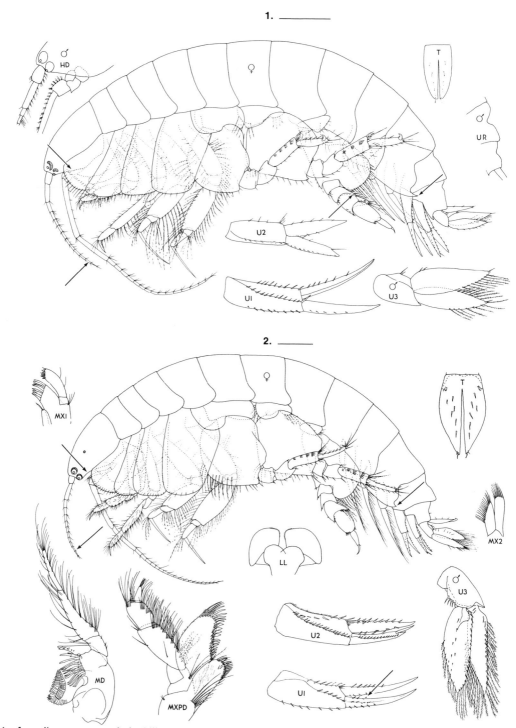

1. *Ampelisca macrocephala* Lilj.

2. *Ampelisca verrilli* Mills

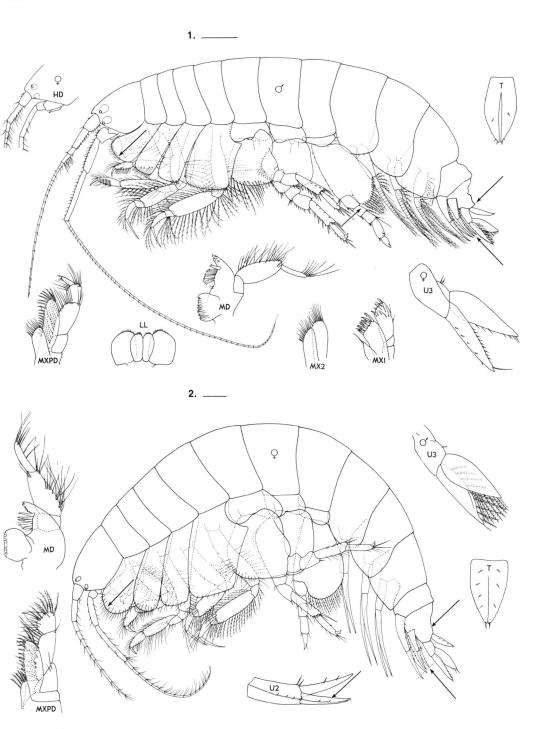

1. *Ampelisca vadorum* Mills   **2.** *Ampelisca abdita* Mills

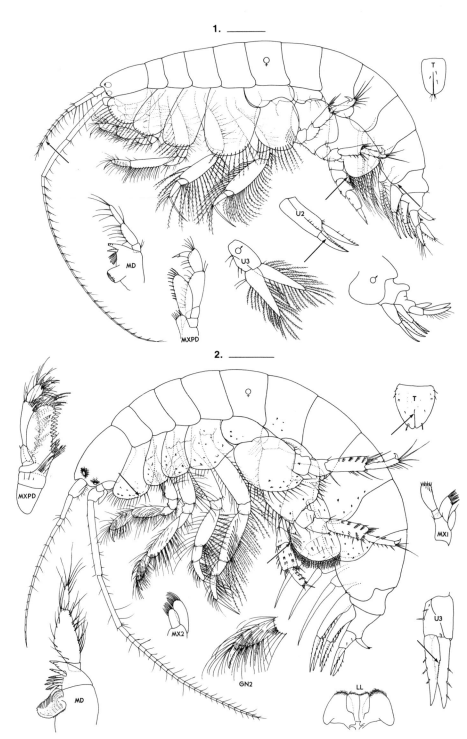

1. *Ampelisca agassizi* (Judd)

2. *Byblis serrata* Smith

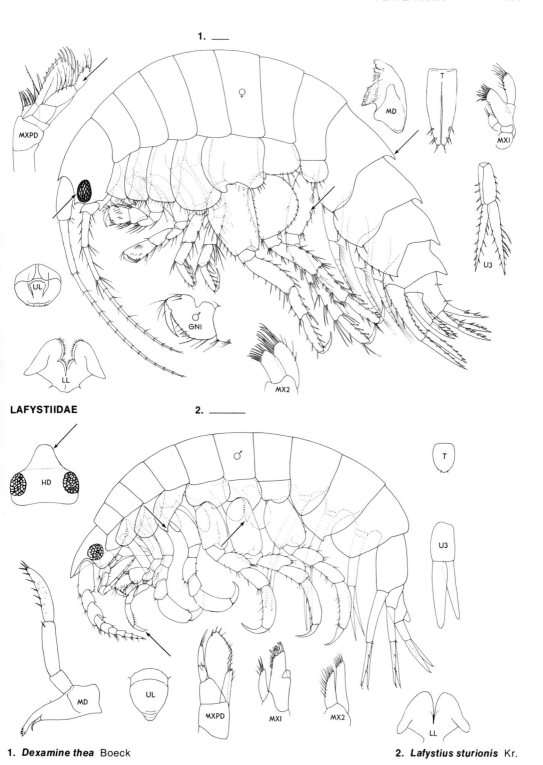

LAFYSTIIDAE

1. *Dexamine thea* Boeck   2. *Lafystius sturionis* Kr.

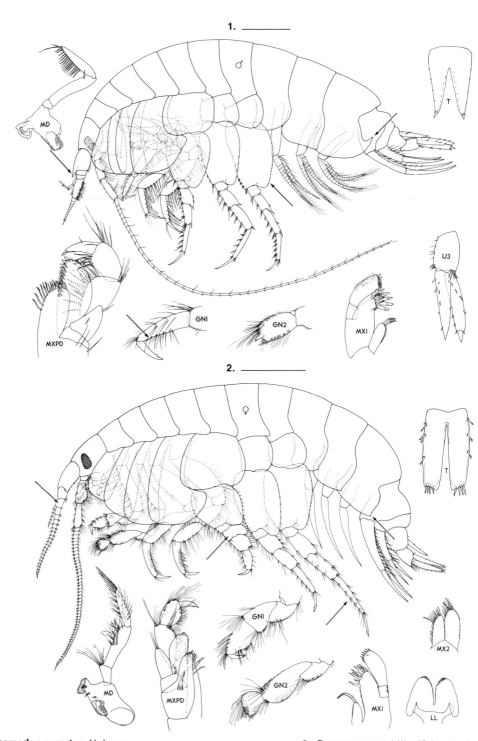

1. *Hippomedon serratus* Holmes   **2.** *Psammonyx nobilis* (Stimpson)

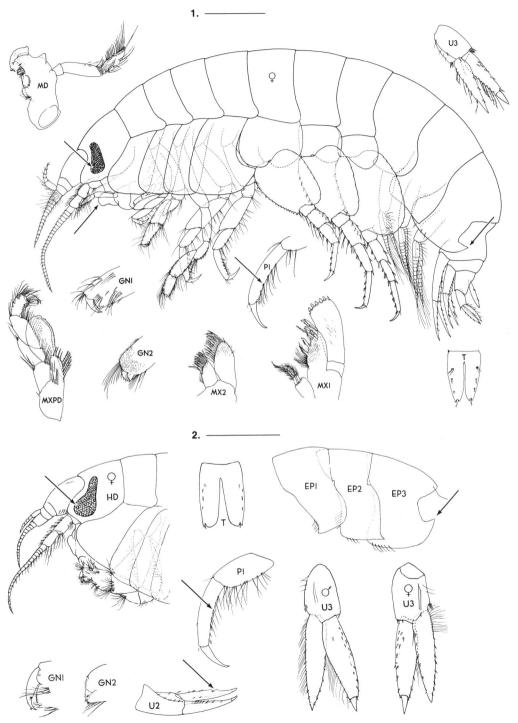

1. MD

1. U3

1. ♀

1. GN1

1. GN2

1. MX2

1. MX1

1. P1

1. MXPD

1. T

2. HD

2. ♀

2. T

2. EP1 EP2 EP3

2. P1

2. GN1

2. GN2

2. U2

2. ♂ U3

2. ♀ U3

1. *Anonyx liljeborgi* Boeck

2. *Anonyx sarsi* Steele and Brunel

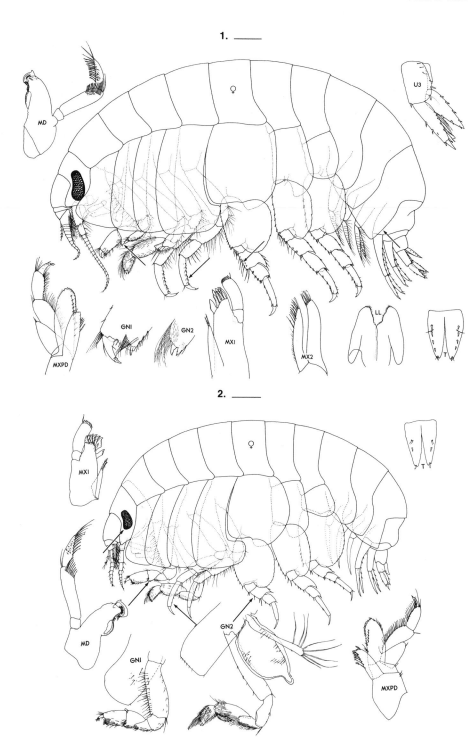

1. *Orchomenella pinguis* (Boeck)   2. *Orchomenella minuta* Kr.

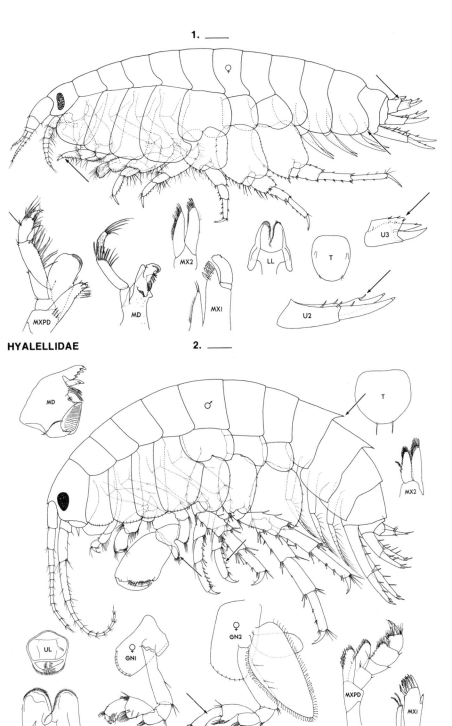

HYALELLIDAE

1. *Lysianopsis alba* Holmes   **2.** *Hyalella azteca* Saussure

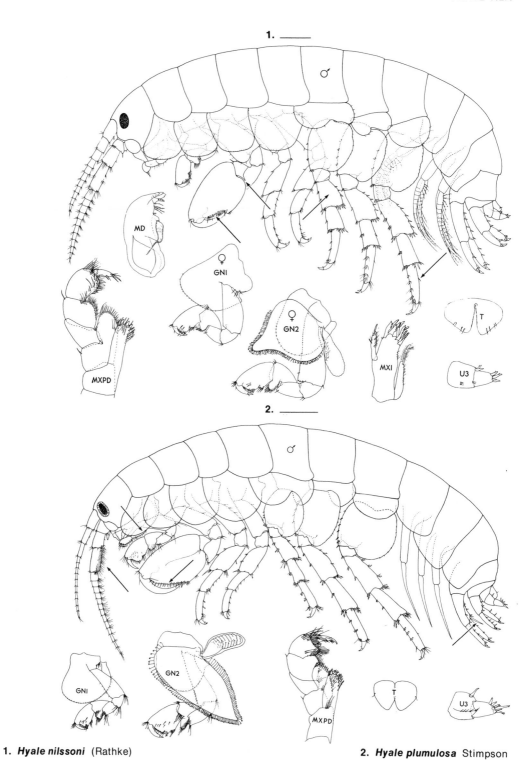

1. Hyale nilssoni (Rathke)   **2. Hyale plumulosa** Stimpson

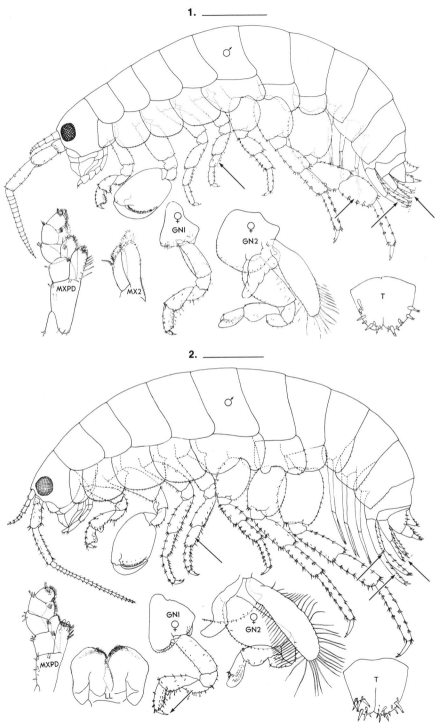

1. *Orchestia gammarella* Pallas

2. *Orchestia grillus* Bosc

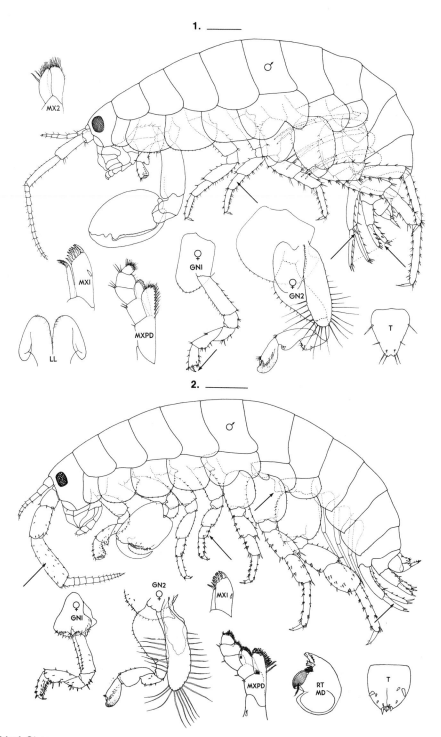

1. *Orchestia uhleri* Shoem.   **2.** *Orchestia platensis* Kr.

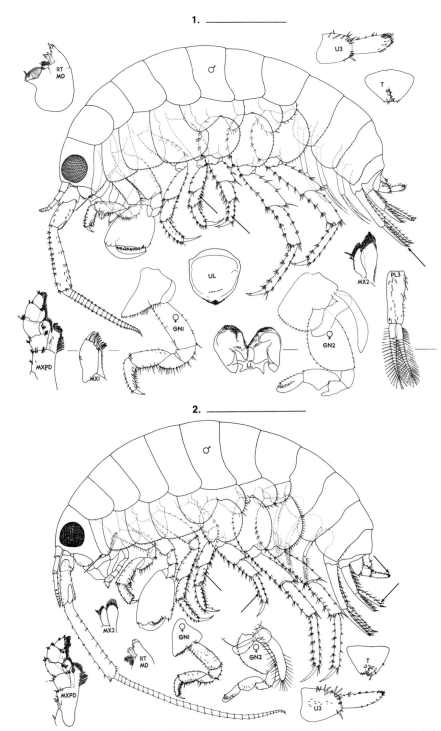

1. *Talorchestia megalophthalma* (Bate)

2. *Talorchestia longicornis* (Say)

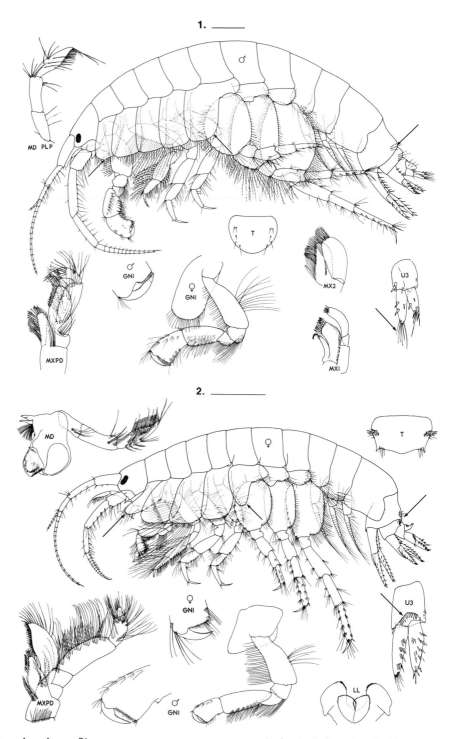

1. *Leptocheirus plumulosus* Shoem.

2. *Leptocheirus pinguis* (Stimpson)

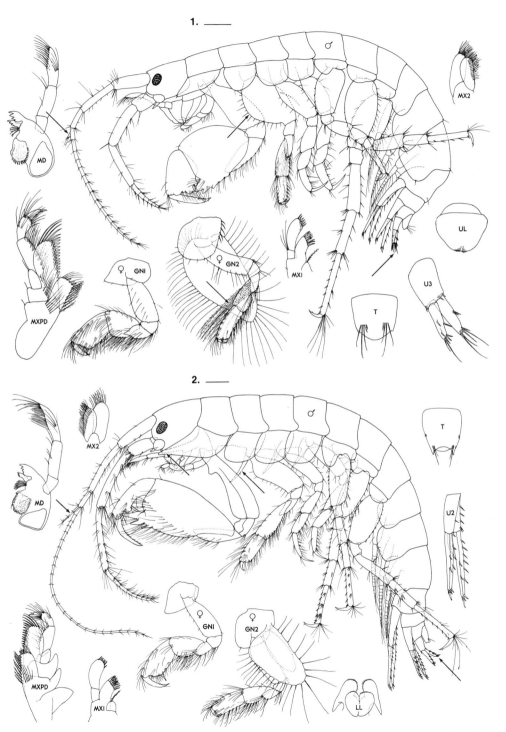

1. *Microdeutopus gryllotalpa* Costa

2. *Microdeutopus anomalus* (Rathke)

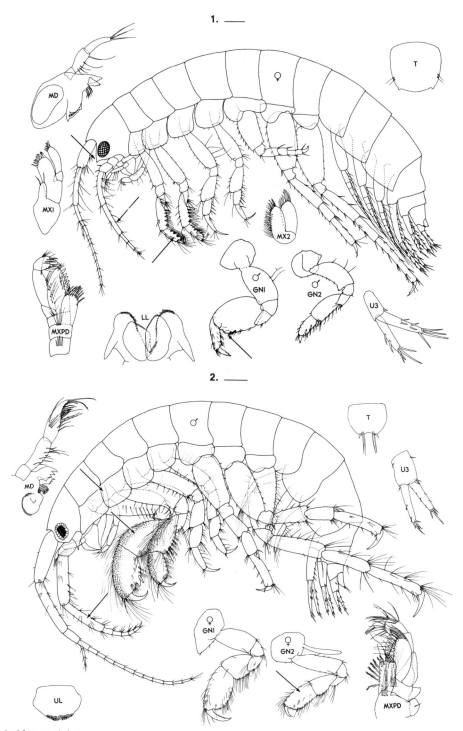

1. *Rudilemboides naglei* n. sp.

2. *Lembos websteri* Bate

PLATE LI   263

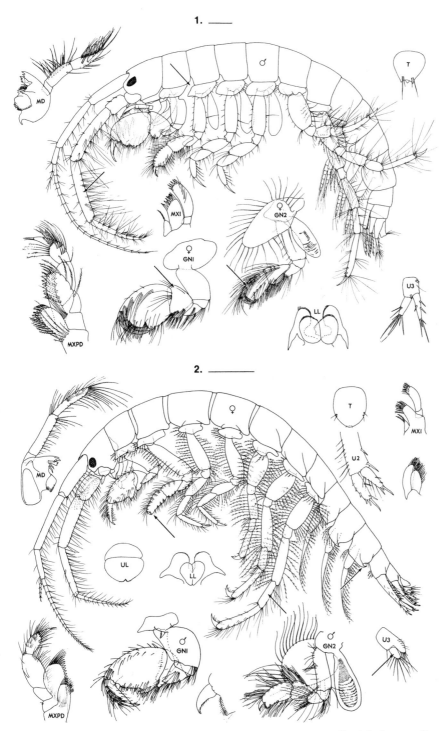

1. _____

2. _____

1. _Lembos smithi_ Holmes   **2. _Unciola irrorata_** Say

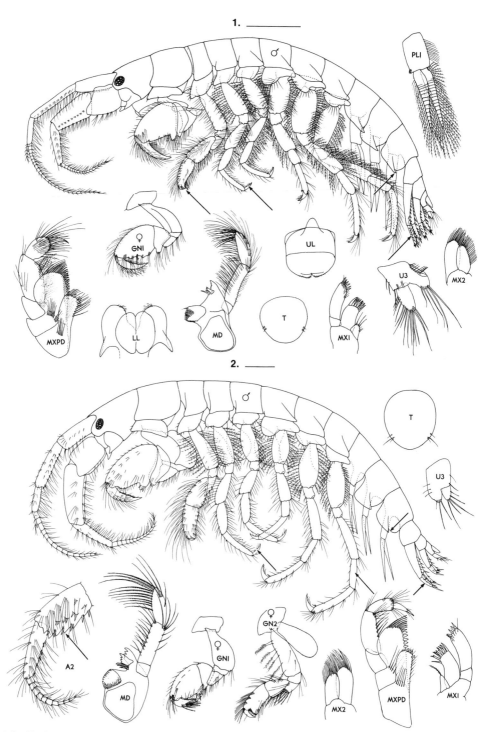

1. *Unciola dissimilis* Shoem.   **2.** *Unciola serrata* Shoem.

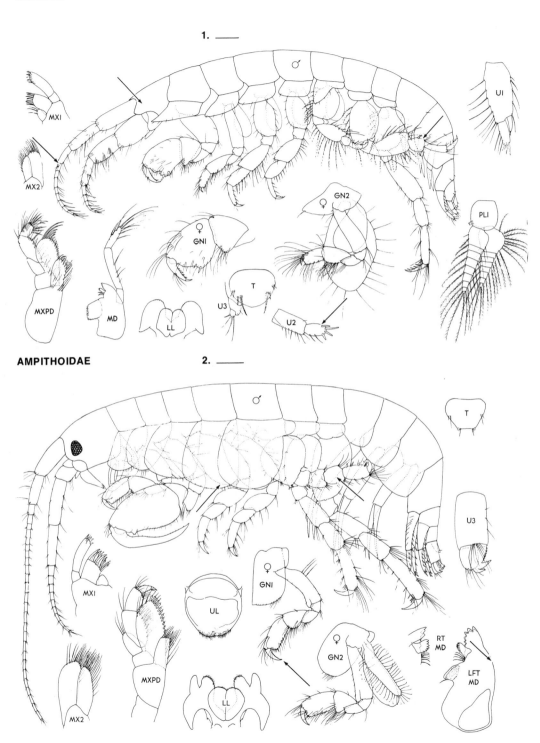

1. *Pseudunciola obliquua* (Shoem.)   **2. *Sunamphitoe pelagica*** Bate

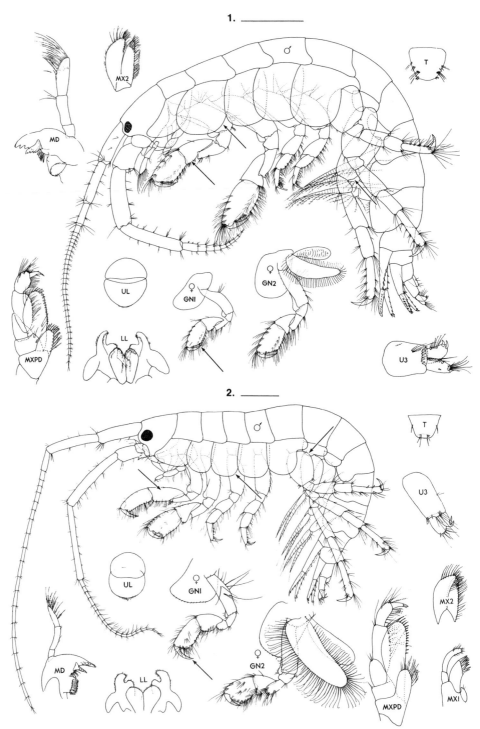

1. *Ampithoe rubricata* Montagu   **2. *Ampithoe longimana*** Smith

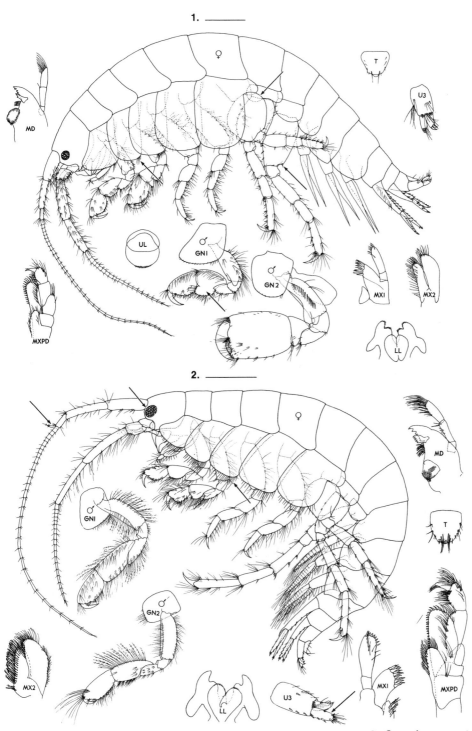

1. *Ampithoe valida* Smith   2. *Cymadusa compta* (Smith)

PLATE LVI

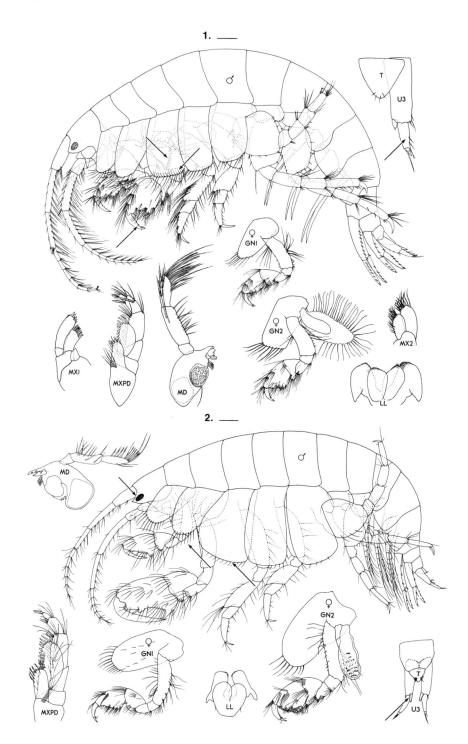

1. *Photis reinhardi* Kr.

2. *Photis macrocoxa* Shoem.

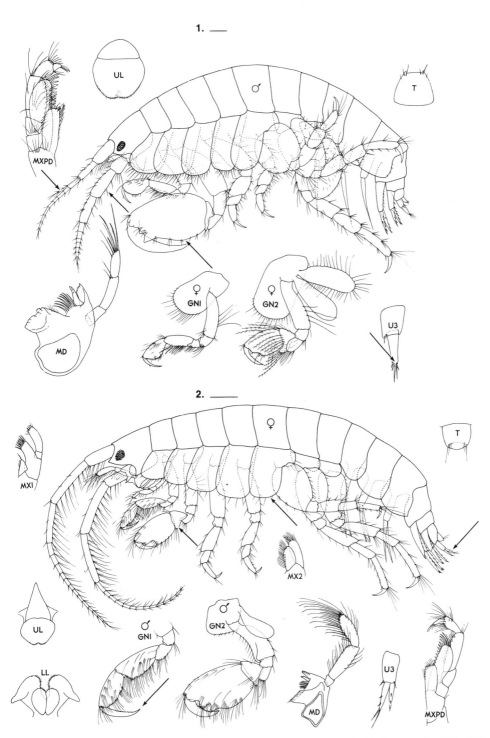

1. *Microprotopus raneyi* Wigley   **2. *Podoceropsis nitida* (Stimpson)**

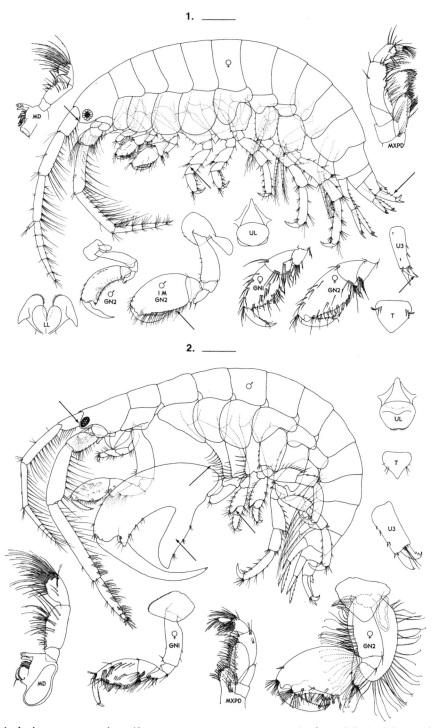

1. *Ischyrocerus anguipes* Kr.　　　　　　　　**2.** *Jassa falcata* (Montagu)

PLATE LIX   271

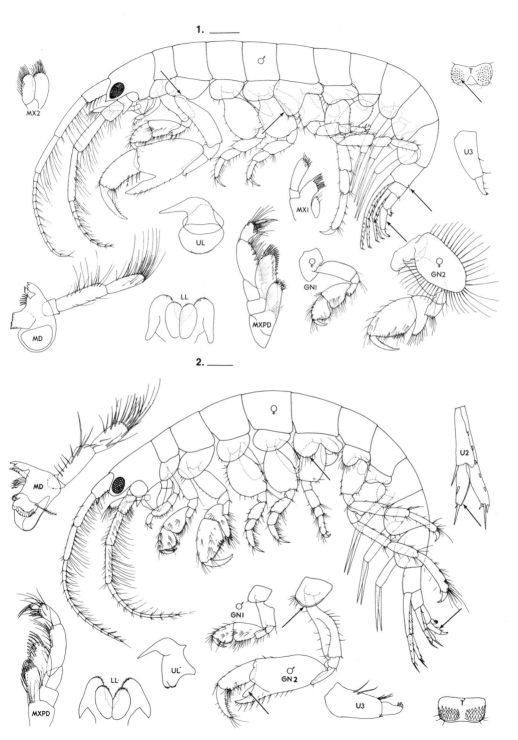

1. *Erichthonius rubricornis* Smith

2. *Erichthonius brasiliensis* (Dana)

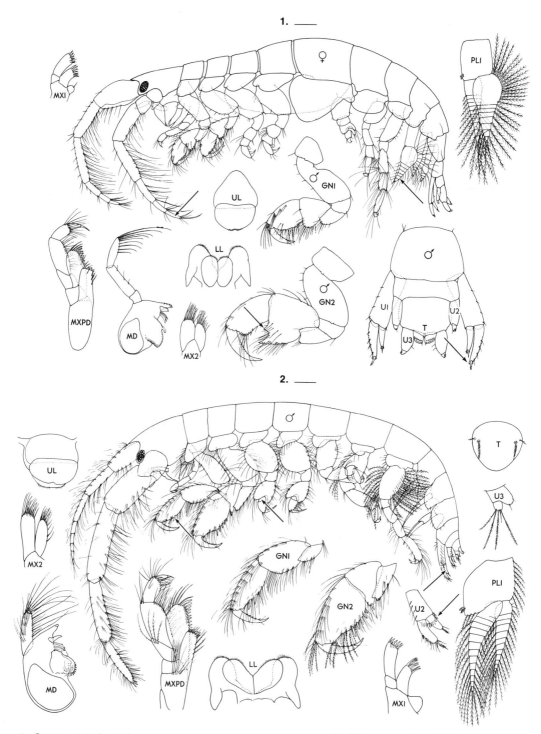

1. *Cerapus tubularis* Say

2. *Siphonoecetes smithianus* Rathbun

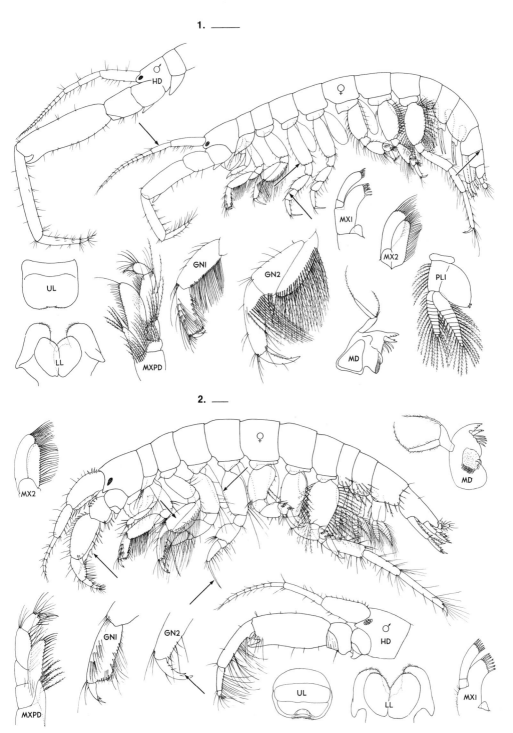

1. *Corophium volutator* (Pallas)

2. *Corophium crassicorne* Bruz.

1. ⎯⎯

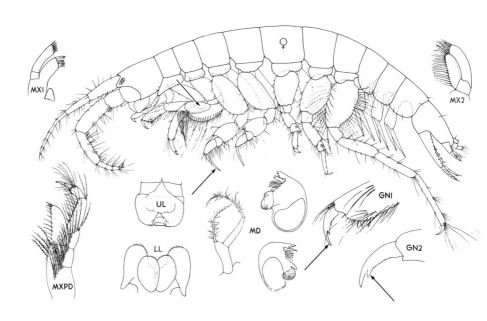

2. ⎯⎯

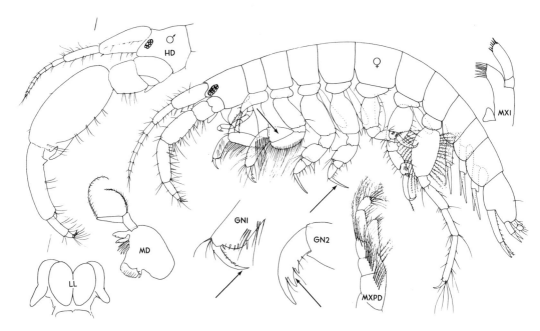

1. _Corophium bonelli_ M.-E.   **2. _Corophium acherusicum_ Costa**

1. ——

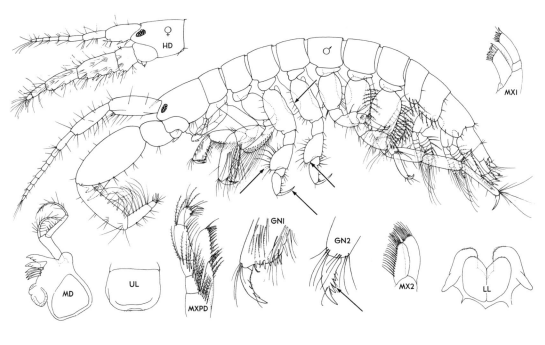

2. ——

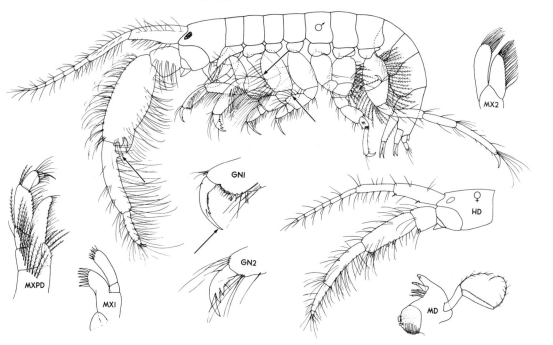

1. *Corophium insidiosum* Crawford

2. *Corophium tuberculatum* Shoem.

1. _____

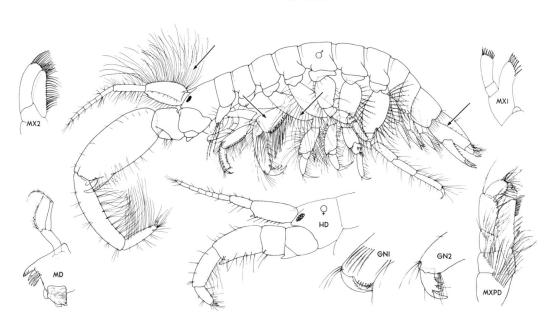

2. _____

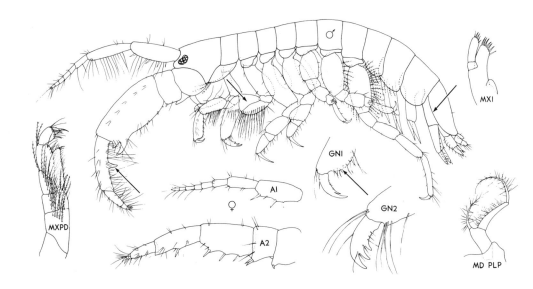

1. _Corophium lacustre_ Vanhoffen

2. _Corophium acutum_ Chevreux

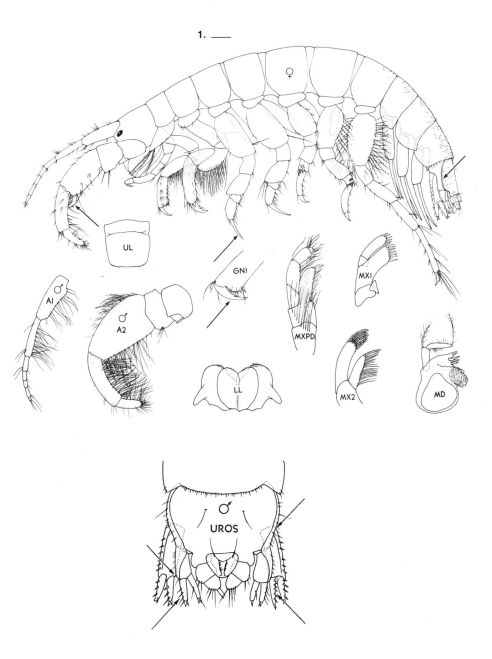

1. *Corophium simile* Shoem.

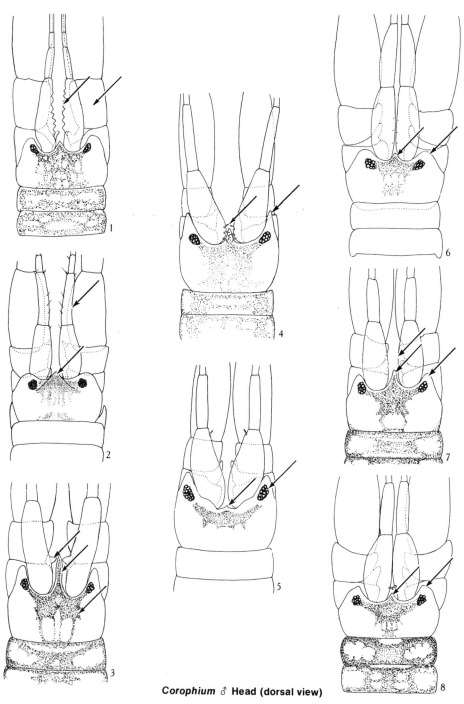

Corophium ♂ **Head (dorsal view)**

1. *Corophium volutator*   **2.** *Corophium simile*   **3.** *Corophium insidiosum*   **4.** *Corophium crassicorne*   **5.** *Corophium acherusicum*   **6.** *Corophium tuberculatum*
7. *Corophium lacustre*   **8.** *Corophium acutum*

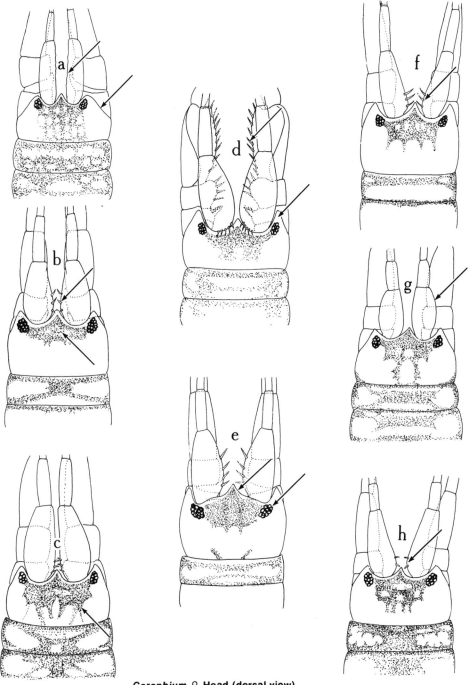

Corophium ♀ Head (dorsal view)

a. *Corophium volutator*   b. *Corophium bonelli*   c. *Corophium insidiosum*   d. *Coro-*
phium crassicorne   e. *Corophium acherusicum*   f. *Corophium tuberculatum*
g. *Corophium lacustre*   h. *Corophium acutum*

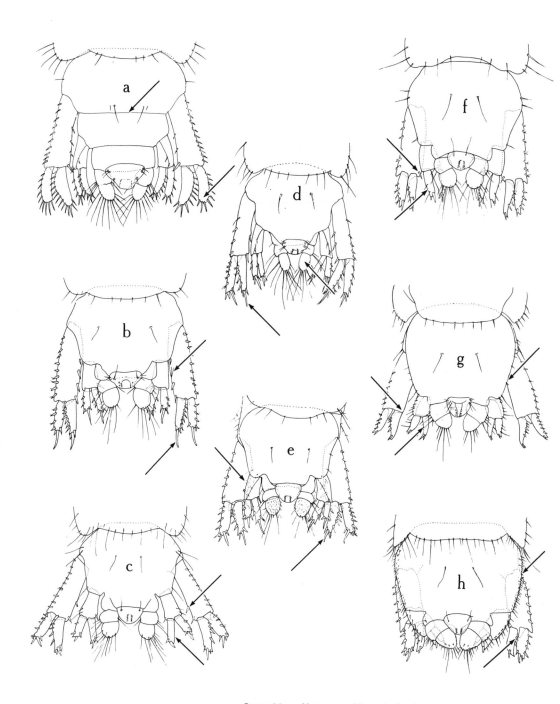

Corophium Urosome (dorsal view)

a. _Corophium volutator_   **b.** _Corophium bonelli_   **c.** _Corophium insidiosum_   **d.** _Corophium crassicorne_   **e.** _Corophium acherusicum_   **f.** _Corophium tuberculatum_
g. _Corophium lacustre_   **h.** _Corophium acutum_

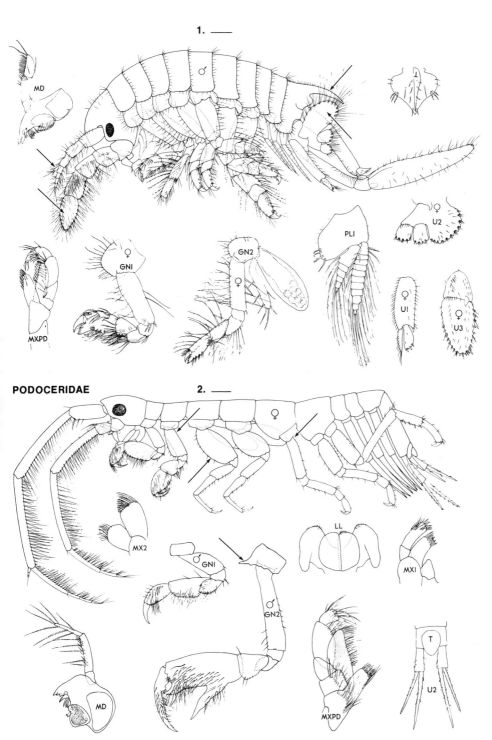

1. *Chelura terebrans* (Philippi)

2. *Dulichia porrecta* (Bate)

LIST OF GAMMARIDEAN AMPHIPODS RECORDED OR PROBABLY OCCURRING IN THE NEW ENGLAND SHELF REGION

SPECIES	KNOWN DISTRIBUTION
ACANTHONOTOZOMATIDAE	
Acanthonotozoma inflatum (Kr.)	Arctic, S. to Gulf of St. Lawrence
A. serratum (Fabr.)	Arctic, S. to Bay of Fundy
AMPELISCIDAE	
Ampelisca abdita Mills	Central Maine to Gulf of Mexico
A. aequicornis Bruz.	Arctic to Cape Cod
A. agassizi (Judd)	S.W. Nova Scotia and central Maine to Caribbean Sea
A. declivitatis Mills	W. Greenland to off Middle Atlantic states (bathyal depths)
A. eschrichti Kr.	Arctic to Gulf of Maine
A. macrocephala Lilj.	Arctic to Cape Cod and Georges Bank
A. vadorum Mills	S.W. Gulf of St. Lawrence to N. Florida
A. verrilli Mills	Cape Cod to North Carolina, prob. to Gulf of Mexico
Byblis gaimardi Kr.	Arctic to New England coast
B. serrata Smith	Cape Cod to Chesapeake Bay
Haploops setosa Boeck	Amphi-Atlantic; Bay of Fundy to Delaware Bay
H. spinosa Shoem.	Bay of Fundy and outer coast Nova Scotia
H. tubicola Lilj.	Arctic to N. Gulf of Maine
AMPHILOCHIDAE	
Amphilocus manudens Bate	Amphi-Atlantic; S. to Gulf of St. Lawrence
Gitanopsis arctica Sars	Amphi-Atlantic; S. to Gulf of St. Lawrence

283

AMPITHOIDAE

Ampithoe longimana Smith   S.W. Gulf St. Lawrence; central Maine to N. Florida
A. rubricata Montagu   Amphi-Atlantic; S. to Cape Cod
A. valida Smith   Central Maine, S. to Chesapeake Bay
Cymadusa compta (Smith)   Cape Cod to S.E. states (Georgia)
Sunamphitoe pelagica (M.-E.)   Gulf Stream N. to Sable Island

AORIDAE

Lembos smithi Holmes   Cape Cod to N. Florida
L. websteri Bate   Amphi-Atlantic; Cape Cod Bay to Georgia
Leptocheirus pinguis (Stimpson)   Labrador to Chesapeake Bay
L. plumulosus Shoem.   Cape Cod to Florida
Microdeutopus anomalus (Rathke)   Amphi-Atlantic; Cape Ann to Chesapeake Bay
M. gryllotalpa Costa   Cape Cod Bay (E.) to North Carolina
Neohela monstrosa (Boeck)   Arctic, to Delaware (deep water)
Pseudunciola obliquua (Shoem.)   Bay of Fundy to New Jersey
Rudilemboides naglei n. sp.   Southern New England to Georgia and eastern Gulf
     of Mexico
Unciola dissimilis Shoem.   Cape Cod to Florida
U. inermis Shoem.   Bay of Fundy, Gulf of Maine
U. irrorata Say   Labrador to Cape Hatteras
U. serrata Shoem.   Cape Cod to Florida

ARGISSIDAE

Argissa hamatipes (Norman)   Amphi-Atlantic; Labrador to Cape Cod Bay

ATYLIDAE

Atylus carinatus (Fabr.)   Arctic, S. to Gulf of St. Lawrence
A. minikoi Walker   Cape Cod to S.E. states and Brazil
A. swammerdami M.-E.   Labrador to off Martha's Vineyard

BATEIDAE

Batea catharinensis Müller   Southern New England to the Gulf of Mexico and
     Brazil

CALLIOPIIDAE

Amphithopsis longicaudata Boeck   Arctic-Atlantic, S. to Gulf of St. Lawrence
Apherusa megalops Buchholz   Arctic-Atlantic, S. to Gulf of St. Lawrence
Calliopius laeviusculus (Kr.)   Labrador to New Jersey
Halirages fulvocinctus (M. Sars)   Arctic, Labrador to Gulf of St. Lawrence and New
     England Coast
Haliragoides inermis (Sars)   Arctic-Atlantic, S. to Gulf of St. Lawrence
Laothoes polylovi Gurjanova   Arctic, S. to Gulf of St. Lawrence
Leptamphopus paripes Steph.   Arctic to Gulf of St. Lawrence
Oradarea longimana (Boeck)   Arctic, S. to Gulf of St. Lawrence
Rozinante fragilis (Goes)   Arctic to Gulf of St. Lawrence

CHELURIDAE

Chelura terebrans (Philippi)   Southern New England and warm temperate waters
     of northern and southern hemispheres

COLOMASTIGIDAE

Colomastix halichondriae n. sp.   Southern New England to Georgia

COROPHIIDAE

Cerapus tubularis Say	Cape Cod to Florida
Corophium acherusicum Costa	Maine to Florida
C. acutum Chevreux	Cape Cod Bay to Florida
C. bonelli (M.-E.)	Arctic to Connecticut
C. crassicorne Bruz.	Arctic to Long Island Sound
C. insidiosum Crawford[a]	Chaleur Bay to Chesapeake Bay
C. lacustre Vanhoffen	Bay of Fundy to Florida
C. simile Shoem.	Cape Cod to Florida
C. tuberculatum Shoem.	Minas Basin to Florida
C. volutator (Pallas)	Bay of Fundy and N. Gulf of Maine
Erichthonius brasiliensis (Dana)	Cape Cod to West Indies; Brazil
E. difformis M.-E.	Arctic to Gulf of Maine
E. ⎱ *hunteri* (Bate)	Amphi-Atlantic; into Gulf of Maine and off New
E. ⎰ *rubricornis* Smith	Jersey
Siphonoecetes smithianus Rathbun	South of Cape Cod to off New Jersey

CRANGONYCIDAE

Crangonyx pseudogracilis Bousf.	S. New England and Hudson River, drainages through central U.S.; Atlantic coastal drainages
C. richmondensis subsp. *richmondensis* Ellis	South Carolina to Newfoundland: fresh waters of Atlantic coastal plain

CRESSIDAE

Cressa abyssicola Sars	Arctic, North Atlantic

DEXAMINIDAE

Dexamine thea Boeck	Amphi-Atlantic; Labrador to Connecticut, a variety to Chesapeake Bay
Guernea nordenskioldi (Hansen)	Arctic and North Atlantic S. to Gulf of St. Lawrence, and outer Nova Scotia (St. Margaret's Bay)

EUSIRIDAE

Eusirogenes deflexifrons Shoem.	Cabot Strait, 387 m
Eusirus cuspidatus Kr.	Arctic to Newfoundland and Bay of Fundy
Rhachotropis aculeata Lepechin	Arctic to off Cape Ann, Mass.
R. distincta (Holmes)	Pan-Arctic, S. to Gulf of St. Lawrence
R. inflata (Sars)	Arctic, S. to Gulf of St. Lawrence
R. lobata (Shoem.)	Arctic, S. to Gulf of St. Lawrence
R. oculata (Hansen)	Greenland to Cape Cod Bay

GAMMARIDAE

Gammarellus angulosus (Rathke)	Amphi-Atlantic; Newfoundland to Long Island Sound
Gammarus annulatus Smith	Sable I; New Hampshire to New Jersey
G. daiberi Bousf.	Delaware Bay to North Carolina
G. duebeni Lilj.	Amphi-Atlantic, S. to Nahant, Mass.
G. fasciatus Say	Atlantic coastal drainages, S. New England to Florida
G. lawrencianus Bousf.	N. Labrador to Long Island Sound
G. mucronatus Say	Chaleur Bay and W. Newfoundland to Gulf of Mexico

[a] *C. cylindricum* Say.

G. oceanicus Segerstråle | Arctic-boreal, S. to off New Jersey
G. palustris Bousf. | New Hampshire to Florida
G. pseudolimnaeus Bousf. | Hudson River drainage and interior continental drainages
G. setosus Dementieva | Pan-Arctic; S. to Penobscot Bay
Marinogammarus finmarchicus Dahl | N. shore Gulf of St. Lawrence to Florida

Amphi-Atlantic; E. Nova Scotia S. to Long Island Sound

M. obtusatus Dahl

Amphi-Atlantic; N.E. Newfoundland, S. to Cape Cod Bay

M. stoerensis (Reid)

Amphi-Atlantic; E. Nova Scotia, S. to Cape Cod Bay

HAUSTORIIDAE

  HAUSTORIINAE (subfamily)

Acanthohaustorius intermedius Bousf. | Cape Cod Bay to Cape Kennedy, Fla.
A. millsi Bousf. | Central Maine to N. Florida
A. shoemakeri Bousf. | Cape Cod Bay and off Georges Bank to Georgia
A. spinosus (Bousf.) | Outer coast of Nova Scotia to Buzzards Bay, Mass.
Haustorius canadensis Bousf. | S.W. Gulf of St. Lawrence; central Maine to Chesapeake Bay

Neohaustorius biarticulatus Bousf. | Vineyard Sound to Chesapeake Bay
N. schmitzi Bousf. | Cape Cod Bay to N. Florida
Parahaustorius attenuatus Bousf. | Georges Bank to Chesapeake Bay
P. holmesi Bousf. | Georges Bank to Chesapeake Bay
P. longimerus Bousf. | Massachusetts Bay to N. Florida
Protohaustorius deichmannae Bousf. | Central Maine to Cape Kennedy, Fla.
P. wigleyi Bousf. | Central Maine to North Carolina
Pseudohaustorius borealis Bousf. | S. side of Georges Bank to Virginia
P. caroliniensis Bousf. | Buzzards Bay to N. Florida

  PONTOPOREIINAE (subfamily)

Amphiporeia gigantea n. sp. | Cape Cod Bay to New Jersey
A. lawrenciana Shoem. | Gulf of St. Lawrence to New Jersey
A. virginiana Shoem. | E. Nova Scotia to South Carolina
Bathyporeia parkeri Bousf. | Cape Cod to Florida
B. quoddyensis Shoem. | Outer coast Nova Scotia to Chesapeake Bay
Pontoporeia femorata Kr. | Arctic Ocean to Penobscot Bay
Priscillina armata Boeck | Arctic Ocean to off Nova Scotia

HYALELLIDAE

Hyalella azteca Saussure | Virtually all permanent fresh-water bodies north to tree line

HYALIDAE

Hyale nilssoni (Rathke) | Labrador to Connecticut
H. plumulosa (Stimpson) | Cape Cod Bay to Chesapeake Bay; Pacific Coast of North America

ISCHYROCERIDAE

Ischyrocerus anguipes Kr. | Arctic to off New Jersey
I. commensalis Chevreux | Newfoundland, Gulf of St. Lawrence
Jassa falcata (Montagu)[b] | Cosmopolitan warm-temperate; N. to Gulf of St. Lawrence

[b]*Jassa marmorata* Holmes.

LAFYSTIIDAE

Lafystius sturionis Kr.

Amphi-Atlantic; S. to Cape Cod; parasitic on cod, skate, cottids

LEUCOTHOIDAE

Leucothoe spinicarpa (Abildgaard)

Arctic-boreal, circum-Atlantic; south from Greenland to New England (Connecticut)

LILJEBORGIIDAE

Idunella sp.
I. aequicornis Sars
Listriella barnardi Wigley
L. clymenellae Mills

North Carolina to Delaware Bay
Arctic to Gulf of St. Lawrence
Vineyard Sound to N. Florida
Cape Cod Bay to Georgia

LYSIANASSIDAE

Acidostoma laticorne Sars
Anonyx compactus Gurjanova
A. debruyni Hoek
A. liljeborgi Boeck
A. ochoticus Gurjanova
A. sarsi Steele and Brunel
Aristias microps Sars
Hippomedon abyssi (Goes)
H. denticulatus (Bate)
H. propinquus Sars
H. serratus Holmes
Lysianopsis alba Holmes

Arctic; Atlantic S. to Nova Scotia
Hudson Bay to Cape Cod Bay
St. Lawrence estuary to Cape Cod
Arctic Ocean to off New Jersey
Arctic to off Nova Scotia
Arctic to Long Island Sound
Amphi-Atlantic; S. to Newfoundland
Arctic to Cape Cod Bay
North Atlantic, S. to Newfoundland
Amphi-Atlantic; Gulf of St. Lawrence
Gulf of St. Lawrence to Chesapeake Bay
Cape Cod Bay to N. Florida and the northern Gulf of Mexico

Onesimus edwardsi (Kr.)
O. normani (Sars)
O. plautus (Kr.)
Opisa eschrichti (Kr.)
Orchomene depressa Shoem.
O. macroserrata Shoem.
O. serrata Boeck
Orchomenella minuta Kr.
O. pinguis (Boeck)
Psammonyx nobilis (Stimpson)
Schisturella pulchra (Hansen)
Socarnes bidenticulatus Bate
Tmetonyx cicada (Fabr.)
Tryphosa groenlandica (Hansen)
Tryphosella compressa (Sars)
T. gulosus (Kr.)
Uristes umbonatus (Sars)

Arctic, S. to S.W. Nova Scotia
Arctic to S.W. Nova Scotia
Arctic to Newfoundland, and off Nova Scotia
Arctic boreal, S. to Bay of Fundy; parasitic on fish
Bay of Fundy region
Hudson Strait to N. Gulf of Maine
Arctic boreal to Long Island Sound
Arctic, S. to Cape Hatteras
Arctic and North Atlantic
Labrador to Long Island Sound
Arctic to Gulf of St. Lawrence
Arctic and North Atlantic, S. to Gulf of St. Lawrence
Arctic, S. to Gulf of St. Lawrence
Arctic to Bay of Fundy
Arctic, S. to off Nova Scotia
Arctic, S. to off Nova Scotia
Arctic, S. to Nova Scotia

MELITIDAE

Casco bigelowi (Blake)
Elasmopus levis Smith
Maera danae Stimpson
M. inaequipes Costa
M. loveni (Bruz.)
Melita dentata (Kr.)

N. Gulf of St. Lawrence to New Jersey
Cape Cod to N. Florida
Gulf of St. Lawrence to New Jersey
Arctic, S. to Gulf of St. Lawrence
Amphi-Atlantic; S. to Nova Scotia and Cape Cod Bay
Arctic and North Atlantic, S. to Cape Cod Bay

M. formosa Murdoch	Arctic, S. to Gulf of St. Lawrence
M. nitida Smith	S.W. Gulf of St. Lawrence; S.W. Nova Scotia; central Maine to Florida, Gulf of Mexico
M. palmata (Montagu)	Arctic and North Atlantic
M. quadrispinosa Vosseler	Arctic, S. to Gulf of St. Lawrence

MELPHIDIPPIDAE

Melphidippa goesi Stebb.	Labrador to Gulf of St. Lawrence
Melphidippella macera (Norman)	North Atlantic, S. to Gulf of St. Lawrence

OEDICEROTIDAE

Arrhis phyllonyx (M. Sars)	Arctic to Gulf of St. Lawrence
Bathymedon obtusifrons (Hansen)	Arctic to Gulf of St. Lawrence
Monoculodes borealis Boeck	Arctic to off Nova Scotia
M. edwardsi Holmes	Gulf of St. Lawrence to Florida (Arctic material is probably a different species)
M. intermedius Shoem.	Gulf of St. Lawrence to Cape Cod Bay
M. latimanus (Goes)	Arctic to Gulf of St. Lawrence
M. longirostris (Goes)	Arctic to Gulf of St. Lawrence
M. norvegicus (Boeck)	Arctic to Gulf of St. Lawrence
M. packardi (Boeck)	Amphi-Atlantic; S. to Gulf of St. Lawrence
M. schneideri Sars	Arctic to Gulf of St. Lawrence
M. tesselatus Schneider	Amphi-Atlantic; S. to outer coast Nova Scotia and Cape Cod Bay
M. tuberculatus Boeck	Arctic to Gulf of St. Lawrence
Monoculopsis longicornis (Boeck)	Arctic to Gulf of St. Lawrence
Paroediceros lynceus M. Sars	Arctic to New England coast
Synchelidium americanum n. sp.	Central Maine to Georgia
S. tenuimanum Norman	Arctic and Atlantic S. to New England
Westwoodilla brevicalcar Bate	Arctic to Cape Cod Bay
W. megalops G. O. Sars[c]	Arctic and Atlantic S. to Gulf of St. Lawrence

PARAMPHITHOIDAE

Epimeria loricata Sars	Arctic to off Nova Scotia
Paramphithoe hystrix (Ross)	Arctic to Bay of Fundy

PARDALISCIDAE

Pardalisca cuspidata Kr.	Arctic and Atlantic, S. to New England
P. tenuipes Sars	Arctic, S. to Gulf of St. Lawrence

PHOTIDAE

Gammaropsis maculatus Johnston[d]	Amphi-Atlantic; S. to Gulf of St. Lawrence
Microprotopus raneyi Wigley	Cape Cod Bay and Vineyard Sound to Florida and the Gulf states
Photis dentata Shoem.	Middle Atlantic states to Florida
P. macrocoxa Shoem.	Gulf of St. Lawrence, Chaleur Bay to Virginia
P. reinhardi Kr.	Arctic to Long Island Sound
Podoceropsis inaequistylis Shoem.	Gulf of St. Lawrence
P. nitida (Stimpson)	Amphi-Atlantic; Gulf of St. Lawrence S. to Connecticut
Protomedia fasciata Kr.	Arctic; S. to Gulf of St. Lawrence

[c] Includes *caecula* Bate.
[d] *G. melanops* Sars.

PHOXOCEPHALIDAE

Harpinia cabotensis Shoem.	Gulf of St. Lawrence to Cape Cod Bay
H. crenulata Boeck	Arctic to Gulf of St. Lawrence
H. neglecta Sars	Arctic to North Atlantic
H. plumosa (Kr.)	Arctic to Gulf of St. Lawrence
H. propinqua Sars	Arctic to Gulf of St. Lawrence
H. serrata Sars	Arctic to Gulf of St.Lawrence
H. truncata Sars	Arctic to Gulf of St. Lawrence
Paraphoxus spinosus Holmes	Vineyard Sound to N. Florida
Phoxocephalus holbolli Kr.	Arctic boreal, S. to Long Island Sound
Trichophoxus epistomus (Shoem.)	American-Atlantic; southern Maine to North Carolina

PLEUSTIDAE

Neopleustes pulchellus Kr.	Arctic to New England coast and Georges Bank
Pleustes medius (Goes)	Arctic to Gulf of St. Lawrence
P. panoplus Kr.	Arctic to New England coast
Pleusymtes glaber (Boeck)	Amphi-Atlantic; Labrador S. to Chesapeake Bay
Stenopleustes inermis Shoem.	Bay of Fundy to Cape Cod Bay
S. gracilis (Holmes)	Vineyard Sound to New Jersey; Chesapeake Bay
Sympleustes latipes (Sars)	Arctic to Gulf of St. Lawrence

PODOCERIDAE

Dulichia falcata Bate	Arctic to Gulf of Maine
D. monacantha Metzger	Arctic to Bay of Fundy and Cape Cod Bay
D. porrecta (Bate)	Arctic-boreal, S. to Cape Cod Bay
D. spinosissima Kr.	Arctic-boreal to Long Island Sound
D. tuberculata Boeck	Arctic to Gulf of St. Lawrence

PONTOGENEIIDAE

Pontogeneia inermis (Kr.)	Arctic-boreal to Long Island Sound

STEGOCEPHALIDAE

Stegocephalus inflatus Kr.	Arctic, S. to Block Island, R.I.

STENOTHOIDAE

Metopa alderi (Bate)	Arctic-boreal, S. to N. Gulf of Maine
M. boecki Sars	Arctic and Atlantic, S. to Gulf of St. Lawrence
M. borealis Sars	Arctic to Gulf of St. Lawrence
M. bruzeli (Goes)	Arctic and Atlantic, S. to Gulf of St. Lawrence
M. clypeata (Kr.)	Arctic-boreal to Gulf of St. Lawrence
M. propinqua Sars	Arctic to Atlantic, S. to Gulf of St. Lawrence
M. solsbergi Schneider	Arctic to coast of Nova Scotia
M. tenuimana Sars	Arctic to Gulf of St. Lawrence
Metopella angusta Shoem.	Bay of Fundy to off New Jersey
M. carinata (Hansen)	Arctic to Gulf of St. Lawrence
M. longimana (Boeck)	Arctic to Labrador and Gulf of St. Lawrence
M. nasuta (Boeck)	Arctic-boreal, S. to Gulf of St. Lawrence
Metopelloides micropalpa (Shoem.)	Gulf of St. Lawrence
Parametopella cypris (Holmes)	Vineyard Sound to N. Florida
Proboloides holmesi n. sp.	Vineyard Sound to Buzzards Bay
P. nordmanni Steph.	Arctic, S. to Gulf of St. Lawrence
Stenothoe gallensis Walker	Warm-temperate, N. to Chesapeake Bay

S. minuta Holmes   Vineyard Sound to Long Island Sound
Stenula peltata (Smith)   Arctic to Gulf of St. Lawrence

TALITRIDAE

Orchestia gammarella (Pallas)   Amphi-Atlantic; Newfoundland to N. Gulf of Maine
O. grillus (Bosc)   Chaleur Bay to Gulf of Mexico
O. platensis Kr.   Anticosti Island, Gulf of St. Lawrence to Argentina;
                                      cosmopolitan-temperate
O. uhleri Shoem.   Central Maine to Gulf of Mexico
Talorchestia longicornis (Say)   Chaleur Bay to N. Florida
T. megalophthalma (Bate)   N. shore Gulf of St. Lawrence to Georgia

TIRONIDAE

Syrrhoe crenulata (Goes)   Arctic Ocean to Cape Cod Bay
Tiron spiniferum (Stimpson)   Arctic to New England coast

GUIDE TO PRONUNCIATION AND ROOT MEANINGS OF SCIENTIFIC NAMES

At present, amphipod crustaceans are known only by binomial scientific names, usually based on Latin and Greek words. To quote R. S. Woods * (*The Naturalist's Lexicon,* 1944): "Since Greek and Latin are seldom included in the studies of those who specialize in any branch of biology, scientific names are to most students of natural history simply meaningless and easily forgotten combinations of letters whose oral use is further hindered by uncertainty as to pronunciation. A knowledge of the derivation and significance of these terms should not only make them less difficult to remember, but may afford some clue to the characteristics of the species to which they are applied, and altogether could contribute to the satisfaction to be found in the study of nature." The following table is provided, therefore, in an attempt to standardize the pronunciation of amphipod names among layman and specialist alike, and to provide the benefits and enjoyment so well described by Professor Woods.

Guide To Pronunciation †

In all Latin or Latinized Greek words, the principal accent falls on the penultimate or next-to-last syllable if that syllable is "long" in quantity, that is, if it contains a "long" vowel or diphthong (except ū) or if the vowel is followed by certain combinations of consonants; also if the word has only two syllables. Otherwise the next previous or antepenultimate syllable is accented.

KEY TO PRONUNCIATION SYMBOLS

ā	as in	rate	ŏ	as in	not
ă		rat	ô		form
â		far	ö		anatomy
ch		church	oi		toy
ē		he	oo		good
ĕ		hen	ōo		fool
ë		her	ow		cow
g		go	s		moss
gw		guano	sh		fish
ī		pine	th		thin
ĭ		pin	ts		Schmitz
j		gem	ū		pure
k		cat	ŭ		nut
kw		queen	y		yard
ng		sing	z		maize
ō		no	zh		vision

* By courtesy of R. S. Woods, the Abbey Garden Press, Pasadena, and rancho Santa Ana Botanic Gardens, Claremont, California.
† From R. S. Woods, *The Naturalist's Lexicon* (Pasadena: Abbey Garden Press, 1944), pp. 1–282.

Scientific name	Pronunciation	Root meaning	Derivative language
abdita	ăb dĭt' ă	hidden, secret	L
Acanthohaustorius	ă kănth ō hôst ōr'ĭ ŭs	spiny Haustorius	Gr,L
acherusicum	ā chĕr ōōs'ĭk ŭm	not bereft	Gr
acutum	ă kūt' ŭm	sharp, pointed	L
agassizi	ăg ă sēz'ī	Agassiz's	
alba	ăl' bă	white	L
alderi	ăl' dër ī	Alder's	
americana (um)	ă mĕr ĭ kăn'ŭm	American	
Ampelisca	ăm pĕl ĭsk' ă	sea plant	Gr
Amphiporeia	ăm fī pör ī' ă	both modes of walking	Gr
Ampithoe	ăm pĭ thö'ē	Nereid (mythology)	Gr
anguipes	ăng' wĭ pās	snake foot	L
angulosus	ăng ŭl ōs' ŭs	cornered, angled	L
angustus (a)	ăng ŭst' ŭs	narrow	L
annulatus	ăn ūl āt' ŭs	ringed	L
anomalus	ă nŏm' ăl ŭs	not even	Gr
Anonyx	ăn ŏn' ĭks	no claw	Gr
Aoridae	ā ôr' ĭ dē	pendulous, hanging	Gr
Argissa	ăr jĭs' ă	located in Thessaly	Gr
attenuatus	ă tĕn ū āt ' ŭs	thin, lessened	L
Atylus	ă tĭl' ŭs	no lumps	Gr
azteca	az' tĕk ă	Aztec (Indian race)	
barnardi	bârn' ârd ī	Barnard's	
Batea	bāt'ē ă	named after Bate	
Bathyporeia	băth ĭ pör ī' ă	deep mode of walking	Gr
biarticulatus	bī ăr tĭk ū lāt' ŭs	two-segmented	L
bigelowi	bĭg ĕl ō 'ī	Bigelow's	
bonelli	bŏn ĕl'ī	Bonell's	
borealis	bör ē ăl' ĭs	northern	L
brasiliensis	bră sĭl ĭ ĕn' sĭs	Brazilian	
Byblis	bĭb' lĭs	Greek nymph	Gr
Calliopius	kăl ĭ ōp' ĭ ŭs	beautiful eye, from muse of epic poetry	Gr
canadensis	kăn ă dĕn' sĭs	Canadian	
caroliniensis	kăr ö lĭn ĭ ĕn' sĭs	Carolinian	
Casco	kăs' kō	named after Casco	
catharinensis	kăth âr ĭn ĕn' sĭs	from Estado de Santa Catarina, Brazil	
Cerapus	sĕr'ă pŭs	horn foot	Gr
Chelura	kēl ūr' ă	claw tail	Gr
clymenellae	klī mĕn ĕl' ē	associated with polychaete Clymenella	Gr
Colomastix	kōl ŏ măst' ĭks	shortened whip	Gr
compta	kŏmpt' ă	elegant, ornamented	L
Corophium	kŏr öf' ĭ ŭm	living with a sea serpent	Gr
Crangonyx	krăng ŏn'ĭks	shrimp claw	Gr
crassicorne	krăs ĭ kôrn'ē	thick horn	Gr
Cymadusa	sī mă dūs'ă	wave diver	Gr
cypris	sĭp' rĭs	like Cypris (Ostracoda) from Venus	L
daiberi	da' bër ī	Daiber's	
danae	dā' nē	Dana's	
deichmannae	dīk' măn ē	Deichmann's	
dentata	dĕn tāt' ă	toothed	L
Dexamine	dĕks ă mĭn' ĕ	sea nymph	Gr
difformis	dĭ fŏrm' ĭs	of two forms	L
dissimilis	dĭ sĭm'ĭl ĭs	unlike	L

Scientific name	Pronunciation	Root meaning	Derivative language
duebeni	dū'bĕn ī	Dueben's	
Dulichia	dū lĭk' ĭ ă	servile; long	Gr
edwardsi	ĕd' wârdz ī	Edward's	
Elasmopus	ĕ lăs mō' pŭs	metal plate foot	Gr
epistomus	ĕp ĭ stōm' ŭs	above the mouth	Gr
Erichthonius	ĕr ĭk thŏn' ĭ ŭs	character in Greek myth	Gr
Eusiridae	ū sĭr' ĭ dē	true sea nymph	Gr
falcata	făl kāt' ă	sicklelike	L
fasciatus	făs ē āt' ŭs	bundled	L
femorata	fĕm ö rāt' ă	with a femur	L
finmarchicus	fĭn mârch' ĭk ŭs	of Finmark	
gammarella (Gam-marellus)	găm âr ĕl' ă	little lobster	Gr
Gammarus	găm' âr ŭs	lobster	Gr
gigantea	jī găn tē' ă	giant	Gr
glaber	glā' bĕr	bare, without setae	L
gracilis	gră' sĭl ĭs	slender	L
grillus	grĭl' ŭs	like a cricket	L
gryllotalpa	grĭl ö tălp' ă	mole cricket	L
halichondriae	hăl ĭ kŏn' drĭ ē	of a sponge	Gr
hamatipes	hăm āt' ĭ pās	hook foot	L
Haploops	hăp' lö ŏps	single eye	Gr
Harpinia	hâr pĭn' ĭ ă	little sea bird	Gr
Haustorius	hôst ör' ĭ ŭs	a drawer of water	L
Hippomedon	hĭp ō mēd' ŏn	figure of Greek mythology	Gr
holbølli	hôl' böl ī	Holboll's	
holmesi	hōmz' ī	Holmes'	
hunteri	hŭnt' ĕr ī	Hunter's	
Hyale	hī'ăl ĕ	glass	Gr
Hyalella	hī ăl ĕl' ă	little *Hyale*	Gr
Idunella	ī dūn ĕl' ă	Greek goddess	Gr
inermis	ĭn ĕrm' ĭs	unarmed	L
insidiosum	ĭn sĭd ĭ ōs' ŭm	lying in wait	L
intermedius	ĭn tĕr mēd' ĭ ŭs	intermediate	L
irrorata	ĭr ör āt' ă	with little colored markings	L
Isaeidae	ī sē' ĭ dē	Greek orator, teacher	Gr
Ischyrocerus	ĭs kĭr ŏ sēr' ŭs	strong horn	Gr
Jassa	yas'ă	Greek mythological character	Gr
lacustre	lă cŭst' ĕr	found in lakes	L
laeviusculus	lēv ĭ ŭs' kūl ŭs	smooth rump	L
Lafystius	lă fĭst' ĭ ŭs	to swallow greedily	L
lawrencianus (a)	lâr ĕnts ĭ ān' ă	Laurentian, of St. Lawrence gulf	
Lembos	lĕm' bŏs	small sailing vessel	Gr
Leptocheirus	lĕp tö kĭr' ŭs	slender hand	Gr
Leucothoe	lo͞ok ö thö' ē	white nimble shrimp	Gr
levis	lēv'ĭs	smooth	L
Liljeborgiidae	lilj bôrg ē' ĭ dē	Liljeborg's (family)	
Listriella	lĭst rē ĕl' ă	small shovel, plough	Gr
longicornis	lŏnj ĭ kôrn' ĭs	long-horned	L
longimana	lŏnj ĭ măn' ă	long-handed	L
longimerus	lŏnj ĭ mĕr' ŭs	long-thighed	L,Gr
Lysianassidae	līs ĭ ăn ăs'ĭ dē	daughter of Nereus and Doris (Greek mythology)	Gr

Scientific name	Pronunciation	Root meaning	Derivative language
Lysianopsis	līs ĭ ăn ŏp′ sĭs	like Lysianassa	Gr
macrocephala	mă krō sĕf′ ăl ă	large headed	Gr
macrocoxa	mă krö kŏks′ ă	large hip	Gr
Maera	mēr′ ă	priestess of Venus	L
Marinogammarus	măr ēn ō găm′ âr ŭs	sea lobster	L
megalophthalma	mĕg âl ŏf thălm′ ă	large eyed	Gr
Melita	mĕl′ĭt ă	honey bee	Gr
Metopa	mĕt ōp ′ ă	brow	Gr
Metopella	mĕt ōp ĕl′ ă	little brow	Gr
microdeutopus	mī krö dūt′ ö pŭs	small second foot	Gr
microprotopus	mī krö prōt′ ō pŭs	small first foot	Gr
millsi	mĭlz′ī	Mills'	
minuta (us)	mĭn ūt′ ă	very small	L
Monoculodes	mŏn ŏk ū lōd′ ēz	one eye	Gr
mucronatus	mū krö nāt ŭs	pointed	L
naglei	nä′ gĕl ī	Nagle's	
Neohaustorius	nē ō hôst ör′ī ŭs	new Haustorius	Gr,L
nilssoni	nĭls′ŏn ī	Nilsson's	
nitida	nĭt′ĭd ă	shining	L
nobilis	nō ′bĭl ĭs	well-known	L
obliquua	ŏb lēk′ wä	slanted	Gr
obtusatus	ŏb tūs āt′ ŭs	blunt, obtuse-angled	L
oceanicus	ō shē ăn′ĭk ŭs	living in the ocean	L
oculata	ŏk ū lāt′ă	eyed	L
Oedicerotidae	ē dĭ sĕr ŏt′ī dē	swollen horn	Gr
Orchestia	ôr kĕs ′ tĭ ă	dancer, leaping sea fish	Gr
Orchomene	ôr kō mēn′ĕ	unknown fish	Gr
Orchomenella	ôr kō mĕn ĕl′ ă	little Orchomene	Gr
palustris	pă lŭs′ trĭs	of the marsh	L
Parahaustorius	păr ă hôst ör′ī ŭs	near Haustorius	Gr,L
Parametopella	păr ă mĕt ōp ĕl′ă	near Metopa	Gr
Paraphoxus	păr ă fŏks′ ŭs	near Phoxus	Gr
parkeri	pârk′ ĕr ī	Parker's	
pelagica	pĕl ăj′ ĭk ă	frequenting the open sea	Gr
Photis	fōt′ĭs	light	Gr
Phoxocephalus	fŏks ō sĕf′ ăl ŭs	peaked head	Gr
pinguis	pĭng′ wĭs	fat	L
platensis	plāt ĕn′ sĭs	of the Plata River	
Pleustes	plē ŭst′ ēz	sail user	Gr
Pleusymtes	plē ū sĭm′ tēz	anagram of Sympleustes	Gr
plumulosus (a)	plŏŏm ūl ōs′ ă	fine downy	L
Podoceridae	pöd ö sĕr′ī dē	foot horn	Gr
Podoceropsis	pöd ö sĕr ŏp′ sĭs	like Podocerus	Gr
Pontogeneia	pŏnt ō jĕn ē′ă	born in the sea	L
Pontoporeia	pŏnt ō pör ī′ ă	walking in the sea	L
porrecta	pö rĕkt′ ă	straight channel	L
Proboloides	prō böl oī dēz	like Probolium	Gr
propinqua (us)	prō pĭnk′ wä	near	L
Protohaustorius	prōt ō hôst ör′ ī ŭs	first Haustorius	Gr,L
Psammonyx	săm ŏn′ ĭks	sand-dwelling little claw	Gr
pseudogracilis	sū dö grä′ sĭl ĭs	false gracilis	Gr,L
Pseudohaustorius	sū dö hôst ör′ī ŭs	false Haustorius	Gr,L

Scientific name	Pronunciation	Root meaning	Derivative language
Pseudunciola	sūd ŭns ĭ ōl′ ă	false *Unciola*	Gr,L
quoddyensis	kwŏd ē ĕn ′sĭs	inhabiting Quoddy	
raneyi	rān ′ ē ī	Raney's	
reinhardi	rīn′ hârd ī	Reinhard's	
Rhachotropis	răk ö trōp′ ĭs	thorny ridge	Gr
richmondensis	rĭch mŏnd ĕn′ sĭs	inhabiting Richmond	
Rivulogammarus	rĭv ūl ō găm ′ âr ŭs	brook lobster	L
rubricata	rōͦ brĭ kāt′ ă	red ochre	L
rubricornis	rōͦ brĭ kôrn′ ĭs	red horn	L
Rudilemboides	rōͦ dĭ lĕm boïd′ ēz	imperfect *Lemboides*	Gr
sarsi	sârs′ ī	Sars's	
schmitzi	shmĭts′ī	Schmitz's	
serrata (us)	sĕr āt′ ă	saw-toothed	L
setosus (a)	sē tōs′ ŭs	setaceous	L
shoemakeri	shōͦ māk′ër ī	Shoemaker's	
simile	sĭm′ĭl ē	similar, like	L
Siphonoecetes	sĭ fŏn ē kēt′ ēz	tube inhabitant	Gr
smithi	smĭth′ ī	Smith's	
smithianus	smĭth ĭ ān′ŭs	relating to Smith	
spinicarpa	spīn ĭ kârp′ ă	spinose carpus	L
spinosissima	spīn ōs ĭs′ĭm ă	most spinose	L
spinosus (a)	spīn ōs′ ŭs	spinose	L
Stenopleustes	stĕn ō plē ŭst′ ēz	narrow *Pleustes*	Gr
Stenothoe	stĕn ō thō′ē	narrow nimble shrimp	Gr
stoerensis	stôr ĕn′ sĭs	living in Stoerr	
sturionis	stĕr ĭ ōn ′ ĭs	of the sturgeon	L
Sunampithoe	sŭn ăm pĭ thö′ ē	with *Ampithoe*	Gr
swammerdami	swäm ër dăm′ ī	Swammerdam's	
Sympleustes	sĭm plē ŭst′ ēz	with *Pleustes*	Gr
Synchelidium	sĭn kēl ĭd′ ĭ ŭm	with small chela	Gr
Talitridae	tăl ĭ ′ trĭ dē	a rap with the finger	L
Talorchestia	tăl ör kĕs′ tĭ ă	like *Orchestia*	Gr
terebrans	tĕr′ĕ bränz	boring	L
tesselatus	tĕs ĕl āt′ ŭs	checkered	L
Thea	thē′ ă	Greek goddess	Gr
tigrinus	tī′grĭn ŭs	striped like a tiger	L
Tmetonyx	mĕt ŏn′ĭks	part little claw	Gr
Trichophoxus	trĭk ō fŏks′ ŭs	narrow peak	Gr
tuberculatum	tū bër kū lāt′ ŭm	having tubercles	L
tubularis	tū bū lâr′ĭs	living in a tube	L
uhleri	ūl′ ër ī	Uhler's	
Unciola	ŭns ĭ ōl ′ ă	little hook	L
vadorum	vă dôr′ ŭm	of shallow places	L
valida	văl′ĭd ă	strong	L
verrilli	vĕr′ĭl ī	Verrill's	
virginiana	vër jĭn ĭ ān′ ă	Virginian	
volutator	vŏ lūt āt′ ôr	coiled up	L
websteri	wĕb′ stĕr ī	Webster's	
wigleyi	wĭg′ lē ī	Wigley's	

GLOSSARY OF SCIENTIFIC TERMS

abdomen. The posterior six body segments, consisting of anterior three segments (pleon) bearing pleopods, and posterior three segments (urosome) bearing uropods and telson (Fig. 1).

accessory flagellum. The secondary ramus of antenna 1, sometimes vestigial or lacking (rarely equal to or longer than the main flagellum), attached to the inner distal margin of peduncular segment 3 (Fig. 4A).

accessory gill. A secondary lobe of a coxal gill arising near its base.

acuminate. Produced in form of a small sharp tooth.

aesthetasc. A flattened, nontapering sensory seta of the antennal flagellae.

antenna. One of two paired segmented appendages arising from the anterior part of the head, anterodorsal to the buccal mass (Fig. 4).

antennal sinus. Emargination of head to accommodate rotation of peduncular segments of antennae: superior—antenna 1, inferior—antenna 2 (Fig. 4A).

anterior. Front, toward head end.

apex. Distal end, terminus, tip.

approximate. Close together, touching or nearly so.

arcuate. Arching, curved.

baler lobe. A flat, marginally fringed expansion of the (base of the) outer margin of the basis of maxilla 1 in some haustoriid amphipods (Fig. 9F).

basis, basipodite. The second segment of a peraeopod (Fig. 4B, C).

bifid. Two-pronged, two-lobed.

biramous. Two-branched; normal, basic, or "primitive" condition of limbs of Crustacea.

branchia(e). Coxal gill; a saclike, platelike, foliate, or dendritic respiratory organ attached to the posterior inner face of the coxal segment of peraeopods 2–7 (6) (Fig. 4B). Sternal gill arises directly from midventral body surface of peraeon segments.

brood lamella (e), *plate.* Thin, expanded, chitinous plate arising from posterior inner margin of coxa of peraeopods 2–5 (in females) that together form a pouch or

296

marsupium for retention of ova and newly hatched young. In mature females, margins (especially distally) are fringed with setae that interlock with those of adjacent plates (Fig. 4B).

calceolus(i). A small globular, plate-shaped, or clavate articulated sense organ attached to segments of one or both antennae; frequently on the posterior margin of antenna 1 and anterior margin of antenna 2, of Gammaridae, Eusiridae, Calliopiidae, Pontogeneiidae, Crangonycidae, Phoxocephalidae, Pontoporeiinae, etc., in which precopula or pelagic mating occurs.

carinate. With middorsal ridge that is compressed and elevated above the dorsum (Fig. 12D, G).

carpochelate. Prehensile appendage formed by segment 6 and parallel to immovable distal finger of the carpus (segment 5), as in *Leucothoe*, gnathopod 1 (Fig. 10F).

carpus, carpopodite. The fifth segment of a peraeopod.

cephalon. Head region, consisting of five true head segments to which is fused the first (anterior) true thoracic segment (Fig. 4A).

chela. A terminal pincer formed by a movable and immovable finger, parallel to the axis of the appendage (Fig. 10G).

chelate. Having a form of a chela.

clavate. Club-shaped.

complexly subchelate. A prehensile condition of the gnathopods in which segments 6 and 7 close against the distal palm and/or fixed tooth of segment 5 (fig. 10E).

compressed. Flattened from side to side.

corneal lens. A biconvex light receptive body of the surface cuticle of the head (contrast with subcuticular ommatidium) in family Ampeliscidae (Fig. 6J).

coupling spines. Small, hooklike spines, usually paired, located on the distal inner margin of peduncle of pleopods; spines mesh with opposite members and ensure pleopod pair's beating in synchrony (Fig. 4D).

coxal gill. See branchia.

coxal plate. The outer ventrolateral expansion of the coxa (segment 2) of the peraeonal (thoracic) appendages, forming a shield for the gills and brood plates (Figs. 4B, 7A–E).

crenate, crenulate. Sawtoothed, serrate.

dactyl, dactylopodite. Talonlike terminal segment or claw, of peraeopods 1–7 or maxilliped palp (Fig. 4B, C).

dactylate. Possessing a dactyl or clawlike terminal segment.

decurved. Sharply curved downwards and backwards.

dentate. Toothed.

depressed. Flattened dorsoventrally, from top to bottom.

distal. Farthest from point of origin or attachment; in limbs, furthest from body.

emargent. Protruding from body surface or margin.

emarginate. Shallowly concave, excavate, or incised, as in apex of telson (Fig. 8K).

entire. Not bilobed; forming one piece (Fig. 8L. M).

epimeron, epimeral plate. Ventrolateral expansion of pleon segments 1–3 (Fig. 7G–L).

epipodite. External process of limb base, especially of coxa and basis, usually serving as gill or branchia.

epistome. The anterior surface of the head immediately above the upper lip, often coming free with the latter on dissection; may be produced forward (Fig. 8C, D).

excavate. Incised, emarginate, usually somewhat deeply so.

facet. One unit of the subcutaneous compound eye; usually pigmented (see *ommatidium*).

facial. Occurring on face or flat side, as opposed to marginal.

falcate, falciform. Sickle-shaped; talonlike; curve-tapering (Figs. 9C, 13E).

filiform. Very slender, linear, elongate (Fig. 10B).

flagellum. The main distal portion of antenna 1 or 2; the portion beyond the peduncle which, in antenna 1, commences with the fourth segment, and in antenna 2, with the sixth segment (Figs. 1, 4A).

foliaceous. With marginal setae, usually plumose (Fig. 13H).

fossorial. Adapted for burrowing, digging, or tunneling; frequently applied to the excessively broadened, spinose, or setose appendages typical of the Haustoriidae, Phoxocephalidae, Oedicerotidae, some Talitridae, and many Lysianassidae.

fusiform. Spindle-shaped; widest in middle, tapering toward the ends (Fig. 3A).

geniculate. Sharply bent in a kneelike fixed position; applied to peduncular segments of antenna 1, or to distal segments of maxilliped palp, or some peraeopods of certain Haustoriidae (Fig. 9J).

gill. See branchia.

gland cone. Conical posterodistal process of segment 2 or antenna 2, through which antennal gland discharges to exterior (Fig. 4A).

gnathopod, gamopod. One of the first two appendages of the peraeon, usually subchelate, and usually differing in form and function from peraeopods 3–7 (Fig. 4B).

head lobe. Anterior head lobe or interantennal head lobe—lateral head process between antennal sinuses.

head lobe angle. Portion of head at junction of antennal sinus and head margin; anterior—upper, inferior—lower.

hook spine(s). Short curved spines, with bosses, set at ends of uropod rami (Fig. 13G).

incisor. The distal or apical portion of the mandible, usually forming a toothlike process or sharp cutting plate; well developed or modified in carnivorous, omnivorous, and parasitic species; usually weak or lacking in herbivorous or filter-feeding species (Fig. 5C).

incrassate. Much thickened, heavy, powerful.

ischium, ischiopodite. The third segment of a peraeopod.

labrum. See *upper lip.*

lacinia mobilis. An articulated toothlike plate attached to base of incisor and distal to spine row of mandible; usually differs in left and right mandibles (Fig. 5C).

laminar. Platelike, flat, nonarticulated.

lanceolate. Lance-shaped, broad proximally, tapering distally.

lappet. A small lobe covering anteroproximal part of peduncle of uropod 1 in some Haustoriinae.

linguiform. Tongue-shaped.

lobe. A flat, rounded process or piece (see *lower lip*).

lower lip (labium). A nonarticulated fleshy plate on the posterior margin of the mouth, consisting of at least one pair of lobes (outer) and often with a medioproximal pair of inner lobes (Fig. 5B); proximal ends of outer lobes (mandibular processes) may be attenuated (Aoridae, Fig. 8H) or lacking (Haustoriinae, Fig. 8F).

mandible. The most anterior of the paired movable mouthparts, originating on either side of the mouth; composed usually of a body bearing a distal incisor, lacinia mobilis, spine row, molar, and 3-segmented palp (Fig. 5C).

mandibular process. See *lower lip.*

marsupium. See *brood lamellae.*

maxilla 1. The second most anterior paired or movable mouthpart, located immediately posterior to the lower lip; consists of a basal segment to which are attached a median (inner) plate bearing marginal or apical setae, an outer plate bearing terminal spine-teeth, and a palp that is usually two-segmented but may be one-segmented, vestigial, or lacking (Fig. 5D); occasionally bearing a basal baler lobe (Fig. 9F).

maxilla 2. Paired mouthpart immediately posterior to maxilla 1 and anterior to maxilliped; consists of a basal segment to which are attached the distally setose inner (medial) and outer (lateral) plates (Fig. 5E).

maxilliped. The hindmost mouthpart and hindmost paired appendage of the head region or cephalon (Fig. 5F); each member of the basally fused pair consists of an inner (proximal) plate, an outer (distal) plate, and a palp of four segments that may be reduced to three (Haustoriidae, some Talitridae, etc., Fig. 9J), rarely two (Lafystiidae, some Dexaminidae), or lacking (all Hyperiidea).

merus, meropodite. Fourth segment of a peraeopod.

mesosome. The peraeon, consisting of the posterior 7 true thoracic segments.

metasome. Pleon segments (pleonites) 1–3.

molar. A medial process of the mandible (Fig. 5C); the normal form is a large subcylindrical body with a terminal surface or ridges or teeth for grinding or trituration (Fig. 9A).

mucronate. Sharply toothed; bearing acute processes that are an integral part of the integument, not movably hinged (Fig. 12A) nor medially ridged as in a carina.

ommatidium(a). A facet of the sessile subintegumentary compound eye.

oostegite. Brood plate.

palm. A distal surface or margin of segment 6 (propod) of a peraeopod, especially a gnathopod, against which segment 7 (dactyl) closes (Fig. 4B); defined posteriorly by a sharp change of marginal slope or occurrence of special spine(s).

palp. The segmented appendage attached laterally to the basal segments of certain mouth parts; in amphipods, found only on the mandible, maxilla 1, and maxilliped (Fig. 5).

parachelate. A linear chela (not subchela) in which the immovable finger is distinct but short, with anteriorly oblique (not horizontal) palm, the apex of which is usually overlapped by the dactyl, as in gnathopod 2 of some species of *Unciola* (Plate LII.1).

pectinate. Bearing fine comblike setae or teeth (Fig. 5C, F).

peduncle. The combined basal segments of certain paired (usually biramous) appendages, consisting of three segments in antennae 1, five segments in antenna 2, one or two segments in pleopods and one segment in uropods.

pedunculate. Attached to body by means of a stalk or peduncle; also applied to gills.

penis papillae. Small, paired processes arising from the sternum just medial to coxa of peraeon 7 in males, through which the sperm strings are ejaculated.

peraeon. The posterior 7 true thoracic segments behind the head, bearing the uniramous peraeopods (Fig. 1).

peraeonite, pereonite. A single segment of the peraeon.

peraeopod, pereiopod. One of the seven paired uniramous limbs of the peraeon, the first two of which are usually subchelate and termed gnathopods or gamopods (Fig. 1). Each limb consists of seven segments, including the coxa (Fig. 4B, C).

plate. A flattened, well-defined, specially armed lobe, attached medially to basal segments of certain mouthparts, or laterally to coxal and pleonal segments.

pleon, pleosome. The first three segments of the abdomen, bearing paired pleopods (Fig. 1). Some authors apply this term to the entire abdomen of six segments.

pleopod. One of three anterior paired biramous appendages of the abdomen, consisting of basal peduncle and marginally setose, multisegmented rami (Fig. 4D).

plicate. Folded.

plumose. Covered with feathery setae, or with fine hairs or microsetae (Fig. 4D).

posterior. Toward tail or rear.

propod, propodus, propopodite. The sixth segment of a peraeopod, used especially to denote the palmar segment of a gnathopod (Fig. 4B).

proximal. Nearest to point of origin or attachment (opp. *distal*).

rastellate. With fine teeth or spinelike projections.

recurved. Curved back upon itself (Fig. 12B).

reniform. Kidney-shaped (Fig. 6F).

rostrum. The dorsal median anterior projection of the head (Fig. 6A–E).

scale. A small plate; usually applied to a vestigial or one-segmented accessory flagellum or the inner ramus of uropod 3 in some Crangonycidae, Melitidae, etc.

serrate. Regularly toothed; crenate, as a margin (Fig. 7L).

seta. A slender, flexible, chitinous outgrowth of the body or limb surface.

simple. Not chelate or subchelate; lacking a true palm distally on segment 6, as in a gnathopod (Fig. 10A).

sinuous. S-shaped, as a margin or limb, (Fig. 7H, I).

spatulate. Shaped like a flat spoon or spatula (Fig. 13J).

spine. A stout, stiff, narrow outgrowth (with basal "boss") of body or limb surface.

spine-tooth. Low conical or short spine on medial and apical margins of certain mouthparts (Fig. 6F), especially maxilla 1, and maxilliped plates.

squamiform. Scalelike; broad-elongate and flattened.

sternal process. Stiff midventral outgrowth of the sternum.

sternum. The ventral surface of the peraeon and pleon, between the coxae and epimeral plates.

stiliform. Very slender and pointed (Fig. 13K).

sub. Prefix meaning "nearly," "almost," as in subcylindrical, subparallel, subquadrate (Fig. 7J); also meaning "below," "under," as in subcuticular.

subchelate. A prehensile condition of a peraeopod or palp in which the palm of the subterminal segment is not produced to form an immovable finger, but is at right angles to (or posteriorly oblique to) the axis of that segment (usually the propod) (Fig. 10C, D) (see also chelate, complexly subchelate, parachelate).

telson. A terminal flap of the urosome attached to segment 6 dorsal to the anus, normally consisting of two nearly separated lobes, but often fused into a single plate, and always present in Amphipoda (Figs. 1, 5G).

triturating. Strongly ridged for grinding, as in medial surface of mandibular molar (Fig. 5G).

truncate. Blunt, sharply cut off, either squarely or obliquely, as in apex of terminal segments and plates of mouthparts.

tumescent. With soft pellucid lobe.

uncinate. Bearing small hooklike ectodermal processes (Fig. 8N, 13I).

unguiform. Bearing a terminal nail, as in the normal dactyl.

uniramous. One-branched; opposite to biramous or two-branched.

upper lip (labrum). Nonarticulated fleshy plate on the anterior margin of the mouth; lower or apical margin may be incised, bilobed, or truncate, and is usually pilose or fringed with minute setae (Fig. 8A–D).

uropod. One member of three most posterior paired biarmous appendages of the abdomen; a paired appendage of the urosome, consisting of a peduncle and paired rami (Figs. 1, 4E, F).

urosome. The posterior three segments of the abdomen, some or all of which may be fused together (Fig. 1).

venter. The ventral body surface or sternum.

$\mathcal{S}$ELECTED REFERENCES

The following list of publications is intended to provide at least one reference to each species and other taxa, including the original description of all species described since Stebbing (1906). Basic reference works are marked with an asterisk.

* Barnard, J. L., 1958. Index to families, genera, and species of Gammaridean Amphipoda (Crustacea), *Oc. Pap. Allan Hancock Found.*, No. 19, 145 pp.

————, 1959a. Generic partition in the amphipod family Cheluridae, marine wood borers, *Pacif. Nat., 1*(3), 12 pp., 5 figs.

————, 1959b. The number of species of Gammaridean Amphipoda (Crustacea), *Bull. So. Calif. Acad. Sci., 58*(1):76.

————, 1960. The amphipod family Phoxocephalidae in the eastern Pacific Ocean with analysis of other species and notes for a revision of the family, *Allan Hancock Pacific Exped., 18:*175–368, 75 pls.

————, 1964. Revision of some families, genera, and species of Gammaridean Amphipoda, *Crustaceana, 7*(1):4–74.

* ————, 1969. The families and genera of marine Gammaridean Amphipoda, *Bull. U.S. Nat. Mus.*, No. 271, 535 pp., 173 figs.

————, and W. S. Gray, 1968. Introduction of an Amphipod Crustacean into the Salton Sea, California. *Bull. So. Calif. Acad. Sci., 67*(4):219–232.

Barrett, B. E., 1966. A contribution to the knowledge of the amphipodous crustacean *Amphithoe valida* Smith 1873. Ph.D. Thesis, University of New Hampshire.

Bate, C. S., 1858. On some new genera and species of Crustacea Amphipoda, *Ann. Mag. Nat. Hist.*, Ser. 3, *1:*361–362.

————, 1862. Catalogue of the specimens of amphipodous Crustacea in the collection of the British Museum, London. 399 pp., pls. 1–58.

————, and J. O. Westwood, 1863. *A History of the British Sessile-eyed Crustacea*, Vol. 1. London: John Van Voorst. 2 vols.

Blake, C. H., 1929. New Crustacea from the Mount Desert Region, *Wistar Inst. Anat. Biol., Biol. Survey Mt. Desert Region*, Pt. 3, 1–34, Figs. 1–15.

Boeck, A., 1871. Crustacea Amphipoda borealia et arctica. Forhandl. Vidensk.-Selsk. Christiana, 1870, pp. 83–280, i–viii (index).

Bousfield, E. L., 1955. Malacostracan Crustacea from the shores of western Nova Scotia, *Proc. N. S. Inst. Sci., 24*(1):25–38.

————, 1956. Studies on shore crustaceans collected in eastern Nova Scotia and Newfoundland in 1954, *Bull. Nat. Mus. Canada,* No. 142, pp. 127–152.

————, 1958a. Distributional ecology of the terrestrial Talitridae (Crustacea: Amphipoda) of Canada, *Tenth Intern. Congr. Entomol. Proc. for 1956, 1:*883–898.

* ————, 1958b. Fresh-water amphipod crustaceans of glaciated North America, *Canadian Field-Naturalist, 72*(2):55–113.

————, 1962. Studies on littoral marine arthropods from the Bay of Fundy region, *Bull. Nat. Mus. Canada,* No. 183, pp. 42–62.

* ————, 1965. The Haustoriidae of New England (Crustacea: Amphipoda), *Proc. U.S. Nat. Museum, 117*(3512):159–240.

————, 1969. New records of *Gammarus* (Crustacea: Amphipoda) from the Middle Atlantic Region, *Chesapeake Science, 10*(1):1–17.

————, 1970. Adaptive radiation in sand-burrowing amphipod crustaceans, *Chesapeake Science, 11*(3):143–154.

————, and L. B. Holthuis, 1969. Proposed use of the Plenary Powers for the suppression of the names proposed between 1814 and 1820 by C. S. Rafinesque for two genera and four species belonging to the Order Amphipoda. Bull. Zool. Nomencl., *26*(2):105–112.

————, and A. H. Leim, 1960. The fauna of Minas Basin and Minas Channel, *Bull. Nat. Mus. Canada,* No. 166, pp. 1–30.

————, and M. L. H. Thomas, 1973 in press. Post-glacial dispersal of littoral marine invertebrates of the Canadian Atlantic region, *J. Fish. Res. Bd. Can.*

Bowman, T. E., and L. W. Peterson, 1965. Bibliography and list of new genera and species of amphipod crustaceans described by Clarence R. Shoemaker, *Crustaceana, 9*(3):309–316.

Briggs, J. C., 1970. A faunal history of the North Atlantic ocean, *Systematic Zoology, 19*(1):19–34.

Brunel, P., 1970. Catalogue d'Invertébrés benthiques du golfe Saint-Laurent recueillis de 1951 à 1966 par la station de biologie marine de Grande-Rivière, *Trav. Pêch. Québec,* No. 32, 54 pp.

* Bulycheva, A. I., 1957. The sea fleas of the seas of the USSR and adjacent waters (Amphipoda-Talitroidea) (in Russian), *Keys to the Fauna of the USSR,* Acad. Sci., USSR, *65:*1–185, 66 figs.

Chevreux, E., and L. Fage, 1925. Amphipodes, *Faune de France,* Vol. 9, 488 pp., illus.

Cole, G. A., 1970. The Epimera of North American fresh-water species of *Gammarus* (Crustacea: Amphipoda), *Proc. Biol. Soc. Wash., 83*(31):333–348.

Crane, J. M., 1969. Mimicry of the gastropod *Mitrella carinata* by the amphipod *Pleustes platypa, Veliger, 12*(2):200, Pl. 36.

Croker, R. A., 1967a. Niche diversity in five sympatric species of intertidal amphipods (Crustacea: Haustoriidae), *Ecol. Monogr., 37*(3):173–200.

————, 1967b. Niche specificity of *Neohaustorius schmitzi* and *Haustorius* sp. (Crustacea: Amphipoda) in North Carolina, *Ecology 48*(6):971–975.

————, 1968. Return of Juveniles to the marsupium in the amphipod *Neohaustorius schmitzi* Bousfield. Crustaceana, *14*(2):1 p.

Cronin, L. E., J. C. Daiber, and E. M. Hulbert, 1962. Quantitative seasonal aspects of zooplankton in the Delaware River Estuary, *Chesapeake Science, 3*(2):63–93.

Dahl, E., 1938. Two new amphipoda of the genus *Gammarus* from Finmarck, *Kong. Norske Vid. Selsk. Forh. Trondheim, 10:*125–128.

————, H. Emanuelsson, and C. Von Meckenburg, 1970. Pheromone transport and reception in an amphipod, *Science, 170* (3959):739–740.

Dana, J. D., 1852–1853. Crustacea, Part II, *U.S. Exploring Exped., 14:*689–1618, atlas.

DeKay, J. E., 1844. *Crustacea,* in *Zoology of New York, or the New York fauna.* New York (State) Nat. Hist. Surv., Pt. 6. 70 pp., 13 pls.

Dennel, R. B., 1933. Habits and feeding mechanism of the amphipod *Haustorius arenarius* Slabber, *J. Linn. Soc. London, 38:*363–388.

Dexter, D. M., 1967. Distribution and niche diversity of haustoriid amphipods in North Carolina, *Chesapeake Science, 8*(3):187–192.

———, 1971. Life history of the sandy-beach amphipod *Neohaustorius schmitzi* (Crustacea: Haustoriidae). Mar. Biol. *8*(3):232–237.

Dexter, R. W., 1944. The bottom community of Ipswich Bay, Mass., *Ecol., 25*(3):352–359.

* Dunbar, M. J., 1954. The amphipod Crustacea of Ungava Bay, Canadian Eastern Arctic, *J. Fish. Res. Bd. Canada, 11*(6):709–798.

Enequist, P., 1949. Studies on the soft-bottom amphipods of the Skagerrak, *Zool. Bidrag fran Uppsala, 28:*297–492.

Enright, J. T., 1963. Tidal rhythm of a sand beach amphipod, *Zool. J. Physiol., 46:*276–313.

Feeley, J. B., 1967. The distribution and ecology of the Gammaridea (Crustacea: Amphipoda) of the lower Chesapeake estuaries. M.Sc. Thesis, Faculty of Marine Science, College of William and Mary, Va., 75 pp.

Gould, A. A., 1841. *Report on the Invertebrata of Massachusetts comprising the Mollusca, Crustacea, Annelida and Radiata.* Cambridge, Mass., 373 pp., 213 figs., 15 pls.

* Gurjanova, E. F., 1951. Amphipoda-Gammaridea of the seas of the USSR and adjoining waters (in Russian), *Keys to the Fauna of the USSR,* Zool. Inst. Acad. Sci. USSR, No. 41, 1029 pp., 705 figs.

* ———, 1962. Gammaridean Amphipoda of the northern part of the Pacific Ocean (in Russian), *Keys to the Fauna of the USSR,* Zool. Inst. Acad. Sci. USSR, No. 74, 440 pp. 143 figs.

Heller, S. P., 1968. Some aspects of the biology and development of *Ampithoe lacertosa* (Crustacea: Amphipoda). Master's thesis, U. of Washington.

Hessler, R. R., J. D. Isaacs, and E. L. Mills, 1972. Giant Amphipod from the Abyssal Pacific-Ocean, *Science, 175*(4022):636–637.

Holme, N. A., 1966. Methods of sampling the benthos, *Adv. Mar. Biol., 2:*171–260.

* Holmes, S. J., 1905. The Amphipoda of southern New England, *Bull. U.S. Bur. Fish., 24:*459–529, 13 pls., many figs.

Hurley, D. E., 1954. Studies on the New Zealand amphipodan fauna, No. 2, The family Talitridae: the fresh-water genus *Chiltonia* Stebbing, *Trans. Roy. Soc. New Zealand, 81*(4):563–567.

———, 1954. Studies on the New Zealand amphipodan fauna, No. 6, The family Colomastigidae, with descriptions of two new species of *Colomastix, Trans. Roy. Soc. New Zealand, 82:*803–811, 3 figs.

Hynes, H. B. N., 1954. The ecology of *Gammarus duebeni* Lilljeborg and its occurrence in fresh water in western Britain, *J. Anim. Ecol., 23:*38–84.

———, 1955. The reproductive cycle of some British fresh-water Gammaridae, *J. Anim. Ecol., 24:*352–387.

Just, J., 1970. Amphipoda from Jørgen Brønlund Fjord, North Greenland, *Meddelelser om Grønland, 184*(6):1–39.

* Kaestner, Alfred, 1967–1970. *Lehrbuch der Speziellen Zoologie.* English translation and adaptation, *Invertebrate Biology,* by H. W. and L. R. Levi. Vol. III, *Crustacea.* New York: Wiley, Interscience.

Kanneworffe, E., 1966. On some amphipod species of the genus *Haploops,* with special reference to H. tubicola Lilj. and H. tenuis, *Ophelia, 3:*183–207.

Kinne, O., 1954. Die *Gammarus*—Arten der Kieler Bucht (*G. locusta, G. oceanicus, G. salinus, G. zaddachi, G. duebeni*), *Zool. Jahrb., Syst., 82*(5):405–424.

Krøyer, 1842. Une nordiske Slaegter og Arter af Amfipodernes Orden, henhørende til Familieb *Gammarina*. (Forelobigt Uddrag af et Større Arbejde) Naturh. Tidsskr., vol. 4, pp. 141–166.

* Kunkel, B. W., 1918. The Arthrostraca of Connecticut, State Geol. Nat. Hist. Surv. Bull. No. 26(1), *Amphipoda*, 1–261, 55 figs.

Leach, W. E., 1814. Crustaceology, *Edinburgh Encylopaedia, 7:*402–403, App. 429–437.

Linnaeus, C., 1758. *Systema naturae*, Vol. 1 (10th ed.), Holmiae.

Low, J., 1965. Preliminary studies on *Hyale pugettensis* (Dana), a tidepool amphipod, Zoology 533 MS, University of Washington.

Lowry, James K., 1972. Taxonomy and distribution of *Microprotopus* along the East Coast of the U.S. (Amphipoda, Isaeidae). Crustaceana, Suppl. 3, 1972 pp. 277–286.

* McCain, J. C., 1968. The Caprellidae (Crustacea: Amphipoda) of the western North Atlantic, *Bull. U.S. Nat. Mus., 278,* 147 pp.

McLusky, D. S., 1968. Aspects of osmotic and ionic regulation in *Corophium volutator* (Pallas), *J. Mar. Biol. Assoc. U.K., 448:*769–781.

Michael, A. D., 1972. Numerical Analyses of Marine Survey Data; A Study applied to Amphipoda of Cape Cod Bay, Mass. Ph.D Thesis, Dalhousie University.

Mills, E. L., 1962. Amphipod crustaceans of the Pacific coast of Canada, II: Family Oedicerotidae, *Nat. Mus. Can. Nat. Hist. Paper,* No. 15, 21 pp.

———, 1963. A new species of Liljeborgiid amphipod, with notes on its biology, *Crustaceana, 4:*158–162.

———, 1964. Noteworthy Amphipoda (Crustacea) in the collection of the Yale Peabody Museum, *Postilla, 79,* 41 pp.

———, 1967a. The biology of an ampeliscid amphipod crustacean sibling species pair, *J. Fish. Res. Bd. Canada, 24:*305–355.

———, 1967b. A reexamination of some species of *Ampelisca* (Crustacea: Amphipoda) from the east coast of North America, *Can. J. Zool., 45:*635–652.

Miner, Roy Waldo, 1950. Field Book of Sea Shore Life. Putnam's Sons, New York. 888 pp.

Montagu, G., 1813. Descriptions of several new or rare animals, principally marine, discovered on the south coast of Devonshire, *Trans. Linn. Soc. London, 11:* 1–26.

Mulot, M., 1967. Description d'*Haustorius algeriensis* n. sp. (Amphipoda: Haustoriidae), *Bull. Soc. Zool. France, 92*(4):815–826.

Nagle, S. J., 1968. Distribution of the epibiota of macroepibenthic plants, *Contr. Mar. Sci., Univ. Texas Mar. Inst., 13:*105–144.

Pallas, P. S., 1766. *Miscellanea zoologica*, Hagae Comitum, 224 pp., 14 pls. (amphipods, pp. 190–194, plate 14 partim).

Paulmeier, F. C., 1905. "Higher Crustacea of New York City," *Bull. New York State Mus.,* No. 91, Zool., *8:*117–189.

Pinkster, S., and J. S. Stock, 1970. Western European Species of the presumed Baikal genus *Eulimnogammarus* (Crustacea: Amphipoda) with description of a new species from Spain, *Bull. Zool. Mus. Univ. Amsterdam, 1*(14):211–219, 305–311, 8 figs.

Rathbun, M. J., 1905. Fauna of New England, 5, List of Crustacea, *Occ. Pap. Boston Soc. Nat. Hist., 7:*171.

Rathbun, R., 1851. The littoral marine fauna of Provincetown, Cape Cod, Mass., *Proc. U.S. Nat. Mus., 3:*116–133.

Sameoto, D. D., 1969a. Comparative ecology, life histories and behavior of intertidal sand-burrowing amphipods (Crustacea: Haustoriidae) of Cape Cod, *J. Fish. Res. Bd. Canada, 26*(2):361–388.

———, 1969b. Some aspects of the ecology and life cycle of three species of subtidal sand-burrowing amphipods (Crustacea: Haustoriidae), *J. Fish. Res. Bd. Canada, 26*(5):1321–1345.

————, 1969c. Physiological tolerances and behaviour responses of five species of Haustoriidae (Amphipoda: Crustacea) to five environmental factors. Jour. Fish. Res. Bd. Canada 26(9):2283–2298.

Sanders, H. L., P. C. Manglesdorf, and G. R. Hampson, 1965. Salinity and faunal distribution in the Pocasset River, Massachusetts, Limnol. and Oceanogr., 10(suppl.):R216–R229.

* Sars, G. O., 1895. An Account of the Crustacea of Norway: Amphipoda. Christiana and Copenhagen. Vol. 1, 711 pp., 240 pls., Suppl.

Say, T., 1818. An account of the Crustacea of the United States, J. Acad. Nat. Sci. Philadelphia, 1:37–401.

* Schellenberg, A., 1942. Krebstiere oder Crustacea IV: Flohkrebse oder Amphipoda, Tierwelts Deutschland, Pt. 40, 252 pp., 201 figs.

Segerstråle, S. G., 1947. Distribution and morphology of Gammarus zaddachi, J. Mar. Biol. Assoc. U.K., 27:1–52.

Sexton, E. W., and D. M. Reid, 1951. The life history of the multiform species Jassa falcata (Montagu) (Crustacea: Amphipoda) with a review of the bibliography of the species, J. Linn. Soc. London, 57:29–88.

Sexton, E. W., and G. M. Spooner, 1940. An account of Marinogammarus (Schellenberg) gen. nov. (Amphipoda) with a description of a new species, M. pirloti, J. Mar. Biol. Assoc. U.K., 24:633–682.

Shoemaker, C. R., 1926. Amphipods of the family Bateidae in the collection of the United States National Museum, Proc. U.S. Nat. Mus., 68(2626):1–26.

————, 1930a. The lysianassid crustaceans of Newfoundland, Nova Scotia, and New Brunswick in the United States National Museum, Proc. U.S. Nat. Mus., 77(2827):1–19, 10 figs.

————, 1930b. Descriptions of two new amphipod crustaceans (Talitridae) from the United States, J. Wash. Acad. Sci., 20(6):107–114.

* ————, 1930c. The Amphipoda of the Cheticamp Expedition, Contr. Can. Biol., n.s., 5:221–359.

————, 1931. The stegocephalid and ampeliscid amphipod crustaceans of Newfoundland, Nova Scotia, and New Brunswick in the United States National Museum, Proc. U.S. Nat. Mus., 78(2883):1–18.

————, 1932. A new amphipod of the genus Leptocheirus from Chesapeake Bay, J. Wash. Acad. Sci., 22:548–551.

————, 1933a. Amphipods from Florida and the West Indies, Amer. Mus. Nov., 598:1–24.

————, 1933b. A new amphipod of the genus Amphiporeia from Virginia, J. Wash. Acad. Sci., 23(4):212–216.

————, 1934. The amphipod genus Corophium on the east coast of America, Proc. Biol. Soc. Wash., 47:23–32.

————, 1938. Two new species of amphipod crustaceans from the east coast of the United States, J. Wash. Acad. Sci., 28(7):326–332.

————, 1945a. The amphipod genus Photis on the east coast of North America, Charleston Museum Leaflet, No. 22, pp. 1–17, Figs. 1–5.

————, 1945b. The amphipod genus Unciola on the east coast of America, Amer. Midl. Nat., 34:446–465, 9 figs.

* ————, 1947. Further notes on the Amphipod genus Corophium on the east coast of America, J. Wash. Acad. Sci., 32(2):47–63.

————, 1949. Three new species and one new variety of amphipods from the Bay of Fundy, J. Wash. Acad. Sci., 39(12):389–398.

Smallwood, M. R., 1905. The salt-marsh amphipod: Orchestia palustris. Brooklyn Inst. Arts and Sci., Cold Spring Harbor Monograph, No. 3, pp. 1–27.

* Smith, S. I., 1873. Report upon the invertebrate animals of Vineyard Sound and adjacent waters, with an account of the physical features of the region (Crustacea: Amphipoda), in Verrill, A. E., pp. 1545–1558.

Spooner, G. M., 1949. The distribution of *Gammarus* species in estuaries, Pt. 1, *J. Mar. Biol. Assoc. U.K., 27*:1–52.

* Stebbing, T. R. R., 1906. Amphipoda I: Gammaridea, *Das Tierreich*, Vol. 21, 806 pp.

——, 1914. Crustacea from the Falkland Islands collected by Mr. Rupert Vallentin, F. L. S., Pt. II, *Proc. Zool. Soc. London, 1*:341–378, 9 pls.

* Steele, D. H., and P. Brunel, 1968. Amphipoda of the Atlantic and Arctic coasts of North America: *Anonyx* (Lysianassidae), *J. Fish. Res. Bd. Can., 25*(5):943–1060.

Steele, D. H., and V. J. Steele, 1969, 1970, 1972. The biology of *Gammarus* (Crustacea: Amphipoda) in the northwestern Atlantic: I, *Gammarus duebeni* Lillj., *Can. J. Zool. 47*(2):235–244; II, *Gammarus setosus* Dementieva, *48*(4):659–672; III, *Gammarus obtusatus* Dahl, *48*(5):989–995; IV, *Gammarus lawrencianus* Bousfield, *48*(6):1261–1267; V, *Gammarus oceanicus* Segerstrale, *50*(6):801–813.

Steele, V. J., 1967. Resting stage in the reproductive cycles of *Gammarus, Nature, 214:*1034.

Stephensen, K., 1944. Amphipoda, The Zoology of East Greenland, *Med. Grønland, Komm. Vedens. Undersøgel, 121*(14):1–165.

* Stimpson, W., 1853. Synopsis of the marine invertebrata of Grand Manan: or the region about the mouth of the Bay of Fundy, New Brunswick, *Smiths. Contr. Knowledge, 6*:5–56, 3 pls.

Stock, J. H., 1967. A revision of the European species of the *Gammarus locusta* group (Crustacea: Amphipoda), *Zool. Verhandl., 90*:3–56.

Sumner, F. B., R. C. Osburn, and L. J. Cole, 1911. A biological survey of the water of Woods Hole and vicinity, Pt. 2, Ser. 3: A catalogue of the marine fauna, *Bull. Bur. Fish., 31*:1–739.

Verrill, A. E., and S. I. Smith, 1873. Report upon the invertebrate animals of Vineyard Sound and adjacent waters, *Rep. U.S. Comm. Fish.*, 1871 and 1872. Washington, D.C., 478 pp.

Werntz, H. O., 1963. Osmotic regulation in marine and fresh-water gammarids, *Biol. Bull., 124*:225–239.

Whiteaves, J. F., 1901. Catalogue of the marine Invertebrata of eastern Canada, *Geol. Surv. Canada*. Ottawa. 271 pp.

Wigley, R. L, 1966. Two new marine amphipods from Massachusetts, U.S.A., *Crustaceana, 10*:259–270.

Williamson, D. I., 1951. Visual orientation in *Talitrus saltator, J. Mar. Biol. Assoc. U.K., 30*:91–99.

Yentsch, A. E., M. R. Carriker, R. H. Parker, and V. A. Zullo, 1966. *Marine and Estuarine Environments, Organisms and Geology of the Cape Cod Region* (an indexed biliography, 1665–1965). Plymouth, Mass.: Leyden Press, Inc. 178 pp.

INDEX TO COMMON AND SCIENTIFIC NAMES OF BIOTA

Boldface numerals indicate the page numbers of the main species or group description.

307

**SHALLOW-WATER
GAMMARIDEAN AMPHIPODA OF NEW ENGLAND**

Designed by Margaret Deines.
Composed by Vail-Ballou Press, Inc.,
in 9 point linofilm Helvetica, 3 points leaded,
with display lines in Helvetica Bold.
Printed offset by Vail-Ballou Press
on Warren's Olde Style, 60 pound basis,
with the Cornell University Press Watermark.
Bound by Vail-Ballou Press
in Columbia Riverside Linen
and stamped in All Purpose foil.

Library of Congress Cataloging in Publication Data
(For library cataloging purposes only)

Bousfield, Edward Lloyd.
　Shallow-water gammaridean Amphipoda of New England.

　Bibliography: p.
　1. Gammaridae.　2. Amphipoda—New England.
I. Woods Hole, Mass. Marine Biological Laboratory. Systematics-Ecology Program.　II. Title.
QL444.A5B65　　　595'.371　　　72-4636
ISBN 0-8014-0726-5